The Phonology/Phonetics Interface

The Phonology/ Phonetics Interface

Elizabeth C. Zsiga

The MIT Press
Cambridge, Massachusetts

In memoriam

Debi Shifflet Olshaw
1958–2017

Every phrase and every sentence is an end and a beginning
– T. S. Eliot, Four Quartets

Published in the United States and Canada by the MIT Press, 2020

First published in Great Britain by Edinburgh University Press Ltd

Typeset in 10/12 Minion Pro
by Servis Filmsetting Ltd, Stockport, Cheshire, and
printed and bound in Great Britain

Library of Congress Control Number: 2020945254

ISBN 978-0-262-54264-7

CONTENTS

FIGURES AND TABLES

FIGURES

TABLES

PREFACE

I'm pleased to be able to offer this contribution to the series of Edinburgh Advanced Textbooks in Linguistics. As an advanced textbook, this text assumes readers will have taken introductory courses in phonetics and phonology, so that basic terms such as "spectrogram" or "allophone" are not defined (unless the definition is being called into question). Familiarity with International Phonetic Alphabet (IPA) transcription is assumed.

In general, however, this book begins with the assumptions that classes in phonetics and phonology often make—that phonology and phonetics are separate disciplines, that there is a correct IPA transcription for every utterance, that phonology operates over natural classes defined by distinctive features, and so on—and questions them. We will be examining the history of many of these assumptions, the evidence for (and against) them, and the work of scholars who may have thought about things in a different way. Because our field continues to grow and change, there is always new data, and there are more questions than there are settled answers.

To further the use of this book as a textbook, each chapter is followed by suggested readings and possible questions for use in class discussions. The suggested readings are divided into types:

- Overviews: articles or chapters that survey aspects of the topic in greater detail, but that don't necessarily advocate for a particular approach.
- Exemplary research: favorite articles that illustrate some of the best scholarship on the topic. Unlike the overview articles, these will be original research articles, exemplifying different methods of data collection and analysis, and best practices in reasoning from data to theory. Where possible, I suggest "something old" (which I will arbitrarily define as prior to 1990) and "something new."
- Opposing views: sets of publications that express opposite views on the topic, that should be read and discussed together.

For each publication or pair of publications suggested, a question to guide your reading is proposed. Additional discussion questions for each chapter review the main points. Some more open-ended questions also bring up broader themes, applications, and implications. Class leaders can decide which readings and questions will lead to the most fruitful discussion. Overall, the eleven chapters of this text should fit reasonably well into a semester or quarter.

At the outset, I would like to thank those who helped bring this project to completion. Thank you to Mits, Peter, and the editors at Edinburgh University Press for their patience and feedback. Special thanks are due to Bob Ladd, for his generous and extremely helpful comments on an earlier version of the manuscript. Georgetown University supported this work with two summer grants in aid of research and writing. Alexandra Pfiffner, Bernie O'Connor, and Goldie Ann McQuaid provided valuable research and editorial assistance. And above all, thank you to my family, whose support made all the difference.

EZ
Sabillasville, Maryland
August 2019

CHAPTER 1

INTRODUCTION

1.1 PHONOLOGY VS PHONETICS

Both "phonology" and "phonetics," as is evident from their basis in the Greek word φωνή ([fone], *sound*, or *voice*), are concerned with the sounds of language. While scholars have been interested in studying the sounds of language for millennia, phonology and phonetics, and the areas of study or approaches they designate, have not always been considered to be separate, and the two terms have not always been defined in a consistent way. A number of authors who have aimed to define and distinguish phonology, phonetics, and the relationship between them (e.g., Keating 1990a; Pierrehumbert 1990b; Myers 2000; Scobbie 2007; Kingston 2007; Raimy and Cairns 2009; and Hamann 2011 among others), have used opposing terms such as those in (1):

(1)

Phonology:	Phonetics:
cognitive	physical
psychological	biological
internal	external
abstract	concrete
categorical	gradient
discrete	continuous
symbolic	numerical
formal	functional
social science	natural science

The distinction seems straightforward. Phonology uses the methods of social science to study speech sounds as cognitive or psychological entities, which are abstract symbols that are discrete and categorical. Phonetics uses the methods of natural science to study speech sounds as physical entities, which are concrete and thus exist in actual space and time that are gradient and continuous.

Speech sounds are obviously on the one hand physical noises—disturbances of air pressure created by movements of the tongue, lips, and other body parts—and they can be studied by the sciences of acoustics and biology like any other noises or bodily actions. But speech sounds also convey meaning between two minds, meaning that is carried by means of a complex mental code in which speech sounds are discrete counters that can be arranged in different combinations that correspond to different meanings: /med/, /dem/, /emd/, and so on. Mental representations are studied not by physics but by psychology, philosophy, or cognitive science. An apt analogy is proposed by Smolensky and Legendre (2006: 6–7): "Not just in contemporary cognitive science, but implicitly since antiquity, the fundamental hypothesis has held that the mind is a serial computer that manipulates discrete structures built of abstract symbols" using "logic-like rules." They continue:

> Going well beyond language and reasoning, modern computational work in higher visual perception, action planning, problem solving, and most other higher cognitive domains has similarly gained great explanatory leverage by analyzing cognition as arising from the manipulation of complex, abstract, symbolic structures. Such a symbolic cognitive architecture seems inescapable from the perspective of research on higher mental functions.

Since speech sounds have two different aspects, physical and cognitive, the case is made for two different disciplines, phonetics and phonology, to study them.

Yet linguists disagree over whether this opposition is correct or necessary. One issue is that it's not clear that the lists of properties in (1) always cluster together—some argue (e.g., Pierrehumbert et al. 2000) that it is possible to be both abstract and gradient, for example, and to study cognition with the methods of natural science. Another question is whether the division (assuming it exists) is one of method or of the object of study. Do phonology and phonetics study the same object (speech sounds) with different methodologies (natural science vs social science), or do they study different objects, one that exists in the mind and the other that exists as patterns of air pressure variation produced by wriggling tongues?

In addition, the physical and cognitive aspects of speech sounds are not independent and unrelated. Zsiga (2013a: xv), suggests that phonetics asks questions such as "How are speech sounds made?" and "How does sound travel through the air?" while phonology asks questions such as "How do languages organize sounds to distinguish different words?" and "What sorts of restrictions, or constraints, do languages put on sequences of sounds?" These seem like separate questions, yet it is abundantly clear that patterns of contrast, and constraints on licit sequences, are intimately related to the way speech sounds are produced and perceived.

Scobbie (2007) describes several different metaphors for thinking about the phonology/phonetics interface. The interface might be thought of as a transducer, or translator: phonological symbols must somehow be turned into physical movements, and back again, and to study the interface is to study how this happens. But the interface might also be thought of as a fence: some speech-related phenomena (tongue movements, formant trajectories) are on one side of the fence, other phenomena (reduplication, plural formation) are on the other, and to study the interface is to attempt to determine, for any given pattern or fact about speech, whether it properly belongs to one side or the other. "Border skirmishes" then may ensue, with phonologists and phoneticians claiming the same process, phenomenon, or pattern (deletion, assimilation, syllable structure) by offering competing explanations. For example, the American English phrase "miss you" is usually pronounced [mɪʃju]. Is this fact the result of speakers executing a plan that substitutes [ʃ] for [s] preceding [j] (phonology), or is it the result of the physical tongue movements for [s] and [j] getting blended together (phonetics)? (Zsiga 2000 argues the latter.) Yet Scobbie also suggests that the phonology/phonetics interface might be more like a tidal zone: some creatures clearly belong to the ocean, others clearly to the land, but there is an area in between, whose edges are not clearly demarcated, where the environment is equally comprised of water and of sand, and creatures are amphibious. Where phonology and phonetics are concerned, different researchers may then have different ideas of how large such a tidal zone may be, or whether it exists at all.

In essence, the question is one of modularity. It can make sense to break a big problem or issue or area of study down into smaller modules/units, and study them separately. Myers (2000: 258) argues that a modular division between phonetics and phonology makes for a simpler and more coherent theory of each:

> If we can effectively rid all phonetic patterns from the set of patterns our phonological theory has to account for, we will presumably be left with something that makes more sense . . . [A] pattern that is anomalous as an instance of phonology makes sense as an instance of phonetics, and vice versa.

But we might argue about exactly what divisions make the most sense, and how best to define the different modules. And we must address and try to understand how the different parts fit together and make a whole. We can't simply rake all the difficult problems into our neighbor's yard. Thus, the goal of this book is to present different viewpoints on the phonology/phonetics interface: how, or whether, the two disciplines are separate, and what evidence has been presented in favor of each way of defining and approaching the interface.

1.2 OUTLINE OF THIS BOOK

The chapters that follow address in more detail the issues that have been raised in this introduction.

Chapter 2 sets the stage by discussing how central themes of the phonology/phonetics interface have developed over the past century, and how current theories more or less divide up the world.

Chapters 3 and 4 address the topics of contrast and categories. Some of the central concerns of both phonology and phonetics are "What is a possible speech sound?" "What is a possible inventory of speech sounds?" and "Why are inventories shaped the way they are?" Chapter 3 looks at the structure of phonological inventories, focusing on the segment, or "phone," as a basic unit. While division of utterances into segments is *almost* universally assumed, segmentation is not always phonetically clear, and not all linguists accept the segment as a basic unit. If not the segment, then what? Chapter 3 also looks at orthography as a source of evidence and discusses the benefits and drawbacks of segment-based transcription systems. Chapter 3 also considers the question of what the units of non-spoken, signed, languages might consist of. Chapter 4 turns to theories of distinctive features, both as category labels on segments and as basic units in their own right. Are features universal? Are they innate? If not, how do phonological categories emerge from articulation, perception or diachronic change?

Chapter 5 discusses the idea of a derivation in phonology and phonetics. Once the contrastive categories are set, how does the phonetic detail get filled in? In a derivational approach, phonology begins with sparsely-specified underlying representations, with more and more detail filled in step by step, until a fully-specified phonetic representation is formed. In this approach, the main question to be answered is "Where is the dividing line?" What kinds of information belong in a phonological representation, and what kinds in a phonetic representation? This chapter examines derivational approaches to the phonology/phonetics interface, considering in particular the range of data that different theories account for, and arguments for and against sequential ordering of the phonological and phonetic models.

Chapter 6 delves more deeply into questions of markedness, naturalness, and abstraction. A key objection to the derivational approaches discussed in Chapter 5 is that they allow for too much abstraction: that phonology and phonetics became too far removed. Competing theories, both classic and current, emphasize naturalness and markedness, particularly the role of "markedness constraints" in determining phonological outcomes. In a constraint-based approach, the central question of the phonology/phonetics interface becomes the interacting roles of phonological and phonetic information, or the tension between "formal" and "functional" pressures. Exactly what kinds of phonetic information does phonology have access to? Must phonology translate phonetic parameters into more abstract and formal units, or does it have direct access to phonetic information like ease of articulation and perceptibility? What can a phonology based on "naturalness" have to say about "unnatural" processes? (Are they perhaps just diachronic residue?) Can phonological and phonetic forces act and interact simultaneously? If there is no sequential derivation, to what extent are two separate linguistic modules even needed?

Chapter 7 considers the issue of boundaries and domains in the prosodic hierarchy. Important work at the phonology/phonetics interface has often examined hierarchical units larger than the segment, including syllables, stress feet, and phrasing. Phonetic studies have been crucial in arguing for the existence of these larger, hierarchical units, but it is unclear the extent to which they have a "life of their own" as abstract units, beyond their physical instantiations in segmental strings, durational patterns, or pitch tracks. How does the hierarchical organization of speech units work at both the phonological and phonetic levels? Chapter 7 reviews influential papers and current issues in the study of prosodic patterning, including syllable structure, stress, and higher levels of the prosodic hierarchy.

Pitch patterns in both tone and intonation, the topic of Chapter 8, have received a great deal of attention in the literature on the phonology/phonetics interface, as a special case of the interaction of the categorical and the continuous. Acoustic pitch tracks show continuous changes in fundamental frequency over time, and some linguists have argued that a complex continuously-changing "tune" can be the basis of tonal or intonational contrast. Others, however, argue for a more abstract, compositional analysis. Chapter 8 reviews these different approaches, and the successes and challenges of each.

Chapters 9 and 10 examine in more depth two particular theories of the phonology/phonetics interface, both the subject of much discussion in the current literature, one based in articulation and the other in perception. Chapter 9 discusses the theory of Articulatory Phonology, which takes the "articulatory gesture" as the basic unit of both phonological contrast and phonetic implementation. Chapter 10 turns to Exemplar Theory. Exemplar Theory takes perception as its starting point, and examines how phonological categories and relations can be formed out of detailed memory traces. This approach leads to new theories of the organization of the lexicon, the representation of phonological alternations, variation, and the role of historical change in explaining synchronic patterns.

Chapter 11 concludes with consensus, controversies, and directions for future research. But we begin, in Chapter 2, with some historical perspective.

RECOMMENDED READING

Overview

Scobbie, J. M. (2007), "Interface and overlap in phonetics and phonology," in G. Ramchand and C. Reiss (eds), *The Oxford Handbook of Linguistic Interfaces*. Oxford: Oxford University Press, pp. 17–52.

- At this point, which of Scobbie's metaphors do you find most convincing? Keep these metaphors in mind as you continue reading in this text, and see if the additional arguments and data presented change your opinion.

Cohn, A. C. and M. Huffman (2015), "Interface between phonology and phonetics," in M. Aronoff (ed.), *Oxford Bibliographies in Linguistics*. New York: Oxford University Press.

- A very helpful recent overview and annotated bibliography. Guided by Cohn and Huffman's suggestions, choose an article or set of articles to read and comment on.

Exemplary research

Cohn, A. C. (1993), "Nasalisation in English: phonology or phonetics," *Phonology*, *10*(1): 43–81.

- How does Cohn define the phonology/phonetics interface? How does this paper use phonetic data to make a phonological argument?

Opposing views

Smolensky, P. and G. Legendre (2006), *The Harmonic Mind: From Neural Computation to Optimality-theoretic Grammar*. Cambridge, MA: MIT Press, Chapter 1.

- Do you agree that the mind is a serial computer?

Pierrehumbert, J., M. E. Beckman, and D. R. Ladd (2000), "Conceptual foundations of phonology as a laboratory science," in N. Burton-Roberts, P. Carr, and G. Docherty (eds), *Phonological Knowledge: Conceptual and Empirical Issues*. Oxford: Oxford University Press, pp. 273–303

- Would the authors of this paper agree with Smolensky and Legendre that the mind is a serial computer? At what points would they agree and disagree?

QUESTIONS FOR FURTHER DISCUSSION

1. Look again at the list of terms on page 1. How would you define each term? Can you think of ways in which the terms might or might not be exclusive? For example, something probably can't be both internal and external at the same time, but the methods of social science and physical science might overlap. From what you've learned from any previous phonology and phonetics courses you've taken, would you add any terms to the list in (1)?
2. Based on what you've learned in previous courses, would you have said that the difference between phonology and phonetics is one of method, or of the object of study?
3. Again looking at the list of terms on page 1, how do you think the mapping between phonetics and phonology differs from, for example, the mapping between phonology and morphology, or morphology and syntax? Do you think this list of terms suggests there is something unique about the phonology/phonetics interface?
4. A modular view advocates breaking up a larger problem into parts that can be addressed separately. In general, what are the advantages and disadvantages of a modular approach? In terms of the phonology/phonetics interface specifically, how might a modular approach be helpful, and how might it be unhelpful?

CHAPTER 2

APPROACHES TO THE INTERFACE

2.1 THE BEGINNINGS OF THE PHONOLOGY/ PHONETICS DICHOTOMY

The idea of a division of labor between two different disciplines in the study of speech sounds, disciplines that more or less resemble what is taught in university courses on phonetics and phonology in Departments of Linguistics at the beginning of the twenty-first century, became embedded in linguistic theory at the beginning of the twentieth century, with scholars such as Ferdinand de Saussure (1857–1913) in Switzerland, and Jan Baudouin de Courtenay (1845–1929) in Russia. The accepted way to study linguistics at the turn of the twentieth century was historical reconstruction—comparing contemporary languages for the purpose of discerning common ancestry and familial relationships, and discovering the properties of the "mother tongue." The study of modern languages was the concern of literature, rhetoric, or pedagogy. Both Saussure and Baudouin de Courtenay were trained as historical linguists and published papers on historical topics, but both became interested in another question—what is the nature of language itself? What is actually going on when people talk to each other? Both sought to define a "general linguistics" that would take this question, particularly as it relates to living languages, as its main object, rather than the question of language history and origins. Then, if understanding language per se was going to be the proper goal of linguistics, and if contemporary languages were to be the proper objects of study, how should the linguist go about doing his work? How would linguistics be different on the one hand from poetics and rhetoric, and on other hand from physics and biology?

Saussure's "Course in General Linguistics" began as a series of lectures first given at the University of Geneva in 1906. It was published posthumously in 1915 in book form based on students' class notes. (In this text, references are cited from the widely-available 1959 edition, translated by Wade Baskin.) The book opens with a section defining "the object of linguistics." This section contains the diagram in Figure 2.1, illustrating the "speech circuit" between two interlocutors.

In this circuit, he defines five stages: two "psychological," two "physiological," and one "physical."

- **Stage 1** is psychological. Interlocutor *A* thinks of something to say, and this mental concept "unlocks a corresponding sound image in the brain" (p. 11).

Figure 2.1 Saussure's speech circuit
Source: Saussure (1959: 11).

- **Stage 2** is physiological. The brain sends a signal to the speech organs to implement the sound image.
- **Stage 3** is physical. The speech organs move and send acoustic waves from the mouth of *A* to the ear of *B*.
- **Stage 4** is physiological again. Having received the acoustic information, the ear of *B* sends a signal corresponding to the sound image to his brain.
- **Stage 5** is psychological again: *B* associates the received sound image to the concept.

Crucially, Saussure insists that only Stages 1 and 5, the psychological association of sound image to concept, count as language, and thus as a proper object of linguistics. It is the linguistic *sign*—the combination of *signifier* (the sound image) and *signified* (the concept)—that constitutes the basis of language. Stages 2–4 are part of "speaking" (Fr. *parole*) but are outside the system of language (Fr. *langue*) per se.

In identifying the sign alone as the proper object of linguistic study, Saussure carves out a unique sphere for linguistic inquiry distinct from any other field. Yet in doing so he excludes from linguistics anything physiological or physical, and thereby cuts off what we would call phonetics from the rest of linguistics. "Everything in Language is basically psychological," he says (p. 6); language consists of "the sum of word-images stored in the minds of all individuals" (p. 13). The ears and articulators implement language, but are not part of it, nor do they influence it: "The vocal organs are as external to language as are the electrical devices used in transmitting the Morse code to the code itself; and phonation, i.e., the execution of sound-images, in no way affects the system itself" (p. 18). Note that he doesn't consider the possibility that Morse code was designed so as to be implementable via a particular technology.

Saussure insists on two separate ways of studying speech sounds: studying the relationship of "sound image" to meaning is one thing ("linguistics proper"), studying the physiological and physical implementation of sound images is something entirely different, and not proper linguistics at all. "[T]o consider language [*langue*] and speaking [*parole*] from the same viewpoint would be fanciful," he says (p. 19). One might, he admits, want to call the study of speech a different kind of linguistics, a "linguistics of speaking" but "that science must not be confused with linguistics proper, whose sole object is language" (p. 20).

During the years Saussure was teaching in Geneva, more than 2,000 miles to the northeast in the central Russian city of Kazan, Jan Baudouin de Courtenay, a Polish aristocrat educated in Prague and St Petersburg, was also lecturing on the proper definition of "General Linguistics." According to Stankiewicz (1972a), Baudouin de Courtenay ended up at a university in the middle of the Russian steppes because of his outspoken criticism both of his academic mentors and of the Czarist government. His professors in historical linguistics didn't appreciate being told they were doing linguistics wrong. The government didn't appreciate his criticism either: Baudouin de Courtenay's professorial duties were interrupted for several months in 1915 when his publication of a pamphlet promoting the rights of ethnic minorities got him thrown into jail. Because Saussure and Baudouin de Courtenay were contemporaries, and corresponded, it's difficult to be clear on who influenced whom the most, but most linguists (see Stankiewicz 1972a; Anderson 1985a) agree that Baudouin de Courtenay was probably first with many formative ideas on the definition of modern general linguistics. However, since Baudouin de Courtenay was so geographically isolated and published exclusively in obscure Polish and Russian journals, Saussure became much better known and had the greater influence on later writers and thinkers.

Like Saussure, Baudouin de Courtenay wanted to shift the focus of linguistics from historical reconstruction to "pure linguistics, linguistics *per se*, whose subject is language itself" (1870; p. 60 in Stankiewicz 1972b). He also insisted on separation of the physical and the psychological in the study of speech sounds, which he described in terms very similar to Saussure's (though without the clever diagrams).

In 1870, Baudouin de Courtenay wrote, "two elements are inseparably linked in language: a physical and a psychological" (1870; Stankiewicz 1972b: 59). In later work (1927; Stankiewicz 1972b: 280), he argued that these two aspects must be investigated separately: "We must distinguish two separate sciences: 1) a natural science . . . anthropo-phonetics, closely related to mechanics (dynamics, kinematics) and physics (acoustics, optics), and 2) psycho-phonetics, which is a 'humanistic' science closely related to psychology and sociology." To anthropo-phonetics he assigned the study of articulatory, auditory, and "cerebral" structures and functions, noting that these were not "truly psychological or linguistic" (Stankiewicz 1972b: 278). To psycho-phonetics he assigned the study of all psychological aspects of language, including abstraction, generalization, and particularly of the *phoneme* (a term which he was the first to use in its modern sense), which he defined as "the psychological equivalent of physical 'sound,' the actual and reproducible unit of linguistic thought" (p. 279).

Thus, the present dichotomy between two separate sub-disciplines of linguistics, one basically psychological (focusing on Saussure's Stages 1 and 5, the relationship between sound and meaning) and the other basically physical (focusing on Saussure's stages 2–4), dates back to Saussure and Baudouin de Courtenay at the beginning of the twentieth century. It was these two scholars who defined the two sub-disciplines in the process of defining the modern study of linguistics as a whole. What Baudouin de Courtenay called "anthropo-phonetics" became known as simply "phonetics" and what he called "psycho-phonetics" became "phonology." (Unfortunately, Saussure used the term "phonology" in exactly the opposite way, to refer to "the physiology of sounds" (e.g., Saussure 1959: 33), so one needs to be careful in reading and understanding Saussure's writing on the topic!) They thus set the stage for discussion of the phonology/phonetics interface for the next hundred years.

2.2 WHAT IS "REAL"? PHONOLOGICAL STRUCTURE VS PHONETIC SUBSTANCE

For the most part, scholars in the first half of the twentieth century who followed Saussure and Baudouin de Courtenay in pursuing "general" linguistics accepted without question the dichotomy between phonetics and phonology, including Saussure's assertion that phonetics was outside linguistics proper. While it's not fair to say that these linguists ignored phonetics, they considered phonetic substance to be the non-linguistic raw material out of which the (truly interesting, truly linguistic) phonological structure was built. Thus, in the first half of the twentieth century, the separation between phonology and phonetics, and their characteristics as separate disciplines, was further elaborated.

Two leading figures who took important stands on the phonology/phonetics dichotomy in the early decades of the twentieth century were Nikolai Trubetzkoy in Europe and Edward Sapir in the United States. They worked within an approach to linguistics that became known as "Structuralism." The name comes from the central idea (due to Saussure) that language (or any human activity or social/cultural construction) can only be understood as a complex set of structured relationships, such as contrast, correlation, parallelism, and so on. The important thing is to study the relationships between things, the way they function as part of a system, their psychological "value," not the physical composition of the things themselves. Saussure defined language as "a *system of signs* that express ideas" (1959: 17) and famously stated that "in language there are only differences" (p. 118). He compares speech sounds to coins, whose value is not really determined by the amount of precious metal they contain, but by the value assigned by the political and economic system (pp. 118–19):

> A coin nominally worth five francs may contain less than half its worth of silver. Its value will vary according to the amount stamped upon it and according to its use inside or outside a political boundary. This is even more true of the linguistic signifier, which is not phonic but incorporeal—constituted not by its material substance but by the differences that separate its sound image from all others.

Thus /b/ exists as a point in the system because "bad" is different from both "pad" and "dad." Another analogy (p. 110) is to a chess piece. A knight is an element in the game, but not by virtue of its physical makeup. It doesn't matter if it's made of wood or ivory. If the piece were lost, you could substitute a coin or any other marker, designate it as "knight," and it would work just as well because it would have the same "value" in the system.

One of the first and most influential structuralist phonologists was Nikolai Trubetzkoy (1890–1938). His life history has some parallels to Baudouin de Courtenay's (see Anderson 1985a for details). A member of the Russian nobility whose father was a professor of philosophy at Moscow University, Trubetzkoy was publishing papers on Finno-Ugric and Chukchee as a teenager. He fled Moscow in the revolution of 1917, and he eventually became a Professor of Slavic Philology at the University of Vienna. He became, with Roman Jakobson (1896–1982), one of the founders of the influential "Prague Linguistics Circle," an association of European linguists in whose meetings and publications the principles of structural linguistics were hashed out. Though he was not as overtly political as Baudouin de Courtenay, Trubetzkoy came under suspicion when the Nazis took power in Vienna, as his work on minority languages was seen as being critical of claims of Aryan supremacy. In the spring of 1938 the Gestapo raided and trashed Trubetzkoy's apartment, confiscating his papers. The raid apparently brought on a heart attack—his health had never been robust—and Trubetzkoy died in June 1938, just as he was finishing a draft of the first volume of his monograph, *Grundzüge der Phonologie* (*Foundations of Phonology*), which was posthumously published the next year by other members of the Circle. (Citations in this text are from the 1969 translation by Christiane Baltaxe.)

In *Foundations of Phonology*, Trubetzkoy reinforced and elaborated on the phonology/phonetics distinction established by Saussure and Baudouin de Courtenay, making the case (as though to clear phonetics out of the way) in the Introduction. Citing both Saussure and Baudouin de Courtenay, Trubetzkoy carefully distinguishes the concrete and specific "speech event" from the abstract and general "system of language." "They are," he writes, "inseparably linked and should be considered two aspects of the same phenomenon 'language.' Still, they are quite different in nature and must, therefore, be studied separately" (1969: 1). There must be, he argues, "two 'studies of sound,'" one directed toward the speech event and using the methods of natural science, and one directed toward the system of language and using "only the methods of linguistics, or the humanities, or the social sciences" (1969: 3–4). Note again that speech science is not considered part of "linguistics."

Trubetzkoy explicitly argues for many of the defining aspects of phonology that were listed in (1) above. He repeatedly uses the term "concrete" to refer to phonetics, and "abstract" to refer to phonology. He argues (1969: 11) that all the details of acoustics and articulation are of interest to the phonetician, but most of these details are "quite unimportant" for the phonologist, who only needs to consider "that aspect of sound which fulfills a specific function in the system of language." A phonological analysis will begin with a phonetic transcription as data, and with consideration of phonetic factors as a starting point, but "the limit of what is absolutely necessary should not be overstepped" and the linguist must realize that "higher levels of the phonological description, that is, the systemic study and the study of combinations, are quite independent of phonetics" (p. 14). Trubetzkoy dismisses the arguments of linguists who don't want to recognize the abstract nature of phonological analysis, and in fact opines that anyone who continues to cling to phonetic realism must be guilty, in his words, of "inertia, mental lethargy, and stubborn rejection of any new thought" (p. 6).

Phonology must make reference to lexical distinctions—the "function" that tells a phonologist what aspects of sound to study—but reference to lexical or grammatical function is prohibited to the phonetician. Trubetzkoy states that phonetics requires "the complete exclusion of any reference to the lexical meaning of the sound complexes under study," while

> it is the task of phonology to study which differences in sound are related to differences in meaning in a given language, in which way the discriminative elements (or marks) are related to each other, and the rules according to which they may be combined into words and sentences. (p. 10)

His definition which still largely holds of the field of phonology today. He argues that phonology is discrete and phonetics is continuous: speech events and the articulations that create them are continuous and "infinitely varied," but the "phonic norms" that constitute the units of the language system are discrete and enumerable (p. 3). "The system of language," he states, "is outside the scope of 'measurement and number'" (p. 8). The view that gradience vs discreteness defines the distinction between phonetics and phonology is reiterated by more recent scholars including Keating (1990a), Pierrehumbert (1990b), and Myers (2000). (But see below for Pierrehumbert's more recent views, especially Pierrehumbert et al. 2000, which plays off Trubetzkoy's *Foundations of Phonology* title and argues against pretty much every one of Trubetzkoy's claims on this issue.)

A different aspect of the phonology/phonetics distinction was emphasized by Edward Sapir, an American structuralist: phonology is more "psychologically real" than phonetics. Like the phonologist, the "naïve speaker/hearer" also discounts certain sound details, and creates an abstract representation in the process of language use. Sapir conducted his research on Athabaskan languages, and taught linguistics at the University of Chicago and later at Yale, and he drew on these experiences for his 1933 essay, "The psychological reality of phonemes." (In this text, citations are from the 1949 issue of Sapir's collected works.) Sapir's thesis was that speakers routinely ignore or downplay physical aspects of speech sounds in favor of a more abstract representation based on systemic relationships. The essay (1949: 45) begins:

> The concept of the "phoneme" (a functionally significant unit in the rigidly defined pattern or configuration of sounds peculiar to a language), as distinct from that of the "sound" or "phonetic element" as such (an objectively defined entity in the articulate and perceived totality of speech) is becoming more and more familiar to linguists. The difficulty that many still seem to feel in distinguishing between the two must eventually disappear as the realization grows that no entity in human experience can be adequately defined as the mechanical sum or product of its physical properties. These physical properties are needed of course to give us the signal, as it were, for the identification of the given entity as a functionally significant point in a complex system of relatednesses; but for any given context it is notorious how many of these physical properties are, or may be, overlooked as irrelevant, how one particular

> property, possessing for the moment or by social understanding an unusual sign value, may have a determinedness in the definition of the entity that is out of all proportion to its "physical weight."

Phonetics is needed as a starting point, but "physical properties" are then either ignored or disproportionately enlarged, in service to the more important abstract "system of relatednesses."

Sapir goes on to describe several "phonetic illusions." He writes of working with a Paiute speaker on the task of developing an orthography, and found that the speaker would often write "p" where [β] was clearly pronounced, ignoring the phonetic facts of voicing and lenition because they are predictable and non-contrastive. Sapir describes non-contrastive differences as "purely mechanical phonetic variations" (p. 47) or "absolutely mechanical phonetic laws" (p. 49) which don't rise to the level of the native speaker's consciousness. He describes a Sarcee speaker who insisted that a word ended in a [t] for which there was no acoustic or articulatory warrant in the actual form being pronounced, but that surfaced in other morphologically-related forms: evidence that the abstract mental representation was more "real" to the speaker than what his vocal apparatus was actually doing. Speakers of English also come under scrutiny: Sapir describes his English-speaking phonetics students mistakenly transcribing glottal stops where none exist, because the rules of American English require them.

Thus, according to Sapir, the native speaker processes speech sounds as distinct categories defined by systemic function (contrast and alternation), and pays no attention to phonetic differences that are not relevant to determining the different categories. It is the linguist who comes to language description "prepared with a gamut of absolute phonetic possibilities" who will introduce a false account of the system. His analogy is to compare an average person who groups wooden objects into the categories "club" and "pole" based on their functions, rather than their physical characteristics (p. 46).

> Just as it takes a physicist or philosopher to define an object in terms of such abstract concepts as mass, volume, chemical structure, and location, so it takes very much of a linguistic abstractionist, a phonetician pure and simple, to reduce articulate speech to simple physical processes. To the physicist, the three wooden objects are equally distinct from each other, "clubs" are romantic intrusions into the austere continuities of nature. But the naïve human being is much surer of his clubs and poles. . . . To the naïve speaker and hearer, sounds (i.e., phonemes) do not differ as five-inch or six-inch entities differ, but as clubs and poles differ. If the phonetician discovers in the actual flow of speech something that is neither "club" nor "pole," he, as phonetician, has the right to set up a "halfway between club and pole" entity. Functionally, however, such an entity is a fiction, and the naïve speaker or hearer is not only driven by its relational behavior to classify it as a "club" or "pole," but actually hears and feels it to be such.

Sapir equates the phonological with what is psychologically real (i.e., "felt"), discrete, and categorical, and the phonetic with physically real, continuous, and gradient, reinforcing the dichotomy, and downplaying the phonetic as less interesting and important than the phonology. What scholar of language would choose to focus on its "purely mechanical" aspects?

A more recent experiment demonstrating a similar phonetic illusion was carried out by Dupoux and colleagues (1999). When Japanese speakers were played stimuli of the form VCCV, they reported hearing "illusory vowels" in the middle of the consonant cluster. That is, they perceived the sequence [ebzo] as [ebuzo]. Dupoux et al. argue that because the rules of Japanese forbid consonant clusters, listeners mentally transformed the illicit sequence into a licit one. French listeners, whose language allows clusters, did not experience the illusion. Dupoux et al. reach a similar conclusion to that of Sapir: "speech perception is heavily influenced by phonotactic knowledge" (p. 1577). An important difference, however, is that Dupoux and colleagues are working within a much more sophisticated phonetic theory. While the role of language structure in shaping the perception of phonetic substance, as well as the role of perception in shaping language structure, remain active areas of research (see especially Chapter 10), the idea that phonetics is automatic and unimportant has been shown to be false.

2.3 PHONETICS AS LINGUISTICS: MODULARITY

Even as the structuralists were arguing for the crucial role in phonology of system, contrast, and mental representation, and for the peripheral role of phonetics, phoneticians were carrying on important and interesting work. Of course, phonetic description, especially of articulation, has been going on for millennia, as in the work of Pāṇini and the "Indian Grammarians" in 500 BCE, or the anatomist Eustachius in Renaissance Italy. Throughout the nineteenth century CE the work of missionary linguists devising scripts and grammars for Bible translation expanded, as did the sets of meticulous sound correspondences being worked out by historical linguists. But the end of the nineteenth century brought a number of new inventions that moved the science of phonetics rapidly forward. The invention of the phonograph (1877) and the magnetic recorder (1898) allowed the ephemeral sounds of speech to be captured and replayed again and again for careful study. Other devices allowed speech vibrations to be made visible. The kymograph ("wave-writer," invented in the 1840s and adapted for speech around 1910 (Collins and Mees 1999)) worked on the same principal as the phonograph, using a stylus to etch patterns of vibration into a medium, except that the kymograph etched visible lines onto smoked paper, creating an image from which durations could be measured. The X-ray (1895) made visible the exact positions of the full tongue body inside the mouth. Phoneticians such as Daniel Jones at University College London organized the first phonetics laboratories in the early years of the twentieth century (Figure 2.2), and started publishing descriptions supported by X-ray images and quantitative measurements on the detailed pronunciation of numerous different languages. Jones's monograph, *The Pronunciation of English*, was published in 1909, followed soon after by phonetic descriptions of Albanian (1910), Cantonese (1911), Mono (1912), Manx (1914), Setswana (with Plaatje, 1916), Sinhalese (with Perera, 1919), Russian (with Trofimov, 1923), Korean

Figure 2.2 Daniel Jones using a kymograph, c.1911
Source: Collins and Mees (1999: 256).

(1924), Gã (1925), and more. Later in the century, the oscilloscope and spectrograph made possible spectral analysis and the quantitative representation of sound quality.

One thing that quickly became apparent with the beginnings of detailed phonetic measurement, coupled with the practical need for devising orthographies for previously-unwritten languages (or languages written in scripts most European linguists couldn't read), was that phoneticians could not comply with Trubetzkoy's demand for "the complete exclusion of any reference to the lexical meaning of the sound complexes under study." Once it became possible to describe speech sounds in great detail, the phonetician had to choose which details to represent in a transcription. The only reasonable choice would be to represent the details that mattered in making speech comprehensible: the differences that distinguish word meanings.

Alexander Melville Bell (whose father and son were also named Alexander and were also speech scientists, the son being the most famous one) published the first "universal alphabet" that could represent any speech sound in any language: "Visible Speech" (Bell 1867; described in more detail in Chapter 3). Bell recognized that he could never represent all the "infinitesimal differences" (1867: 15) between sounds, but by concentrating on the more limited "elementary relations" that distinguished the meanings of words, a manageable set of symbols could be devised. Henry Sweet's (1877) *Handbook of Phonetics*, which laid the foundation for the science of phonetics as it is understood today, with an emphasis on accurately describing the pronunciation of the sounds of all languages as they are actually spoken, says (p. 86):

> [O]ur ordinary word division is a logical and not a phonetic analysis. No amount of study of the sounds only of a sentence will enable us to recognise the

> individual words of which it consists. We may write down every sound, every shade of synthesis, but we shall never be able to analyse it into separate words till we know its meaning.

Lexical meaning could not be excluded from phonetic study.

A simpler and more successful universal phonetic alphabet (the International Phonetic Alphabet, or IPA) was proposed by the International Phonetic Association in 1888, with a revised edition published in 1912 (again, see Chapter 3 for details). Among the foundational "principles" defining the IPA was the admonition that "it is in general sufficient to represent the distinctive sounds only" (#65; 1912: 15), where "distinctive sounds" were defined as "those which if confused might conceivably alter the meanings of words" (#70; p. 16). In 1950, in fact, Daniel Jones published a book, *The Phoneme: Its Nature and Use.* Though clearly dealing with phonological principles, Jones did stick closely to the idea of the phoneme as a label for a group of actual, pronounceable, sounds, which he described as a "family" (p. 7); and he took a dim view of "mentalistic" and "functional" approaches, such as those taken by Baudouin de Courtenay, Sapir and Trubetzkoy, that saw the phoneme as an abstract, non-physical ideal (Jones 1950: 212–17).

From the other side, by the middle of the twentieth century, phonologists were finding the new phonetic data to be more and more relevant to patterns of phonological structure. Trubetzkoy's colleague Roman Jakobson, and Jakobson's student Morris Halle, pursued the idea that the all-important "differences" and "relationships" between contrastive sounds were far from "incorporeal," as Saussure had claimed, but were in fact instantiated in a limited set of phonetic parameters. In 1952, Jakobson and Halle, together with Gunnar Fant, a speech scientist who was trained as an electrical engineer, published a treatise, *Preliminaries to Speech Analysis: The Distinctive Features and Their Correlates*, reducing Trubetzkoy's complex webs of correspondences to a list of twelve binary features with explicit acoustic and articulatory definitions, and including a tutorial on how to use a sound spectrograph and interpret speech spectrograms (Jakobson et al. 1952; see Chapter 4 for more details).

In an address entitled "Why and how do we study the sounds of speech?" Halle (1954: 18) elaborated on the relation between phonetics and phonology:

> It is my purpose to show that a sharp distinction between phonetics and phonemics cannot usefully be maintained; that phonetics actually makes use of considerations which are strictly phonemic; that physical criteria are an integral part of phonemics; and that a description of language of any level, from phonetics to stylistics, cannot be properly evaluated without considering its consequences on all other levels.

Phoneticians, Halle argued, cannot deal with speech as "a physical phenomenon pure and simple" (1954: 76), but have to take account of the purpose of speech—conveying information—in deciding which aspects of the signal to encode/represent. Similarly, phonologists don't just ignore phonetic substance; for example, they require that allophones should be phonetically similar. Halle writes (1954: 80):

> A science of phonetics that disregards the social function of language is as trivial and barren as a science of phonemics that refuses to take into consideration the material aspect of language, its sounds. Neither phonetics, nor phonemics can get along without the other: phoneticians must have an understanding of phonemics, phonemicists must understand the methods of phonetics.

We have to judge our analyses of language as whole, Halle argued, and resist the temptation to radically simplify one area at the expense of radically complicating another. Syntax, morphology, phonology, and phonetics should all work together in a coherent system.

In *The Sound Pattern of Russian* (1959) and then *The Sound Pattern of English* (SPE; 1968) co-authored with his MIT colleague Noam Chomsky, Halle set out to provide a comprehensive account of the "sound patterns" of these two languages, from abstract mental representation to actual pronunciation. In these publications (together with Chomsky's publications in syntax, e.g., Chomsky 1957), Halle and Chomsky broke with the structuralist idea of strict separation of levels, and instead proposed a series of step-by-step "transformations" that convert one level into the next. (See further extensive discussion in Chapter 5.) In so doing, Halle and Chomsky introduced new ways of thinking about what a formal "grammar" should look like, but many structuralist ideas remained. The phonological representation was still a cognitive construct, à la Saussure (in fact à la Descartes according to Chomsky (1966), *Cartesian Linguistics*), and phonetics was still considered to be outside linguistics.

In the SPE model, phonology begins with an abstract underlying representation, in which lexical items that are represented as partly-specified matrices of binary features (mostly following Jakobson et al. 1952) are inserted into the structural description provided by the syntax. Then transformational rules, applying step by step like a computer program (Halle 1959: 71), insert or change features, or permute segments, until the surface structure, in which each segment of an utterance is fully specified for all of the features that will define its pronunciation, is obtained. As a final step in the derivation, the + and - values of the binary features are replaced, according to language-specific rules, with numbers that represent a scalar realization (Chomsky and Halle 1968: 165–70). For example, while /b/ would be [+voice] in both Russian and English, it might be [8 voice] in Russian and [2 voice] in English, capturing the stronger vocal fold vibration typical of Russian voiced stops. Because the set of distinctive features is claimed to represent all of the "independently controllable" (p. 298) phonetic dimensions of language, the specification of features with numbers sufficed to provide all of the language-particular, and thus phonological or linguistic, information required for speech. The actual phonetic realization, when the numerically-specified features are realized in actual articulator movement, was assumed to be universal, automatic, mechanical, and therefore of no linguistic interest.

Thus, while Chomsky and Halle's transformational grammar created a new formalism for representing phonological contrast and alternation, and new ways of thinking about the relationships between syntax, morphology and phonology, the SPE conception of the place of phonetics was straight out of Saussure and Trubetzkoy.

> But even if the phonetic transcription were as faithful a record of speech as one could desire, there is still some question whether such a record would be of much interest to linguists, who are primarily concerned with the structure of language rather than with the acoustics and physiology of speech. (Chomsky and Halle 1968: 293)

That is, acoustics and physiology might be important starting points (in defining the distinctive features) and eventual endpoints (as the features were realized), but all of the interesting work on the structure of language came in-between, where the phonetic content of features was completely irrelevant. In fact, Chomsky and Halle themselves point out (1968: 400) that if every feature value in their account of English were reversed from plus to minus or vice versa, the result would be, if not English, still a perfectly well-formed language according to the formal rules they set out. While they are apparently troubled by this fact, and agree that a more "satisfactory" account would not have this property, the problem does not actually cause them to change their analysis, and the fact that this drawback isn't even mentioned until page 400 is telling. In the end, however, other linguists over the next fifty years would pursue more aggressively the problem of SPE being "overly formal," and many would end up discarding the derivational theory of phonology for others—notably Optimality Theory (Prince and Smolensky 2004), but others as well—in which the phonetic content of features and the phonetic conditioning of alternations would play a much more central role. (See Chapter 6 for further discussion, in addition to the summary below.)

An important next step was the assertion, coming from the increasingly sophisticated research of phoneticians in the tradition of Daniel Jones, that phonetics was *not* mechanical or universal, and that simple numerical specification of feature values was not enough to determine actual articulatory output. Phonetic implementation, too, had to be learned and language-specific, and thus, contra Saussure, part of linguistics. A important reference here is Keating (1985), who showed that even small phonetic details, such as intrinsic vowel length, which was assumed to be an automatic mechanical consequence of differences in vowel height, were language specific. The point was made even more dramatically in Kingston and Diehl (1994), who argued for active speaker control of all the small articulatory adjustments needed to maintain and enhance vocal fold vibration in voiced stops. Keating concludes (1985: 128):

> I am suggesting here that we consider all phonetic processes, even the most low-level, to be phonologized (or grammaticized) in the sense that they are cognitively represented, under explicit control by the speaker, and once-removed from (that is, not automatic consequences of) the physical speaking machine.

If both phonology and phonetics are learned, language-specific and cognitively represented, then the dividing line is not between linguistics and non-linguistics, but between different parts of linguistics.

Following Fodor (1983, *The Modularity of Mind*), phonology and phonetics could be conceived of as different *modules* of linguistics—equal in importance, but each dealing with separate phenomena, with different fundamental concepts and operations. In SPE-type theories of linguistics, the modules were linked over the course of a derivation: the output of the syntax was the input to the phonology, and the output of the phonology was the input to the phonetics. This was essentially the consensus view by the end of the twentieth century.

In a modular approach to phonetics and phonology, a basic definitional question then becomes, "Where is the dividing line?" What characteristics define the phonological module vs the phonetic module? To return to the metaphors proposed by Scobbie (2007), discussed in Chapter 1, in a modular theory the interface between phonology and phonetics is a fence or boundary, and every fact about sound patterns needs to be placed on one side or the other, depending on how its characteristics match up with the essential properties of that module. Delineating the definitional properties of phonology and phonetics, and discovering the exact characteristics of particular alternations or rules, become important research questions.

By 1990, consensus (e.g., Pierrehumbert 1990b; Keating 1990a, 1990b; Cohn 1990; Zsiga 1993) had settled on the criteria in (1) in Chapter 1. A 1990 special issue of the *Journal of Phonetics* on "phonological and phonetic representation" specifically asked for contributions on the topic. Pierrehumbert's contribution to that volume lays out clear criteria: phonology is qualitative and mental, concerned with the sound/meaning correspondence. Phonological rules are formal/syntactic/algebraic, and accessible to introspection. In contrast, phonetics deals with data that is measurable and physical, described by physics and continuous mathematics, and is not accessible to introspection. While this sounds a lot like the distinction that was originally made by Saussure and Baudouin de Courtenay, an important difference is that Pierrehumbert argues that phonetic representations also involve a type of abstraction, as when we model the vocal tract as a hard-walled tube or interpret formants from a spectrogram, and crucially, that both phonology and phonetics are indispensable parts of linguistics. To say that one or the other is the whole story, Pierrehumbert argues, amounts to "intellectual imperialism." The result of a century of phonetic work was to establish phonetics as a module of linguistics equally important to the phonological module, but in the consensus view, still separate from it.

2.4 PHONOLOGY WITHOUT PHONETICS AND PHONETICS WITHOUT PHONOLOGY

Some researchers have taken the dividing line between phonology and phonetics to be absolute, and have argued that while phonetics may be interesting, it still has nothing to do with phonology per se. In the 1950s, just as Halle and colleagues were arguing for a greater role for phonetics in linguistics, the Danish linguist Louis Hjelmslev was arguing against any role for linguistic phonetics at all. Hjelmslev pushed the phonology/phonetics dichotomy to its most logical extreme in the theory of Glossematics, which he laid out in the (dense, nearly impenetrable) *Prolegomena to a Theory of Language*, first published in Danish in 1943 and in English in 1953. The term "prolegomena" refers to a preface that deals with methodological issues that have to be settled before the real work can begin. Hjelmslev's prolegomena focused on the procedure that should be used in linguistic investigation. Hjelmslev argued that, far from forming the basic elements of linguistic theory, phonetic "substance" should be completely irrelevant. Like Saussure who argued that a "knight" was a knight only by virtue of its role in the game of chess, Hjelmslev argued in pure structuralist fashion that his basic unit, the glosseme, was defined entirely by virtue of its relationships to other glossemes: "A totality," Hjelmslev argued, "does not consist of things but of relationships . . . not substance but only its internal and external relationships have scientific existence" (1953: 14). We must

> recognize that language is a form and that outside that form . . . is present a nonlinguistic stuff, the so-called substance. While it is the business of linguistics to analyze the linguistic form, it will just as inevitably fall to the lot of other sciences to analyze the substance. (1953: 49)

Again, phonetics is defined as being outside linguistics. Thus, Hjelmslev argued, linguistic analysis should not start with physical sounds as building blocks which are combined into words and words into sentences. Rather, a linguistic analysis should proceed in a completely top-down manner. Beginning with an unanalyzed text, and using logical procedures including substitution, mutation, commutation, and so on, the text should be broken down into smaller and smaller parts, until an unanalyzable element (the glosseme) is reached. Linguistic analyses were to be judged in terms of consistency, exhaustiveness, and most importantly, simplicity. Linguistics was to be "an algebra of language, operating with unnamed entities, i.e., arbitrarily named entities without natural designation" (p. 50). If, in the end, the glosseme turned out to be equivalent to a phonetic segment (as was the case more often than not), this was purely accidental.

A modern counterpart to Hjelmslev's theory is the approach of "Substance-free Phonology" (Hale and Reiss 2008; Reiss 2018). Hale and Reiss argue that phonology should be completely formal: "all form and no substance" (2008: 162), a computational system operating solely on arbitrary symbols (p. 167). That is, even Chomsky and Halle were too phonetic. Everything about language that can be explained in other ways, such as perceptual salience, ease of articulation, historical change, or general learning mechanisms should be so explained, leaving true phonological theory to deal only with symbolic, cognitive manipulations. (See also Anderson 1981, and the discussion in Chapter 6.) The problem with such an approach, other linguists would argue, is that when all the explaining from other domains is done, there is no phonology left.

This argument, that symbolic phonology adds no value to explanations from other domains, is made by John Ohala in his 1990 article, "There is no interface between phonology and phonetics," which appeared in the same themed issue of the *Journal of Phonetics* as Pierrehumbert's article discussed above. Other contributors to that volume argued that Saussure and the structuralists were wrong in underestimating the linguistic value of phonetics. Ohala argued that the structuralists were wrong in conceptualizing phonology as separate from phonetics at all: there is no interface between phonology and phonetics because there is no separate formal phonological module.

Ohala argues that there are two different ways to define phonology and phonetics as separate modules: either as focusing on two separate objects of study, or as two different ways of studying the same object. Both views, he argues, are incorrect. Phonetics cannot be pursued independent of phonology, because (following the argument of Halle from 1954) how speech is implemented crucially depends on conveying lexical and discourse meaning, and how speech is perceived depends on comparing the signal to stored patterns in memory. Nor can phonology be independent of phonetics, because the structures and patterns of the phonology "are what they are largely due to the physical and physiological structure of the speech mechanism" (p. 155). Doing phonology without phonetics, Ohala argued, is like doing medicine without physiology and chemistry.

Continuing in the tradition of Trubetzkoy's barb that linguists who insist on phonetics in phonology are guilty of "inertia, mental lethargy, and stubborn rejection of any new thought" (1969: 6), Ohala argues that the proponents of "autonomous phonology" are guilty of circular reasoning, reification, projection, and myopia. Such phonologists are guilty of circular reasoning because they make up phonological "constraints" that simply restate phonetic facts, and then treat the constraints as explanations for the facts; guilty of reification because they treat notations such as features as "real entities," and then treat these "elements of the notation for phonological events as somehow having more interest than the events themselves" (1990: 162); guilty of projection because they assume that the linguist's knowledge is necessarily present in the mind of a speaker; and guilty of myopia because a focus on symbols and abstractions causes them to miss real physical explanations. The questions asked by phonology, on typology, alternation, contrast, and the structure of language, are indeed interesting questions, Ohala states, but the answers are to be found in the empirically-testable, physical science of phonetics. Trubetzkoy (perhaps handicapped by the under-developed techniques available at the time he was writing) was wrong in thinking that the methods of natural science can't be used to answer the questions of phonology. Turning the argument of Saussure, Trubetzkoy and Sapir on its head, Ohala argues that phonology as a separate module is worthless to real linguistic study, and that the only fruitful way to study linguistic sound patterns is to study phonetics.

This argument for phonology as a natural science, albeit in less strident form, is found in Pierrehumbert et al. (2000), "Conceptual foundations of phonology as a laboratory science," the title of which speaks for itself. Modern phonologists, the authors argue, have "abandoned the doctrine of dualism" (p. 274): the assumption that the mind is separate from the body, and thus that phonology must to be separate from phonetics. "The modularization of phonetics and phonology that was still assumed by most laboratory phonologists up through the early 1990's is no longer universally accepted, and we ourselves believe that the cutting edge of research has moved beyond it" (pp. 284–5). Because the study of speech articulation and perception requires reference to meaning and categories (as Halle argued in 1954), and because the study of contrast and alternation requires reference to physical dimensions, sometimes in great detail, "there is no place the line [between phonology and phonetics] can be cogently drawn" (p. 287). With Ohala, the authors believe that categories, inventories, alternations, and other manifestations of linguistic/phonological structure should be studied, but with the quantitative tools of phonetics, not the traditional phonological methods of "symbolic records of morphological alternations" (p. 274). Phonetic representations may be quantitative and gradient, but mathematics is also a type of abstraction and formality, and quantitative representations are not thereby automatically unsuited to represent cognition. Mathematical formalism and gradient representations are in fact more accurate than representing speech as a string of symbols.

As Keating and others who studied the phonetic implementation of features and "surface structure" found, Pierrehumbert et al. point out that even the detailed symbols of narrow IPA transcription cannot be equated across languages. Categories, the authors argue, arise out of discontinuities or non-linearities in acoustics and articulation (see the discussion of features in Chapters 4 and 6). These non-linear relationships should be studied directly: nothing is gained by imposing a completely separate system of features or symbols on top of or instead of the physical dimensions. For example, we can study the independent movement of lips, tongue tip, and tongue body; nothing further is gained by writing algebraic rules manipulating features like [labial], [coronal], and [dorsal]. In fact, Pierrehumbert et al. argue (as Ohala did) that focusing on the algebraic representation obscures the real phonetic processes that give rise to categorical alternations. Instead of assuming two separate systems, there is only one system, where categories arise out of discontinuities, and linguistic knowledge is both fine-grained and categorical.

The view of Pierrehumbert et al., however, should be contrasted with that of Smolensky and Legendre (2006), which was quoted at the beginning of Chapter 1. They and others (e.g., Berent 2013; Reiss 2018) continue to argue that the best way to represent cognition is with discrete, symbolic, algebraic computation, not continuous mathematics. There is also the case made by Myers (2000), that trying to solve both phonological and phonetic problems with one big theory just leads to a muddled theory. Instead, each module will solve its problems more easily if the two are *not* conflated.

Thus we have two ends of the continuum represented in current theories: Hale and Reiss who argue that phonology should be completely algebraic, with no phonetic content, and Pierrehumbert et al. who argue that phonology should be completely free of algebraic symbolism, and couched completely in terms of phonetic content. Most current theories lie somewhere in between the two: there is some algebra, and some phonetic content. Phonology is not free of phonetics, but cannot be subsumed by phonetics either. Thus the need for this book. There is still a phonology/phonetics interface. But what does it look like?

Chapter 3 begins the inquiry by looking at the problem of segmentation, and the issue of how to represent the "building blocks" of phonological and phonetic structure.

RECOMMENDED READING

Overviews

Anderson, S. R. (1985a), *Phonology in the Twentieth Century: Theories of Rules and Theories of Representations.* Chicago: University of Chicago Press.

- Choose one of the influential linguists mentioned in this chapter (Ferdinand de Saussure, Jan Baudouin de Courtenay, Nikolai Trubetzkoy, Louis Hjelmslev, Edward Sapir or Morris Halle) and learn more about their life, research, and contributions.

Myers, S. (2000), "Boundary disputes: the distinction between phonetic and phonological sound patterns," in N. Burton-Roberts, P. Carr, and G. Docherty (eds), *Phonological Knowledge: Conceptual and Empirical Issues.* Oxford: Oxford University Press, pp. 245–72.

- What are Myers' arguments in favor of modularity?

Exemplary research

Something old

Sapir, E. [1933] (1949), "The psychological reality of phonemes," in D. G. Mandelbaum (ed.), *Selected Writings of Edward Sapir in Language, Culture, and Personality.* Berkeley and Los Angeles: University of California Press, pp. 45–60.

- Which of Sapir's phonetic illusions (if any) do you find most convincing? Can you think of others?

Something new

Dupoux, E., K. Kakehi, Y. Hirose, C. Pallier, and J. Mehler (1999), "Epenthetic vowels in Japanese: a perceptual illusion?" *Journal of Experimental Psychology: Human Perception and Performance, 25*(6): 1568–78.

- Why did the Japanese participants in the study hear vowels that the French participants did not? What does this study tell us about the "psychological reality" of phonology?

Opposing views

Pierrehumbert, J. (1990b), "Phonological and phonetic representation," *Journal of Phonetics, 18*: 375–94.

vs

Pierrehumbert, J., M. E. Beckman, and D. R. Ladd (2000), "Conceptual foundations of phonology as a laboratory science," in N. Burton-Roberts, P. Carr, and G. Docherty (eds), *Phonological Knowledge: Conceptual and Empirical Issues.* Oxford: Oxford University Press, pp. 273–303.

- How did Pierrehumbert's views change between the two publications?

Reiss, C. (2018), "Substance-free Phonology," in S. J. Hannahs and A. R. K. Bosch (eds), *The Routledge Handbook of Phonological Theory.* London: Routledge, pp. 425–52.

vs

Ohala, J. J. (1990), "There is no interface between phonetics and phonology," *Journal of Phonetics, 18*: 153–71.

- Imagine that you have invited Reiss and Ohala to debate the importance of symbolic phonology. What arguments and counterarguments would each offer?

QUESTIONS FOR FURTHER DISCUSSION

1. Return to Figure 2.1, Saussure's "speech circuit." What were Saussure's reasons for deciding that only Stages 1 and 5 were true linguistics? Keep this figure in mind as you continue reading, as it will come up throughout the book.
2. How did the invention of news ways of recording and analyzing speech at the beginning of the twentieth century change how phonologists thought about phonetics, and how phoneticians thought about phonology? Do you think that any technological inventions at the beginning of the twenty-first century might change the field in similar ways? How closely wedded are theory and technology?
3. Both Baudouin de Courtenay and Trubetzkoy ran afoul of authorities because their descriptive linguistic work was seen as advocating for minority rights. To what extent does linguistic research in and of itself legitimize minority languages? What responsibilities do linguists have toward the groups whose languages and dialects we describe?
4. In addition to the biographies of the men discussed by Anderson (1985a), research the scientific contributions of one or more of the influential women mentioned in this chapter (Mary Beckman (co-author of Pierrehumbert et al. 2000), Patricia Keating or Janet Pierrehumbert) or Ilse Lehiste, a phonetician featured in Chapter 7. Does it matter for linguistics in the twenty-first century that all the linguists discussed by Anderson (1985a) as having made important contributions to phonology in the twentieth century are white men?

CHAPTER 3

ABCS: SEGMENTS, TRANSCRIPTION, AND THE BUILDING BLOCKS OF INVENTORIES

3.1 THE DELIMITATION OF UNITS

What are the basic units, the "primitives" (Cohn 2011: 17) of human speech? When we speak, when we plan an utterance, what are the building blocks we reach for to build it out of? When languages choose an inventory of contrasts from which to build a vocabulary, what is the stock from which they draw?

Saussure, as he was defining modern linguistics at the turn of the twentieth century, wrote that language was, at bottom, a system of "signs," pairings of a chunk of sound with a chunk of meaning. Before this pairing takes place, he wrote, both thought and sound are chaotic, indistinct, and undifferentiated. It is the creation of the linguistic sign that "subdivides" and "orders" each plane, by pairing some distinct "phonic material" with some distinct meaning, thus bringing about "the reciprocal delimitation of units" (1959: 112). He illustrates the idea with the diagram in Figure 3.1.

Language is neither thought nor sound, but the system that links them: "language works out its units while taking shape between two shapeless masses" (1959: 112).

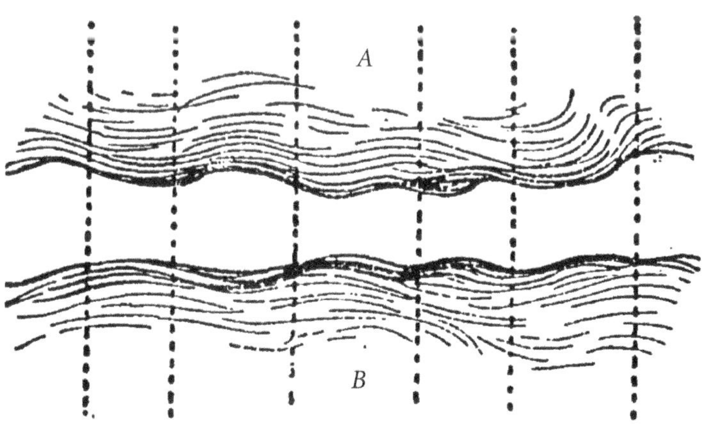

Figure 3.1 Language is "a series of contiguous subdivisions marked off on both the indefinite plane of jumbled ideas (A) and the equally vague plane of sounds (B)"
Source: Saussure (1959: 112).

Phonology and phonetics must deal with the question of how the "phonic material" associated with meanings can be characterized.

All phonology works on the premise that words are not "vocal wholes" that have no subcomponents. Hockett (1960) defined the pairing of sound and meaning with the term "duality of patterning": any utterance can be analyzed simultaneously as a string of meaningful morphemes and as a string of meaningless sounds. "The meaningful elements in any language—'words' in everyday parlance, 'morphemes' to the linguist—constitute an enormous stock. Yet they are represented by small arrangements of a relatively very small stock of distinguishable sounds which are themselves wholly meaningless" (Hockett 1960: 90). Hockett took duality of patterning to be a "basic design feature" of human language, along with arbitrariness, discreteness, productivity, and prevarication, among others.

What, then, are the units that languages use to pair sound with meaning? Of what does this "very small stock" consist? Hockett describes the stock as being composed of strings of "distinguishable sounds." Like Hockett, those of us who were raised on alphabetic writing systems take the segmental nature of speech for granted: just as written words and sentences are made up of sequences of letters, so spoken utterances are assumed to be made up of sequences of sounds to which the letters more or less correspond. Phonologists and phoneticians alike rely on segment-based descriptions of language sound systems, and represent spoken language using letter-based systems of phonetic transcription. Statements like "Hawai'ian has five vowels [i, e, a, o, u] and eight consonants [p, k, s, h, m, n, w, ʔ]" or "the English word [sprɛd] begins with three consonants" are ubiquitous. Hockett took the existence of "strings of sounds" as obvious and axiomatic. Yet phonetic study quickly shows that "segmenting" the acoustic record is not obvious. Saussure's diagram notwithstanding, it is impossible to draw exact lines in a waveform indicating where one segment ends and another begins. In both articulation and acoustics, characteristics of segment sequences overlap and blend. Given this difficultly, can we be so sure that the segment is the basic unit of both phonology and phonetics? Could it be something larger (mora, syllable, or word), smaller (distinctive feature), or different (articulatory gesture)? This chapter considers the evidence.

We begin (Section 3.2) with a discussion of the segment as a phonetic and phonological unit, which leads to a discussion of orthography and alphabetic writing in general in Section 3.3. Writing systems are not always considered to be in the domain of either phonology or phonetics, and may be more or less closely based on spoken language, but writing systems are symbolic representations of inventories and are thus implicit theories of what the basic units of a language are. Section 3.4 turns to orthographies developed for phonetic transcription, including Alexander Melville Bell's "Visible Speech" and the extremely useful but theoretically problematic International Phonetic Alphabet. Does the usefulness of segmental transcription prove that the segment is the basic unit of Language, or do the problems with the IPA prove that the segment is not basic? Section 3.5 changes the focus somewhat, looking not at linear strings, but at the problem of selecting the set of segments for any given language inventory, briefly discussing Dispersion Theory and Quantal Theory. Section 3.6 then begins the consideration of subsegmental "parameters" or features as basic units, a discussion continued in Chapter 4. Concluding the chapter, Section 3.7 considers the question of how all of the previously-described principles might apply to non-spoken signed languages, which also show duality of patterning. What insight do signed languages give us about what the basic units of Language might be?

3.2 SEGMENTATION

As has often been pointed out, due to coarticulation and gestural overlap, it is impossible to perfectly "segment" the stream of speech by indicating exact timepoints where one segment begins and the previous one ends. Thus segmentation is one of the first and most basic problems confronting the phonology/phonetics interface. We assume discrete and categorical segments, but how can we identify them in the continuous acoustic record?

Ladd (2014, in press) emphasizes that reliance on segmental transcription has been a constant through a century of change in phonological theory: theories from the Prague School through SPE to various contemporary versions of Optimality Theory "[take] for granted the scientific validity of a segment-based idealized phonetic representation of speech" (in press: 7) and "all assume that the primary phonetic data can be expressed in terms of a segmented symbolic representation" (in press: 8). Bloch (1948: 13), for example, in laying out his "postulates for phonemic analysis" states in Postulate 12.4 that "Every utterance consists wholly of segments." Ladd (2014: 31–7) further points out that the structuralist ideas of phoneme and allophone presuppose the "phone" whose distribution can be analyzed (e.g., in Bloch 1948 and Hockett 1958), and that most generative analyses, whether rule-based or constraint-based, assume a phonological "output" that consists of a segmental string.

The problem with assuming that segments are obvious, however, can be illustrated with the simplest of utterances, such as [apa], as shown in Figure 3.2. The figure shows the waveform, spectrogram, and a proposed segmentation. The lines are not drawn randomly, and the acoustic representation is not amorphous. Discontinuities can be marked. The vowels show high amplitude, periodicity, and complex resonance structure. The consonant shows low amplitude, lack of periodicity, and no resonance structure. But are the lines in exactly the right place? The discontinuities do not all exactly line up. The voicing of the vowel dies out gradually into the voiceless closure. At exactly what point does vowel end and consonant begin? The voicelessness of the aspiration is a characteristic of [p], but the open vocal tract is a characteristic of [a]. To which segment does the period of aspiration belong? Should the utterance be transcribed as [apʰa], [apha], or perhaps [apḁa]? The answer cannot be simply read off the acoustic record, but requires a phonological analysis to determine the units, as sample problems presented for students in both Hockett (1958) and Kenstowicz and Kisseberth (1977) attest.

The problem only gets worse if one considers perception rather than acoustics. During the voiceless closure for the [p], which is where we might want to say the real consonant segment resides, the hearer perceives only silence. The labiality of the

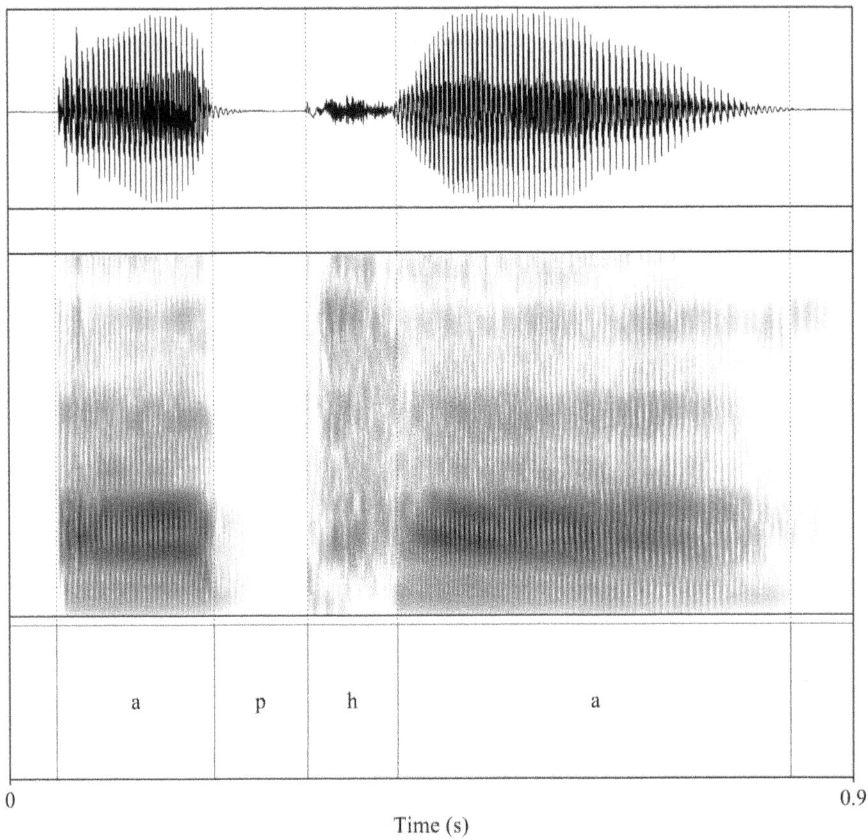

Figure 3.2 Waveform, spectrogram, and segmentation for the utterance "apa"

consonant is actually perceived in the formant transitions into and out of the closure, caused by the gradual closing of the lips, during the periods Figure 3.2 delineates as belonging to the vowel.

Yet, even if the edges are not crisp, phonologists have assumed that evidence for segments resides in perception of the (relatively) steady states. Bloch (1948: 12) writes in Postulate 11 that discrete segments can still be perceived in speech even if they can't be marked off in a spectrogram.

> Phoneticians have long known that the movements of the vocal organs from one "position" to another proceed by continuous, uninterrupted flux . . . Such instrumental data, however, need not be taken as evidence that speech AS PERCEIVED cannot be segmented; every phonetician has had the experience of breaking up the smooth flow of speech into perceptibly discrete successive parts. In Postulate 11 we do not imply that the vocal organs assume static positions or move in unidirectional ways at constant acceleration; rather, we imply that a phonetically trained observer can interpret the auditory fractions of an

> utterance in terms of articulations that seem (to his perception) to be static or unidirectional. (Emphasis original)

Some more recent models of speech perception, notably Stevens (2002), also assume that steady states, interspersed with rapid transitions, serve as "landmarks" that listeners recognize as signaling segmental units. (See further discussion in Chapter 10.)

Another argument for the existence of segments is that phonology needs them. Ohala (1992) argues that segmental organization is crucial to the creation and maintenance of phonological contrast, and that language must have evolved to have segmental structure. Segmental organization, he argues, provides the necessary temporal coordination of features to one another in order to efficiently build contrastive units, and the alternation of steady states and transitions that segments provide make the different parts of the signal more readily perceptible. Ladd (2014), as quoted above, reminds us that most phonological theories are premised on segmental structure. As Anderson (1974: 6) puts it,

> The only justification [for positing segments in the continuous speech stream] that can be given is a pragmatic one; such a description, based on a segmental structure imposed on the event by the analyst, has been the basis of virtually every result of note that has ever been obtained in the field of linguistic phonetics or phonology.

That is, if segments didn't exist, we would have had to invent them.

Not everyone agrees with Anderson's conclusion, however. A number of researchers have argued that processes and patterns that have traditionally been modeled in terms of segments, including phonological alternations (Port and Leary 2005; Silverman 2006; Golston and Kehrien 2015), speech errors (Mowrey and MacKay 1990; Pouplier 2003; Brown 2004; Pouplier and Hardcastle 2005), and language acquisition and processing (Jusczyk 1992; Cheung et al. 2001; Port and Leary 2005; Golston and Kehrein 2015), can be as well or better modeled in terms of larger units such as onsets or codas, or smaller units such as features or articulatory gestures. (See also Raimy and Cairns 2015a for further examples and arguments both for and against the utility of segmental analyses in phonology.) For example, speech errors in which whole segments seem to move from one word to another, such as "toin coss" for "coin toss" and the famous "tasted the whole worm" for "wasted the whole term," seem to provide clear evidence that words consist of segmental sequences that can be permuted in error (Fromkin 1971). The authors cited above, however, argue that the transposed units could equally well be onset constituents or tongue body gestures.

Throughout this book, we'll see many cases of competing analyses that do and do not assume segmental organization. For example, Chapter 9, Articulatory Phonology, is an extended discussion of a phonological system without segments. In the sections that follow in this chapter, we'll look at some further evidence for the identity of basic units of sound structure: from orthography, from systems of phonetic transcription, and from the ways that inventories are organized. Notwithstanding the claims of Bloch (1948), Ohala (1992), and Anderson (1974) for the centrality of the segment to phonological analysis, and the assumption of segmental structure by most phonologists (Ladd 2014), some have argued (e.g., Read et al. 1986; Silverman 2006) that our assumption that the segment is a basic unit of phonology comes solely from exposure to an alphabetic writing system.

3.3 ORTHOGRAPHIES

3.3.1 MORPHEME, SYLLABLE, SEGMENT

Orthographies are relevant to the phonology/phonetics interface because every orthography is an inventory: a set of symbols that represent the units of the language, whether those units are words/morphemes (attested in Egyptian and Chinese logograms), syllables/moras (attested in numerous syllabaries from Korean to Cherokee), or segments (in the Semitic, or Phoenician/Roman alphabets). The way that people write reflects, albeit imperfectly, an analysis of the structure of language (Berent 2013). As Aronoff (1985: 28) puts it, “Written language is a product of linguistic awareness.” Thus, looking at writing systems provides a window on how languages (and thus Language) “mark[s] off . . . both the indefinite plane of jumbled ideas and the equally vague plane of sounds” (Saussure 1959: 112). Important references on the relation of writing to speech include Gelb (1963), DeFrancis (1989), Coulmas (1989, 1999), Daniels and Bright (1996), Rogers (2004), Sproat (2006), Gnanadesikan (2009), and Sampson (2015). (This section is indebted to these sources, but covers only a few of the major points raised therein.)

Orthographies differ in how close they are to spoken language. In principle, an orthography could be completely “semasiographic,” consisting of symbols like ⚠ and 🛇, which are designed by the International Organization for Standardization (https://www.iso.org) to be comprehensible in any language, and stand for ideas rather than specific words. While the inventory of such symbols is large, no semasiographic system exists that is capable of conveying the full semantic range of human language, including utterances as simple and practical as “Haven’t we met before?” (Try it. Or consider the online games that attempt to write passages of famous novels in emojis.)

In practice, all orthographies are used to represent specific utterances in specific languages, though orthographies differ in how directly they represent pronunciation. Some orthographies are closer to logo-graphic, in which every symbol represents a morpheme; others are more phono-graphic, in which every symbol represents a sound. None is perfectly one or the other, however. Even basically logo-graphic systems, like Chinese, have some phonetic elements, and even the most transparently phono-graphic have some non-phonetic symbols, such as 2 or $. (Exceptions are linguistically-designed phonetic alphabets, discussed below.) English is more toward the phono-graphic end of the spectrum, where most words can be “sounded out” based on spelling, but there are many cases where the pronunciation is unpredictable and thus has to be memorized, and many cases where the reader has to know what morpheme is represented by the spelled word in order to know how to pronounce it.

The earliest orthographies (Egyptian and Sumerian, about 3000 BCE) began as logo-graphic. Essentially, writing began as a series of pictures, a drawing representing each morpheme. Through repeated use, the drawings became more and more simplified and stylized, until the symbols became shapes only vaguely reminiscent of the meaning of the morpheme (a circle for "mouth" or a wavy line for "water"). Pictograms, however, are cumbersome. A language has tens of thousands of words, and it is difficult to draw pictures of every morpheme, particularly very specific items (such as "donkey" vs "horse") or anything abstract ("yesterday," "loyalty"). Pictograms undergo extension, both semantic extension (such as using the pictogram for "foot" to also mean "walk" or "go") and phonetic extension (as in a rebus puzzle, where  stands for "I"). If pictograms develop into a completely stylized and abstract system, the result is morpho-graphic writing, as in Chinese characters. If the phonetic "rebus" use takes over, the result is a syllabary, where each symbol stands for a particular chunk of sound, not a particular meaning.

Of the different writing systems in use around the world, the majority are syllabaries. The name implies that in these systems there is a one-to-one correspondence between syllables and symbols, though there is some question as to whether the unit represented in these systems is in fact a syllable or a mora. According to Poser (1992; see also Sproat 2006; Buckley 2006), it is very rare for a "syllabary" to actually include a separate symbol for every possible syllable, which would number at least in the hundreds for most languages. Rather, a set of a few dozen "core" CV syllables are symbolized, with extra symbols used to extend the inventory, for example to indicate vowel length.

Regardless of whether "syllabaries" are based on syllables or moras, there is little question as to the utility of either the syllable or mora as a phonological unit (as evidenced by allophony, stress systems, poetry, reduplication, language games, the ability of naïve speakers to count syllables, patterns of articulatory coordination, etc.). If we follow the claims of Aronoff (1985) and Berent (2013) that orthographies reflect linguistic awareness, then the fact that syllabaries have been independently invented numerous times is evidence for the "core" CV syllable as a basic natural division in the stream of speech.

On the other hand, the alphabet as we know it was invented only once. According to scholars of the ancient Near East (beginning with Gardiner 1916), sometime around 1700 BCE Semitic speakers, the ancestors of the Phoenicians, borrowed Egyptian symbols to write their own language. In a crucial development, however, the symbols were used acro-phonically: the symbol stands not for the meaning nor for the syllable, but for the first sound in the word (a strategy that is also used by children's puzzle-makers today). Since all the words in this ancient variety of Phoenician began with consonants, only consonants could be written. This kind of writing system, where only the consonants are represented, is known as an "abjad." Modern Arabic and Hebrew orthographies are fundamentally abjads (though vowel diacritics can be added), direct descendants of this earliest Semitic writing system.

The abjad became a true alphabet when the Greeks borrowed writing from the Phoenicians, around 700 BCE. (A syllabic writing system, "Linear B," had been used on the Greek peninsula by the Mycenaeans centuries earlier, but was lost with the collapse of the Mycenaean civilization.) The Greek sound system differed from Phoenician in several important ways. First, Greek syllables, and thus words, can begin with vowels. In Greek, vowels and consonants are equally important in conveying lexical contrast, while in Semitic the consonants carry lexical meaning and the vowels indicate grammatical distinctions. For both these reasons, a consonant-only writing system would be problematic for Greek. Finally, Greek doesn't use the pharyngeal and laryngeal consonants that Semitic does. At some point, an ancient Greek writer may have made a conscious choice to use the otherwise un-needed pharyngeal symbols for the needed vowels, or it may have been a mistake. Sampson (2015: 101) imagines the scenario:

> A Greek sees a Phoenician using a mysterious system of written marks and asks for an explanation. The Phoenician . . . begins, "This mark is called ʔalp—no, not 'alp'—ʔalp, ʔalp, can't you hear, ʔʔʔalp!", while the bewildered Greek perceives only the [alp], and ends up calling the letter "alp^{h}-a" and using it for /a/ since by the acrophonic principle that will now seem to him to be its proper value. This would explain the Greek use of Semitic <ʔ h ħ ʕ> as vowel letters, without the need to attribute any special linguistic sophistication to the first Greek user of the alphabet.

Every truly alphabetic system is descended from the Phoenician/Greek system. The Greek alphabet was borrowed by the Romans, and the Roman alphabet was then adapted, with varying degrees of success (as discussed below), for numerous languages around the world.

One might argue that the one-time development of alphabetic writing argues against the idea of the segment as a basic linguistic unit. If segments are in fact the default basic units of speech, we would expect many different alphabets to have been independently created. Alternatively, one could argue that the popularity of the alphabet, once invented, argues in favor of the segment as a basic unit. Most neutrally, one can assume that orthographies can represent different levels of phonological/morphological structure, which exist independent of writing. Syllabaries exist because syllables are basic units, and alphabets exist because segments are basic units.

3.3.2 ALPHABETS AND PRONUNCIATION

The preceding section has described alphabetic writing as "phono-graphic," which implies that each letter transparently represents a specific sound. That is not necessarily (or usually) the case, however, depending on the structure of the language and the depth of time over which the alphabet has been used. Despite its popularity, the Latin alphabet is not a very good fit for most of the languages of the world.

Because of the high regard for Latin among literate Europeans (who for most of the history of writing would have been a very small minority of the population), there was a belief that the way Latin was written was "correct," and all other languages ought to correspond to it. English, for example, has thirteen or so vowels (depending on the dialect and how you count diphthongs and vowel-r combinations), written with only five symbols. We get around this by using digraphs "ea" "oi" "ai," doubled letters "ee" "oo," or other spelling cues (final "e" makes the vowel long), though these are far from perfect (e.g., "read" as [rid] or [rɛd], and the famous triplet "though," "through," "trough"). English readers rely a lot on memorization and context. English spelling was closer to pronunciation 500 years ago, but mismatches between sound systems and orthographies are a perennial cross-linguistic problem.

The "First Grammarian" was an anonymous scribe writing in Iceland about 1150 CE (so called because his was the first "Grammatical Treatise" of four that were appended to the Prose Edda (Haugen 1950)). The Grammarian praises the idea of each country writing its own history and laws in its own language, but complains at length about the inadequacy of the Roman alphabet to capture the more numerous contrasts of the Icelandic language (which he calls "Danish" but which we now call Old Norse). Through extensive demonstrations with minimal pairs, he argues that Icelandic contrasts thirty-six different vowels—nine different qualities, any one of which can be contrastively nasalized and/or contrastively lengthened—far beyond the power of "a, e, i, o, u" to capture. Almost 900 years later, Trudell and Adger (2015: 17) discuss similar problems in teaching Maasai children to read in the same Latin script that the First Grammarian complained about: "Maasai-language text is reported to be difficult to read, even for native speakers of the language. The principal problem seems to be that the orthography marks neither tone nor four of the nine Maasai vowels."

The only orthography in use today (other than those explicitly designed by linguists) that at least somewhat systematically represents units smaller than the segment is Korean Hangul. This writing system is said to have been created *c.*1500 CE by King Seychong, but was not widely used until the nineteenth century. Up until that time, the educated elite wrote Korean using Chinese characters. In Hangul, the shape of the character to some extent corresponds to the articulatory characteristics of the sound represented.

Some of the Hangul consonants are shown in Figure 3.3, with their IPA equivalents. The labials are based on a square shape that (plausibly) indicates the lips, while the dentals and velars are based on a wedge that could correspond to raising the front and back of the tongue respectively. The most basic symbol is the nasal (though /ŋ/ is an exception, probably because it never occurs syllable-initially). The plain (voiceless lenis) stop then adds a horizontal stroke, the aspirated stop two horizontal strokes. The fortis consonants are indicated by doubling the symbol for the lenis, straightforwardly indicating their extra length. Other consonant and vowel symbols are similarly systematic. However, as with English spelling, the phoneme/grapheme correspondence has grown opaque over time, so Hangul cannot be read as simple phonetic symbols today.

In summary, then, cross-linguistically, orthographies can represent different levels of language: the morpheme, the syllable, the segment, or in the case of Hangul even

ㅁ	ㅂ	ㅍ	ㅃ
[m]	[p]	[p^{h}]	[pˈ]
ㄴ	ㄷ	ㅌ	ㄸ
[n]	[t]	[t^{h}]	[tˈ]
ㅇ	ㄱ	ㅋ	ㄲ
[ŋ]	[k]	[k^{h}]	[kˈ]

Figure 3.3 Some consonants in Hangul (Korean)

a subsegmental articulatory configuration. Orthographies thus can provide evidence that each of these levels is a valid and useful way of dividing the speech stream into discrete re-combinable categories. Yet even the most phono-graphic writing systems are not perfectly so. Because written marks are relatively permanent and because they are strongly influenced by conservative social pressures (like respect for Latin), orthographic conventions change much more slowly than spoken language. Conventionalized (even fossilized) writing systems can serve a purpose of communication across time and across diverse speech communities. But for accurately representing the sounds of actual utterances, a different system, one of explicit phonetic transcription, is needed.

3.4 PHONETIC TRANSCRIPTION

Accurate phonetic transcription is an indispensable tool for recording, describing, and analyzing sound systems. Yet even in a "purely phonetic" transcription, the questions of what the units of representation should be, and what level of detail should be represented, must be addressed. As was noted above and as discussed by Ladd (2014), in the Western tradition both phonologists and phoneticians have assumed that transcription should consist of strings of segments. But segments can be described with more or less detail, and the acoustic and articulatory similarities between segments can be emphasized or ignored. Segments can be represented as simple units (as is for the most part the case with the International Phonetic Alphabet) or as composites of more basic constituents, as was assumed by Alexander Melville Bell's "Visible Speech" (Bell 1867).

Visible Speech was the first orthography that attempted an explicit and universally-applicable phonetic transcription. The work was subtitled *The science of universal alphabetics: Or self-interpreting physiological letters, for the writing of all languages in one alphabet.* Figure 3.4 shows the alphabet as Bell presented

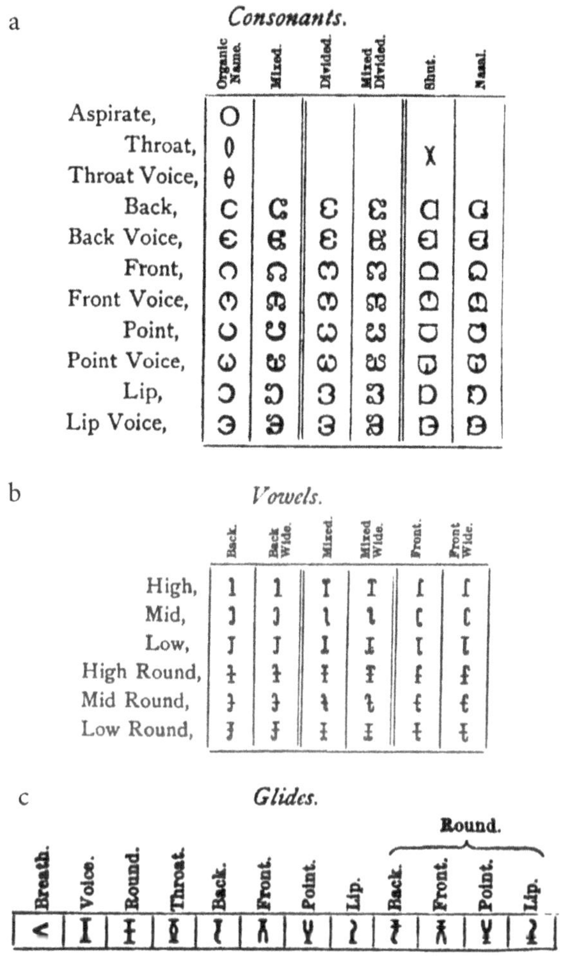

Figure 3.4 *Alexander Melville Bell's "Visible Speech"*
Source: Bell (1867: 37).

it; Figure 3.5 shows modern equivalents for the consonants in the International Phonetic Alphabet.

The letters are "self-interpreting" because each symbol contains explicit instruction for making the sound. As can be seen in Figures 3.4 and 3.5, the basic symbol for most consonants is a rounded horseshoe, with the direction of the opening indicating the place of articulation. Open to the right equals "Back" (a velar), open downward equals "Front" (palatal), upward "Point" (tongue tip), and to the left "Lip" (labial). Without any additional embellishment or diacritic, the horseshoe symbol stands for a voiceless continuant. A line inside the ring adds voicing. A line across the opening indicates a stop, and a curly line, reminiscent of a tilde, indicates a nasal stop. The "mixed" or "modified" consonants allow for intermediate places of articulation such

	Primary	Mixed	Divided	Mixed Divided	Shut	Nasal
Lip	ɸ	ʍ	f	fˠ	p	m̥
Lip voice	β	w	v	vˠ	b	m
Point	r̥	s	l̥	θ	t	n̥
Point voice	r	z	l	ð	d	n
Front	ɕ	ʃ	ʎ̥	ɬ	c	ɲ̊
Front voice	ʝ	ʒ	ʎ	ɮ	ɟ	ɲ
Back	x	xʷ	ʟ̥	ʝ̊ʷ	k	ŋ̊
Back voice	ɣ	ɣʷ	ʟ	ʟʷ	g	ŋ
Throat	breathy				ʔ	
Throat voice	hoarse					
Aspirate	h					

Figure 3.5 "Visible Speech" consonants, with equivalents in the International Phonetic Alphabet

Source: Adapted from omniglot.com/writing/visiblespeech.htm.

as alveopalatal and dental, and the "divided" consonants indicate a "side channel," thus corresponding to laterals for the tongue consonants and labio-dentals for the lip consonants. (Note the velar laterals. Bell confesses, p. 49, that this "back divided" consonant is "perhaps the most difficult of all articulations to unpracticed organs.") For the vowels, Bell's system indicates three degrees of backness—Front, Back, and Mixed (or central)—and six degrees of height—High, Mid, and Low, each of which could further modified as Wide or not. Each of these eighteen vowels could then be either round or unround. Glottal stop (which Bell described as a "cough"), [h], and glides each have their own separate symbols. It is particularly impressive that Bell apparently worked out his system primarily through observation and introspection, without any previous systematic description or phonetic alphabet to rely on.

Bell's goals for his alphabet were very practical—he was primarily interested in teaching Deaf people to speak, and his idea was that reading aloud would be easier if each letter explicitly conveyed the articulatory configuration necessary to make the sound. Bell also predicted that, if popularized, his alphabet would make foreign language pronunciation simple, and would "convert the unlettered millions in all countries into readers" (1867: ix). (He wrote a very indignant preface to his book, criticizing the British government for failing to take him up on his offer of making his system freely available to all if only the government would bear the costs of printing.)

Despite these lofty goals, "Visible Speech" never caught on as a practical orthography (though if you're paying attention you can see it featured in several scenes of Henry Higgins's laboratory in the movie *My Fair Lady*). The system was too cumbersome to print, too different from the alphabets Europeans were used to, and too difficult to learn. It also turned out that Deaf people did not find it helpful. In fact, while Bell was praised during his lifetime for his work, the Deaf community today does not hold the Bell family in high regard (to put it mildly), because the insistence of both father and son on exclusively oral communication was not only unsuccessful, it also prevented Deaf students from becoming proficient in manual sign language. (Signed languages are further discussed in Section 3.7 below.)

Bell's work resulted in practical failure and social harm. What good came of it? Several innovations pioneered by Bell became important to phonetic science, which (as was noted in Chapter 2) was developing rapidly at the end of the nineteenth century. The first was simply the demonstration that a "universal alphabet" was an achievable goal. Henry Sweet, who published *Handbook of Phonetics* in 1877, using a phonetic alphabet of his own, which he called "Romic," wrote (p. viii) that Bell's system "is the first which gives a really adequate and comprehensive view of the whole field of possible sounds . . . applicable to all languages."

Second, as Halle (2009) emphasizes, Bell's system was the first (European) script that expressed the idea that "speech sounds are composite entities." Each sound, and thus each symbol, is a combination of place, manner, voicing, rounding, and nasality.

Third, Bell's work emphasized the need to abstract away from the fine-grained details of shades of articulatory difference and concentrate on differences in sound that produce differences in meaning. As was noted in Chapter 2, in earlier attempts at a phonetic alphabet, focusing on English, Bell was frustrated by the endless variety of possible vowel sounds:

> the plasticity of the organs is so great, that shades of vowel quality are endless, arising from infinitesimal differences in the relative positions of the lips and the tongue. The number of possible varieties can as little be estimated as the number of possible shades of colour. (1867: 15)

However, by concentrating on systematic articulatory differences between contrastive sounds, "the expectation of ultimate success in the construction of a complete Physiological Alphabet, on the principle of Elementary Relations, was now, however, fully entertained" (p. 15). A few decades later a more successful universal alphabet, incorporating the idea of describing sounds in terms of "elementary relations," but based on the familiar letters of the Roman alphabet, was developed by the International Phonetic Association.

As noted above, phonetics was flourishing at the end of the nineteenth century. Phoneticians such as Henry Sweet (1845–1912) and his student Daniel Jones (1881–1967) at University College London sought to put the study of speech on a firmly scientific footing. Their laboratories made full use of new inventions that permitted sounds and articulations to be imaged and permanently recorded. In addition to this increasing technical innovation, there was increasing contact between Europeans and the languages of Africa and Asia, through trade and conquest, but also through missionaries who were hoping to create orthographies and Bible translations. Sweet (1877: Preface page b) records that now that linguists are becoming interested in describing "savage languages" that need to be written down for the first time, they need to be interested in phonetics. Increasing literacy and education in Europe led to a greater need to teach reading, writing, and foreign languages at home as well.

It has to be said at this point that the intellectual and social arrogance of Bell, Sweet, and their contemporaries leaps off the page. They see themselves as "men of science" using their knowledge to save the ignorant, whether the ignorant like it or not, and it falls to us to weigh that attitude and the harm that it did alongside their scientific achievements and the good that resulted. The effects of this Eurocentrism in linguistics have been long-lasting: see the discussions in, for example, Gilmour (2006), Errington (2007), and Zimmerman and Kellermeier-Rehbein (2015). One example is found in the International Phonetic Alphabet, which, while aspiring to be universal, is in its core largely based on Germanic and Romance languages, especially French.

The International Phonetic Association first met in 1886, and first published their International Phonetic Alphabet in 1888. "The principles of the International Phonetic Association," a set of numbered statements that described the alphabet and how it was intended to be used, was published by the association in 1912. Note that in the published principles there is much discussion of the usefulness of phonetic writing, and which symbols should be used for which sounds, but it is taken for granted that an alphabetic system should be the basis of phonetic transcription.

Like Bell's "Visible Speech," the IPA was meant to be practical, a tool for language teaching. The association's first president was Paul Passy (1859–1940), who was both a phonetician and a French language teacher. The association recognized that for "scientific" description a very detailed system of "narrow transcription" was needed, but that such details were not needed for other practical tasks. For most purposes, including language teaching and creating new orthographies, a "broad transcription," including only enough detail as necessary to distinguish words, was sufficient.

Thus, contra the ideas of Saussure on the need for phoneticians to completely avoid any reference to meaning, actual phoneticians recognized that writing down every shade of detail was impossible, and the emphasis had to be on representing differences in sound that created differences in meaning. As stated in the principles of the International Phonetic Association:

> Principle 65: "The general rule for strictly practical phonetic transcription is therefore to leave out everything that is self-evident, and everything that can be explained once for all. In transcribing any given language, it is in general sufficient to represent the distinctive sounds only; for each distinctive sound the typical international [nb: = European] symbol should be chosen; and if necessary, the exact shades of sound used either throughout or in certain positions may be explained (with the use of modifiers) in an introductory note." (1912: 15)
>
> Principle 70: distinctive sounds are "those which if confused might conceivably alter the meanings of words." There is no need to symbolize "shades of sound which are occasioned by proximity to other sounds, absence of stress, and the like." (1912: 16)

As Anderson (1985a) notes, the more phonetic science improved during this time, the clearer it became that most of the detail that phoneticians were able to measure was not relevant to phonology as they understood it. The goal of "practical" transcription is to abstract away from coarticulation and all predictable aspects of speech. However, Anderson also notes that the IPA was designed to accommodate "scientific" description as well: the systematic differences between languages that are neither contrastive within the language nor predictable across languages. For example, because final stops are always released in Georgian, sometimes released in English, and never released in Korean, for the purpose of language description the IPA needs a way to indicate stop release, even if it is never contrastive. The goal, if all systematic aspects of pronunciation are indicated, is to create a "language-neutral phonetic transcription" (Anderson 1974: 8).

Unlike "Visible Speech," the IPA used, insofar as possible, the familiar letters of the Roman alphabet, forgoing Bell's principle of making every symbol "self-interpreting" in favor of simplicity, a strategy that has been successful. But note that there are still some "featural" aspects, where articulation is indicated by an aspect of the symbol, such as hooks (originally under-dots) for retroflex consonants, or the convention of using small caps for uvulars. This is not consistent however: there is no consistent difference between the symbols for voiced and voiceless consonants, for example, or any common visual design element shared by all the labials. Nonetheless, the IPA follows Bell in presenting consonants as organized into a table by place, manner, and voicing, and in so doing, recognizing that speech sounds are "composite entities."

While the IPA has been extremely successful and widely adopted, the goal of creating a phonetic transcription that indicated all non-predictable aspects of pronunciation was never reached. There is no "language-neutral" way of representing all the systematic aspects of pronunciation, since there is no language-neutral way of deciding a priori what is systematic. Even the tiniest details of quality or quantity may be systematic and language-particular, and most transcription is done in the

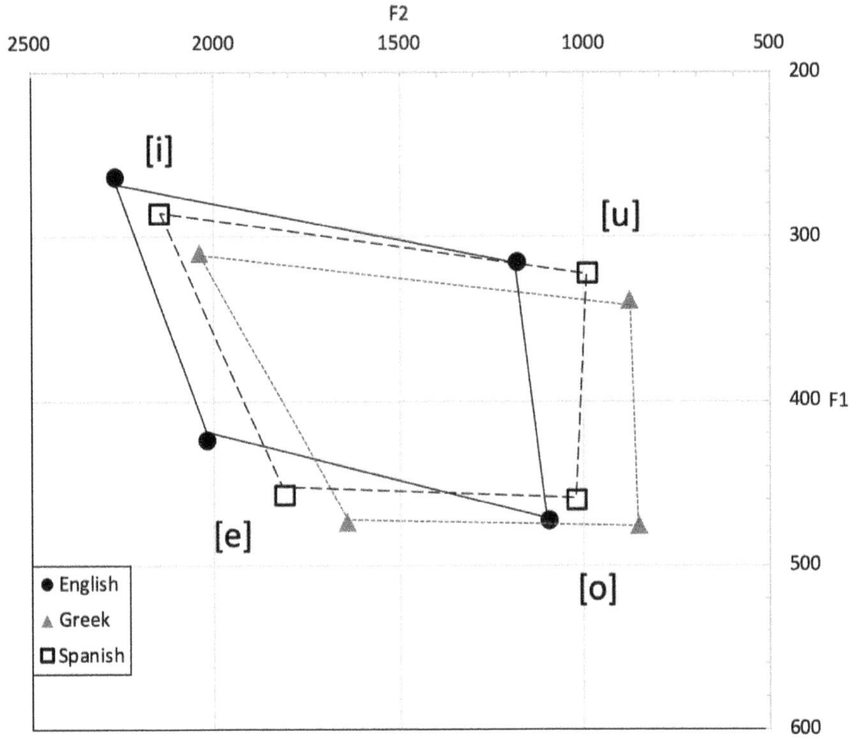

Figure 3.6 The vowel spaces of Greek, Spanish, and English are not identical
Source: Adapted from Bradlow (1993: 57).

absence of phonetic analyses that would reveal these details. (See further discussion of language-specific phonetics in Chapter 5.) In using the IPA, the analyst always had the choice of how much detail to represent, and of how exactly to divide the phonetic space (especially for vowels), and such choices were not consistently made or explained. For example, the actual quality of the front mid vowel of a five vowel-system is quite variable cross-linguistically, and thus the symbol [e] does *not* stand for the exact same sound in different languages. Figure 3.6, adapted from Bradlow (1993: 57), shows that the typical quality of the vowels transcribed /i/, /e/, /o/ and /u/ differ systematically between Greek, Spanish, and English.

To take another example, there is a longstanding problem of deciding whether languages with four vowel heights should be transcribed as [i, ɪ, e, a] or [i, e, ɛ, a]. The transcription of the non-peripheral vowels has to be based on phonological patterning, such as participation in ATR harmony, not phonetic quality (Casali 2008; Rose 2018).

Further, a reader attempting to pronounce an IPA transcription such as [bed] would probably have to guess as to whether there was any dipthongization or not: was [e] chosen over [eɪ] because slight dipthongization is a predictable detail that could be left off, or because dipthongization is absent? How much dipthongization is enough to be worthy of transcribing? The same is true of a consonant symbol such as [b]. Laryngeal configurations and amount of vocal fold vibration for "voiced" consonants differ widely across languages (see e.g., Kingston and Diehl 1994). If there are two stop consonants in the language, one with some prevoicing and the other with some aspiration, should the consonants be transcribed as [b] and [p] or as [p] and [pʰ]? Honeybone (2005) notes that different analysts of the same language often differ in the choice they make. Because such questions are in general answered according to the particular goals and choices of the transcriber, Pierrehumbert (1990b: 388) found it necessary to note of "fine phonetic transcription" that "the representation it is claimed to provide is not a coherent one." Pierrehumbert et al. (2000: 286) further assert that because "it is impossible to equate phonological inventories across languages," an IPA transcription is therefore not "a technically valid level of representation in a scientific model" but rather "a useful method of note-taking and indexing." The same point is made by Lindau and Ladefoged (1986), Bradlow (1993), Cohn (2011), and Ladd (2014), among others.

Port and Leary (2005: 927) take this line of reasoning a step further, arguing that because the IPA fails in the task of being a systematic, language-neutral, phonetic description, and because no other segment-based transcription system is any better, or is likely to be any better, that therefore discrete, symbolic phonology has no empirical basis. They argue that because "languages differ in far more ways than any learnable alphabet could represent," then "an alphabet cannot capture most of the important structure in speech," although it may be "a very useful technology for reading a language one knows well."

> Much of the evidence about the inadequacy of traditional transcription has been available for a long time, but most of us have ignored it or made excuses for it. Since all formal models of language are built on a foundation of discrete, a priori phonetic symbols, the very idea of a formal model of language is rendered impossible, unless someone figures out a way to provide a genuinely discrete universal phonetics. But there is little hope for that. (2005: 927)

The problem with Port and Leary's critique, however, lies in equating "discrete phonetic symbol" with "IPA symbol." As Pierrehumbert et al. (2000) note, if the IPA is taken to be simply a very useful tool for taking notes on the important of aspects of pronunciation, it works, but it is not in itself a phonological theory. The "atoms" or units of phonological organization are not to be found in the symbols of an orthography, no matter how phonetically clear. That is, the IPA symbol [e] should not be taken as referring to a particular set quality, but as a "cover symbol" or short-hand representation for the unit "mid front vowel," however that might be instantiated in a particular system.

As the problem was posed at the beginning of this chapter, does the usefulness of segmental transcription prove that the segment is the basic unit of Language, or do the problems with the IPA prove that it is not? On balance, the success of the IPA and of segment-based phonetic and phonological analyses show that the segment is indeed a *useful* level of analysis. Whether or not it is the ultimately *correct* analysis continues to be debated (see Chapters 9 and 11). Even if the segment is accepted as a unit, analyses at levels both suprasegmental (Chapters 7 and 8), and subsegmental (Chapter 4) also provide important insights. Crucially, when considering how segments are organized into systems of contrast, the nature of the "elementary relations" (Bell 1867: 15) between them must be taken into account. It is to these elementary relations that the next sections now turn.

3.5 SELECTING THE INVENTORY

The International Phonetic Alphabet provides symbols for seventy-nine consonants and twenty-nine vowels, not counting diacritics. How do languages choose which to use in creating inventories? According to the database compiled by Maddieson (2013), the smallest consonant inventory is six (Rotokas, Papua New Guinea) and the largest is 122 (!Xóõ, Botswana). The largest inventories include many secondary and complex articulations, in addition to contrasts in voice quality and airstream mechanism, thus many "segments" use a combination of IPA symbols. In fact, for such complex articulations, it can be difficult to tell whether a combination of constrictions should count as one segment or two, perhaps further evidence against the segment as a basic unit. Most languages, however, select a much simpler inventory. According to Maddieson (2013), the median size for a consonant inventory is twenty-one. English has (about) twenty-four, depending on dialect and how you count. The range for contrastive vowel qualities (not counting tone, voice quality, or diphthongs) is from two (Yimas, Papua New Guinea) to fourteen (German, Western Europe). English is on the high end for vowels, with thirteen (British English). The median number of contrastive vowel qualities is five, and according to Maddieson, fully one-third of the languages in the database have five vowels.

While both very large and very small inventories exist, Maddieson (2013) shows that inventory size is close to normally distributed—most languages are average. Communicative needs keep inventories from getting too small—there have to be enough segment combinations to keep tens of thousands of vocabulary items distinct. (Languages with small segment inventories tend to have long words.) At the other end of the spectrum, acoustic and articulatory pressures keep inventories from becoming too big. It has to be possible, in environments sometimes not perfectly conducive to speaking and hearing, to keep too many words from becoming too confusable.

Within an inventory, the set of sounds is not chosen randomly. It is not accidental that the inventory of Rotokas is [p, t, k, b, d, g] and [i, e, a, o, u], not [b, l, q, h, x, ŋ] and [y, ø, a, ʌ, ɨ]. For a given inventory size, how are contrastive segments chosen from among the indefinitely large number of ways that sounds can vary? How do inventories emerge? Linguists have long argued, beginning with Passy (1890: 227) as well as Martinet (1955) and Jakobson (1968), that if phonology is about contrast, then systems should be organized so that contrasts are *maximized*.

"Dispersion Theory" (Liljencrants and Lindblom 1972; Lindblom 1990; Flemming 1996, 2004) quantifies the idea of "maximizing contrast." According to Dispersion Theory, the members of an inventory are selected so that each segment is as acoustically distinct from each other segment as possible, given the number of segments.

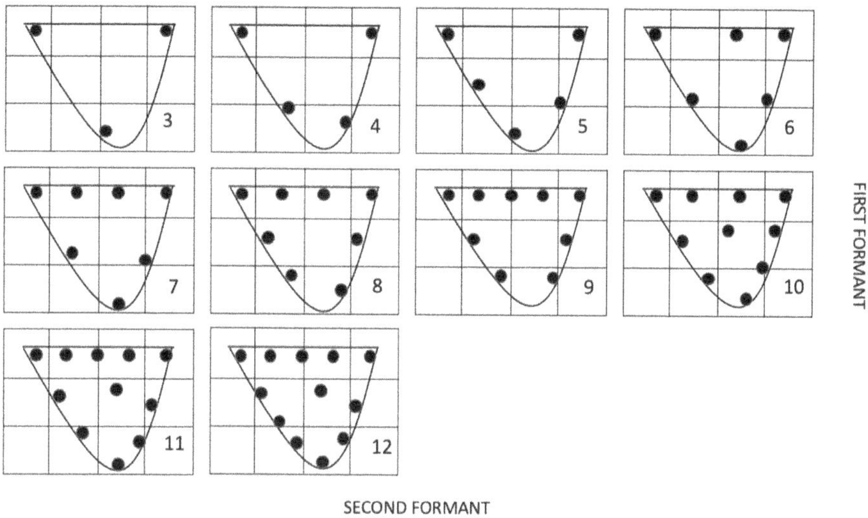

Figure 3.7 Dispersion Theory

Source: Adapted from Liljencrants and Lindblom (1972: 843, Figure 2).

Liljencrants and Lindblom suggest an analogy to magnets attached to corks floating in a tub of water: the magnets will repel each other, resulting in a stable configuration in which the distance between the magnets is as large as possible within the confines of the tub. The positions of the corks will depend on their number: the more there are, the closer together they will be forced.

For vowel systems, Liljencrants and Lindblom propose a formula to maximize the Euclidean distance between points in the F1/F2 space (in mel units), within the limits of possible human articulation/perception, based on the number of points (vowels) in the system. The results of their simulations do not do badly, as shown in Figure 3.7, adapted from Liljencrants and Lindblom's Figure 2 (1972: 843). (The graphs have been rotated from the original to match the orientation of a familiar vowel chart, where /i/ is in the top left corner and /a/ at the bottom.) For a three-vowel system, the results of the simulation produce formant values close to those of /i, a, u/. For a five-vowel system, the simulation produces values close to /i, e, a, o, u/. As the number of vowels increases, however, the predictions become less accurate (see Disner 1984 and Vaux and Samuels 2015 for further discussion). One of the biggest problems is that the model predicts that the high central space is utilized more extensively than is actually the case in attested languages. As seen in Figure 3.7, for example, the theory predicts that a nine-vowel system will have five degrees of backness, but only three vowel heights. Instead, actual nine-vowel systems (such as Akan or Maasai) use only front vs back, and five distinct heights: [i/u, ɪ/ʊ, e/o, ɛ/ɔ, a]. Languages also use the center of the acoustic space (schwa-like vowels) more often than Liljencrants and Lindblom predict. More recent versions of Dispersion Theory (Flemming 1996, 2004) refine the theory by incorporating both acoustic and articulatory information into the predictions. (Flemming's work is described in more detail in Chapter 6.)

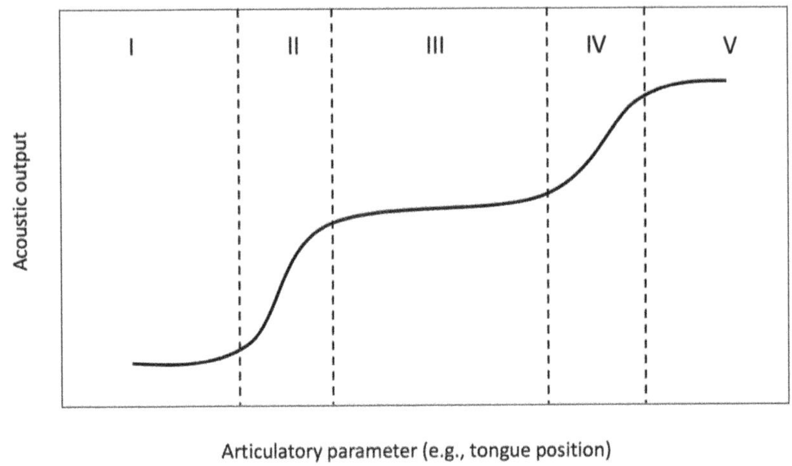

Figure 3.8 Quantal regions in the relations between acoustic and articulatory parameters

Source: Adapted from Stevens (1989: 4, Figure 1).

Stevens's (1989) Quantal Theory was one of the first to propose a quantitative model of the interaction of acoustics and articulation in predicting inventories. Using a schematic model of the vocal tract and calculating predicted resonance frequencies for different configurations, Stevens maps the acoustic output for a range of vocal tract positions. He finds that there are certain regions where a slight change in articulation results in large changes in acoustic output, but other regions where even fairly large changes in articulation result in little change in acoustic output. Stevens argues that languages organize their inventories using these regions of stability.

The idea is graphed in Figure 3.8, a schematic (adapted from Stevens 1989, Figure 1) that could be applied to the relation between any articulatory parameter and acoustic output, such as tongue position and F2, or constriction degree and fricative noise. To take the case of a velar fricative, there would be a large region (I in the diagram) where the tongue approaches the velum, but no fricative noise is produced. There would then be a sudden transition (region II) to fricative noise production, which would in turn not change much as the constriction degree became somewhat tighter (region III), until there was another sudden change (IV) to a region of closure/compression (V).

For vowels, Stevens models the vocal tract as a series of tubes with different resonance frequencies (similar to organ pipes). Changes in tongue position create different tube lengths with higher or lower frequencies. In configurations where the resonance frequencies of two tubes approach one another, the two systems interact in such a way as to produce regions of stability: because of the acoustic linkage between the two tubes, small tongue movements have less of an effect on the overall system (see Stevens 1989 and 2000 for details and formulas). Languages organize their vowel systems, Stevens argues, around such areas of stability, avoiding other areas where

Table 3.1 Relative difficulty of producing a voiced stop based on place of articulation and environment. Values in shaded cells are > 25

Environment	Place of articulation		
	b	d	g
After an obstruent	43	50	52
Word-initial	23	27	35
After a non-nasal sonorant	10	20	30
After a nasal	0	0	0

Source: Adapted from Hayes (1999: 251).

slight changes in tongue position have large acoustic effects. This point that "phonemes are located in well-behaved regions of the articulatory-to-acoustics mapping" is also emphasized by Pierrehumbert (2000: 16). It turns out that the palatal region is an area of instability, where small movements produce large changes. Thus, a great deal of articulatory precision would be required to maintain multiple vowel contrasts in the palatal region, explaining why languages don't use such systems, and why a purely acoustic Dispersion Theory makes the wrong predictions. (Chapter 10 returns to the discussion of the role of perceptual distinctiveness in predicting not only inventories but alternations.)

Pierrehumbert (2000) also notes that overall characteristics of the human vocal tract, such as the fact that the fundamental frequency of vocal fold vibration is lower than the tube resonances of the oral cavity, allow vowel inventories to encode linguistic information. But biology is not enough. Inventories arise, Pierrehumbert argues, from multiple interacting constraints, some of which are physical and biological, and some of which are cognitive.

Implicit in the preceding discussion are two ideas. First is the idea that languages create contrastive units (segments) from acoustic continua. The second is the idea that languages are not choosing segments per se (in which case they might select any random set), but choosing and reusing intersections of phonetic parameters, such as F1 and F2, or place of articulation, nasality, and voicing. Thus, very generally, if a language chooses [p, t, k] for voiceless stops, it will reuse those places with voiced stops [b, d, g] and nasals [m, n, ŋ]. A language will choose three symmetrical series of three, rather than nine randomly related segments, even if those nine are easy to both produce and perceive.

This point was made by Hayes (1999) in his proposal for "Phonetically-driven phonology." (See also the further discussion in Chapter 6.) Hayes argues that systems of contrast emerge through a combination of "phonetic sensibleness" and "good design" (1999: 244). An example is the "difficulty map" in Table 3.1.

The "map" shows the relative difficulty (based on an aerodynamic model of the vocal tract) of producing a voiced stop based on place of articulation and environment, where a higher number means more difficult. In any given environment, a voiced bilabial stop is easier than a voiced coronal which is easier than a voiced velar, because airflow is easier to maintain when there is a larger space between larynx and closure. For the effect of environment, post-nasal position favors voicing (0 difficulty), and then difficulty increases following other sonorants, in word-initial position, and then finally in post-obstruent position, the most difficult environment in which to induce voicing.

Hayes' point is that if languages organized their phonology based solely on articulatory difficulty, you would expect to find phonologies that allowed any voiced stop with degree of difficulty less than twenty-five: that is, [b, d, g] after nasals, [b, d] after non-nasal sonorants, only [b] in word-initial position, and no voiced stops at all after other obstruents. Instead, languages choose the easier over the harder, but in symmetrical ways. To bring in examples from other languages, Dutch (Gussenhoven 1992) bans [g] and allows [b, d] across environments, even though post-obstruent [b] (forty-three) is more difficult than intervocalic [g] (thirty) would be. In Sakha (formerly known as Yakut; Krueger 1962), on the other hand, all stops are voiced in intersonorant position and voiceless in initial position, even though initial [b] (twenty-three) would be easier than intervocalic [g] (thirty). Hayes concludes that "phonological constraints tend to ban phonetic difficulty in simple, formally symmetrical ways" (1999: 252). Phonetic difficulty matters, but is mediated by a formal phonology.

Further, Hayes argues that "The 'boundary lines' that divide the prohibited cases from the legal ones are characteristically statable in rather simple terms, with a small logical conjunction of feature predicates" (1999: 252). That is, possible inventories and possible sequences, on both phonetic and phonological grounds, are not defined by reference to segments per se. A segment, represented by an IPA symbol, is not an atomic whole, but can be understood as a combination of distinctive features that delimit its function within the linguistic system.

3.6 PHONETIC PARAMETERS AND PHONOLOGICAL FEATURES

Are distinctive features, then, rather than segments, the true building blocks of language? Linguists have been classifying segments according to their articulatory characteristics since antiquity (see further discussion in Chapter 4), and in the twentieth century, with the emphasis of phonological theory on systems of contrasts, characterizing the phonetic parameters that differentiate segments took on even greater importance. Saussure wrote that "[t]he important thing in the word is not the sound alone, but the phonic differences that make it possible to distinguish this word from all others, for differences carry signification" (1959: 118). Trubetzkoy took the emphasis on contrast a step further in arguing that an inventory is not a list of sounds, but a list of distinctive parameters. "Phonemes should not be considered as building blocks out of which individual words are assembled," he writes (1969: 35). "The phonemic inventory of a language is actually only a corollary of the system of distinctive oppositions. It should always be remembered that in phonology the major role is played, not by the phonemes, but by the distinctive oppositions" (1969: 67). The inventory, then, is a by-product. The language doesn't choose a list of sounds, but a list of distinctive dimensions, and the intersections of dimensions result in particular sounds.

It remains a question, however, whether features are actual constituents of segments (as were the components of Bell's "Visible Speech") or labels that describe and categorize segments (much as place and manner describe and categorize the set of IPA symbols). As Ladd (2014: 2) put the question, are features "particles or attributes"?

Trubetzkoy saw the phoneme as an abstraction that characterizes all and only the features that all of its allophones share. He writes, "The phoneme is not identical with an actual sound, but only with its phonologically relevant properties. One can say that the phoneme is the sum of the phonologically relevant properties of a sound" (1969: 36). At the allophonic level, features characterize a segment. At the phonemic level, they constitute the contrastive unit.

Later phonologists (including Jakobson et al. 1952 and Chomsky and Halle 1968) would argue explicitly that phonetic segments, not just abstract phonemes, are literally nothing more than "bundles" of features. Ladd quotes phonology textbooks by Harms (1968) and Hyman (1975) as explicitly stating that "the fundamental unit of generative phonology is the distinctive feature" (Harms 1968: 1) and "symbols such as *p, t, k, i, a, u* are used as convenient shortcuts for the feature compositions which combine to produce these segments" (Hyman 1975: 24ff.). In a more recent textbook, Zsiga (2013a: 257) states "Sentences are made up of words, words are made up of segments, and segments are made up of features." If segmental representations, including IPA symbols, are just a convenient notation for sets of features, which are the "real" constituents, that removes the problem of the failure to define IPA transcription as "a technically valid level of representation in a scientific model" (Pierrehumbert, et al. 2000: 286). Instead, it puts the burden on feature theory to define the basic units.

Thus for many phonologists, features, not segments, are the basic building blocks of language. Ladd (2014) goes on to argue, however, that considering all features as "particles" autonomous from the segment turned out to be more confusing than helpful, and others such as Ohala (1990) have argued against the "reification" of features as entities rather than descriptors. The question of whether segments are basic units to which featural labels attach, composite entities composed of features, or just useful fictions to which linguists trained to read the alphabet are predisposed remains a matter of debate. Decades of phonological and phonetic research have shown that the feature, the segment, and larger units such as syllables are all useful ways of breaking down continuous speech into component parts. The question of which level is the most important or basic has not been fully answered.

Chapter 4 delves deeper into distinctive feature theory, considering in detail some of the sets of actual features that have been proposed in the last 100 years or so since Saussure and Trubetzkoy, concentrating on the ways that distinctive feature theory illuminates and illustrates the phonology/phonetics interface, including the question of basic units. However, before digging in to the specifics of feature theories for spoken languages, Section 3.7 turns to the question of the basic units in a different modality: manually signed languages.

3.7 THE UNITS OF SIGNED LANGUAGES

Spoken languages are not the only ones that have phonology and phonetics. The same questions of phonological contrast and phonetic implementation apply to signed languages as well. In fact, studying the phonology and phonetics of languages that use a visual rather than an auditory modality provides a different and important perspective on the general nature of the relationship between the cognitive and physical aspects of language, and allows us to better investigate the ways that the physical modality of expression constrains the structure of language. Overviews of sign language phonology and its relation to spoken phonology include Sandler and Lillo-Martin (2006), Johnson and Liddell (2010), and Brentari (2010). Linguistic studies of signed languages are much more recent than linguistic studies of spoken languages; there is a smaller body of research and arguably less consensus on representation.

What is the "basic unit" of a signed language, the "building blocks" out of which utterances are built? The top-level answer is, of course, that signed languages consist of "signs": gestures of the hands, arms, and face that express a meaning. But just as words are not "vocal wholes" but can be broken down into successive syllables or segments, signs are not "gestural wholes" either. What are the distinctive parameters, to use Trubetzkoy's terminology, that make up the signs?

There are several issues that arise when trying to apply the principles that work for spoken languages to signed languages. First, there are many more degrees of freedom in the visual/spatial domain. Given these many degrees of freedom, there turn out to be few minimal pairs. Thus the top-down principle, of contrasting morphemes/signs to determine the smallest sequential unit, doesn't work well. Third, more "features" can be realized simultaneously in sound than in speech. Every meaningful sign by definition corresponds to a morpheme, and signs definitely combine into phrases within utterances, but it is not at all clear if there are sign correspondents to the spoken segment or syllable (see especially Battison 1980; Sandler and Lillo-Martin 2006). If there are not, we cannot argue that the segment and syllable are necessary and universal components of Language, just convenient ways of organizing *spoken* languages.

Linguists disagree on how signs should be decomposed into more basic units, and there is no generally-accepted transcription system for signed languages, analogous to the IPA for spoken languages. Publications often use pictures or diagrams to reference signs, if they reference the physical form of signs at all, without making any claim or assumption about sign components. The problem faced by all transcription systems for signed languages, as for spoken languages, is choosing what aspects to represent. The greater number of degrees of freedom in signed languages (the hand has many more possibly contrastive shapes than the tongue) makes the problem that much harder. (See Eccarius and Brentari 2008 and Hochgesang 2014 for further examples and discussion.)

Stokoe (1960) offered the first linguistic analysis of a signed language. He proposed that each sign could be specified in terms of three contrastive parameters: handshape, location, and movement. Obviously, each of these parameters had multiple possible values. Eccarius and Brentari (2008) propose a more differentiated system for transcribing handshapes (based on Brentari 1998), specifying the "base" shape of the hand, as well as joint configurations for selected fingers and thumb. The features are arranged in a hierarchical structure similar to the feature geometry proposed for spoken language segments (see the discussion in Chapter 4). Johnson and Liddell (2010) offer a different system (Sign Language Phonetic Annotation, or SLPA), in which handshapes are broken down into simultaneous feature bundles that specify hand configuration, placement, and orientation. The authors explicitly compare these bundles to the segmental feature bundles of spoken language phonology. They argue that signed languages have two types of segments. "Postural" feature bundles are connected by "transforming" segments that specify path and contact, among other aspects of movement.

Both the Eccarius and Brentari system and the Johnson and Liddell system are cumbersome as methods of transcription, however. As Tkachman et al. (2016) point out, a single handshape requires twenty-four to thirty-four characters to transcribe in SLPA, and many feature combinations are either anatomically impossible or non-occurring. Tkachman et al. suggest a number of modifications to SLPA that would make it easier to use, especially for large-scale corpus studies, modifications that include reducing the degrees of freedom in hand configurations, and creating templates for feature combinations that often re-occur. While different systems are still in development, no general consensus has been reached on transcription of signed languages. There simply has not been enough research into the phonology and phonetics of different signed languages to work out a complete theory of what the "distinctive parameters" are, so representation systems err on the side of more detail rather than less.

At the phonology/phonetics interface, we find that there is the same pressure to adapt the message to the medium in signed and spoken languages. There is the pressure of ease of articulation, resulting in assimilation and reduction. There is the pressure of clarity of communication, which results, for example, in an increase in signing space when the interlocutor is more distant, and the familiar effect of less assimilation and reduction in a more formal context. There are physical constraints: no sign requires the ring finger to move independently, for example. There are "markedness" constraints: in signs that require two hands, for example, the non-dominant hand can either mimic the dominant hand, or take on one of a limited set of simplified default shapes, such as a flat palm or fist (Battison 1978). There are dialects of signed languages, and signer-specific individual variation (Crasborn 2012).

A difference is that "iconicity" seems to play a more important role in signed than in spoken languages. Onomatopoeia plays a real but marginal role in spoken language, such that Saussure's concept of the "arbitrariness" of the sound/meaning mapping has become a central tenet of phonology. There is definitely arbitrariness in sign as well, but many signs bear a physical resemblance to the things or concepts they signify. For example, Figure 3.9 (Klima and Belugi 1979; Hamilton 2018) shows the sign for "tree" in three different signed languages. American Sign Language and Danish Sign Language show the full shape of the tree in different ways, while Chinese Sign Language depicts the trunk.

Work continues on the role of iconicity in signed languages (e.g., Hamilton 2018; Becker 2018). We simply don't have the data yet to determine whether iconicity plays a similar role in spoken and signed languages, just to a greater extent in signed because the visual modality provides a greater opportunity, or whether the role of iconicity in the grammar of a signed language is somehow different and deeper.

So for the major questions of signed language phonology, the state of the art is that work continues, but much more data and analysis is needed before consensus can be reached. Given the overwhelming preponderance of data and discussion in the

(a) American Sign Language (b) Danish Sign Language (c) Chinese Sign Language

Figure 3.9 The sign for tree in American, Danish, and Chinese sign languages
Source: Klima and Bellugi (1979: 21).

literature, most of what is discussed in the following chapters of this book is based on spoken language research. Based on the spoken language data, this chapter has argued for the utility of representing the basic building blocks of language as morphemes made up of syllables, syllables made up of segments, and segments made up of features. But it is just not clear if this same breakdown applies to sign, and if it does, exactly how a syllable, segment, or feature in sign should be defined. It is always worth asking, as we go forward, how the questions raised might be applied to sign, and how research into signed languages could help us reach a more truly universal answer.

RECOMMENDED READING

Overviews

Raimy, E. and C. E. Cairns (2015b), "Introduction," in *The Segment in Phonetics and Phonology*. Malden, MA and Oxford: Wiley-Blackwell.

- What are the "contemporary issues concerning the segment" that Raimy and Cairns discuss?

Sampson, G. (2015), *Writing Systems*, 2nd edn. Sheffield: Equinox Publishing.

- According to Sampson, how do writing systems represent linguistic knowledge?

Sandler, W. and D. Lillo-Martin (2006), *Sign Language and Linguistic Universals*. Cambridge, UK and New York: Cambridge University Press.

- Why is studying sign language important for understanding linguistic universals?

Exemplary research

Something old

Liljencrants, J. and B. Lindblom (1972), "Numerical simulation of vowel quality systems: the role of perceptual contrast," *Language, 48*(4): 839–62.

- How do Liljencrants and Lindblom "simulate" perceptual contrast? What are some of the limits of their simulation?

Stevens, K. N. (1989), "On the quantal nature of speech," *Journal of Phonetics, 17*: 3–45.

- Stevens's simulations are a little harder going than those of Liljencrants and Lindblom, but are well worth the effort. How does "Quantal Theory" take advantage of the intersection articulation and acoustics? Make reference to Figure 3.8 in your answer.

Something new

Hayes, B. (1999), "Phonetically-driven phonology: the role of Optimality Theory and inductive grounding," in M. Darnell, E. Moravscik, M. Noonan, F. Newmeyer, and K. Wheatly (eds), *Functionalism and Formalism in Linguistics, Volume I: General Papers*. Amsterdam, The Netherlands: John Benjamins, pp. 243–85.

- According to Hayes, how do systems of contrast emerge through a combination of "phonetic sensibleness" and "good design"?

Hochgesang, J. A. (2014), "Using design principles to consider representation of the hand in some notation systems," *Sign Language Studies*, *14*(4): 488–542.

- What "design principles" does Hochgesang propose? How do they compare to those proposed by Hayes?

Opposing views

Nolan, F. (1992), "The descriptive role of segments: evidence from assimilation," in G. J. Docherty and D. R. Ladd (eds), *Papers in Laboratory Phonology II: Gesture, Segment, Prosody*. Cambridge, UK: Cambridge University Press, pp. 261–79.

- Read Nolan's paper, and then the commentaries by Hayes, Ohala, and Browman that follow. Which of the four accounts of assimilation do you find most convincing?

QUESTIONS FOR FURTHER DISCUSSION

1. Given that orthographic systems only imperfectly represent the structure of language, how do such systems provide evidence concerning units of phonological representation? Investigate a writing system you do not know. Is the system based on the morpheme, syllable, mora, or segment? What evidence did you use to make this determination?
2. Discuss: The IPA is a *useful* level of representation, even if it is not "a technically valid level of representation in a scientific model" (Pierrehumbert et al. 2000: 286). What do Pierrehumbert et al. mean by that phrase, and how can the IPA be useful anyway? What evidence does the IPA provide both for and against the segment as a basic unit of spoken language?
3. How would you redesign the IPA? How might you make it easier to read and remember? How could it be less Eurocentric? What would the IPA look like if it were designed on the principle of "Visible Speech" and Hangul?
4. Consider the vowel inventory of a language other than English. Would Dispersion Theory correctly predict this inventory? If not, would Quantal Theory help? Is the inventory symmetrical? If the answer to all three questions is "no," how do you think this inventory might have come to be?

5. Consider the consonant inventory of a language other than English. Is the inventory symmetrical? What dimensions of contrast are utilized? What tradeoffs do you see between ease of articulation and ease of perception?
6. What insight do signed languages give us about what the basic units of Language might be? Should the basic units of signed and spoken languages be the same? What factors make analyzing the units of signed languages difficult?
7. In what ways did linguists such as Bell and Sweet "see themselves as 'men of science' using their knowledge to save the ignorant, whether the ignorant like it or not"? If you're not sure, read Bell's preface to "Visible Speech." What effects of Eurocentrism in linguistics have you come across? What efforts can we make now to reduce it?

CHAPTER 4

DISTINCTIVE FEATURE THEORY

4.1 FEATURES AND THE GOALS OF PHONOLOGY

Peter Ladefoged, in his 2004 essay on "Phonetics and phonology in the last 50 years," writes that the task of phonology is three-fold: to define the sets of possible lexical contrasts in the languages of the world, to describe the relationship between forms stored in memory and actual pronunciation, and to investigate constraints on possible sequences of sounds. Feature theory, he writes, is the place where "[p]honetics and phonology meet most obviously" (2004: 1). Ladefoged argues that feature theory is most directly concerned with the first task: the set of features defines the set of possible contrasts, and thus the set of possible inventories. But as was discussed in Chapter 3, for many theories of phonology, features are the crucial building blocks of the whole system, what Clements (2009a: 19) calls "the recurrent elementary components." If these elementary components are correctly defined or characterized, they will not only account for the set of possible inventories (every possible combination of features should be a possible phoneme), but will also define the set of possible alternations and constraints, by defining the classes of sounds to which phonological rules or constraints can refer.

According to Clements et al. (2011: 6), evidence for a set of distinctive features must include:

1. phonetic motivation;
2. recurrence across languages;
3. formal simplicity;
4. comprehensiveness (accounts well for the full set of data).

As it turns out, no set of features meets all of these goals perfectly, and the goals may contradict one another: a set of features that is tailor-made for the data in one language (thus winning on "comprehensiveness") would likely not be well-represented cross-linguistically, and a close mirroring of phonetic motivation might well introduce more formal complexity. No perfect set of features has yet been proposed (see the discussion of Mielke 2008 below), and different analysts have emphasized different criteria.

This chapter provides an overview of distinctive feature theory. Section 4.2 discusses some historical background, focusing on the origin of distinctive feature theory in the work of Pāṇini and of Trubetzkoy. Section 4.3 then surveys a number of theories that have been proposed for what the set of distinctive phonological features should be. Each theory reflects a specific view of the relation between phonetic cues and phonological contrasts. Section 4.4 concludes by asking the question of whether phonological theory should be pursuing the definition of a canon of features at all.

4.2 HISTORICAL BACKGROUND

Linguists from antiquity have been describing segments in terms of the characteristics that distinguish them. The *Aṣṭādhyāyī* of Pāṇini (written in India *c.*350 BCE) is a complete grammar of Sanskrit. The grammar consists of about 4,000 short statements (called "sutras"), which lay out, with "rigorous simplicity and beauty" (Kiparsky 1994: 2919), concise rules for the language's phonology, morphology, and syntax. The first set of fourteen sutras are the "Shiva sutras," said to have been directly revealed by Shiva to Pāṇini as the starting point of his grammar. These sutras list the individual sounds of Sanskrit, grouped according to articulatory characteristics such as vowels vs consonants, sonorants vs obstruents, nasality, voicing, and aspiration. (Though these characteristics are not explicitly defined, they can be inferred from the lists.) Later sutras define phonological processes by referring to the classes laid out in the Shiva sutras. They do so in a remarkably concise and elegant way, using a code that Panini defined to refer to contiguous strings of sounds (which are listed together because they share characteristics) by referencing the first and last symbol in the string. See Kiparsky (1994, 2002, in press) for a fuller description of this remarkable work, as well as additional discussion in Chapter 5.

Halle (1959: 91) notes that the work of the Indian Grammarians involved reducing the infinite variability of speech sounds to descriptions in terms of a limited set of "significant variables":

> The description begins with a fixed set of parameters of definite structure. Every sound in the language is classified according to a small number of attributes, e.g., degree of closure, point of articulation, etc. All other attributes which a sound may possess are omitted from consideration: they are not significant variables. Furthermore, each of the significant variables can assume only a very restricted number of values: there are five significant points of articulation, two degrees of aspiration and of voicing, etc.; although physically speaking there are, of course, infinitely many.

Phonologists in the twentieth century further explored the idea of classifying sounds according to sets of "significant" phonetic variables, developing a phonology based on "distinctive features." The discussion in the last 100 years has centered around which phonetic attributes should be considered "significant."

As was noted in Chapter 3, what can be called "distinctive feature theory" began with the idea of contrast. In his "Course on General Linguistics," Saussure emphasized that "everything that has been said up to this point boils down to this: in language there are only differences" (1959: 120). Between two signs, he continues "there is only *opposition*. The entire mechanism of language . . . is based on oppositions of this kind" (p. 121, emphasis original). It was from this basic idea of language being defined by opposition that distinctive feature theory grew.

The next major step in the development of the theory of distinctive features was Trubetzkoy's *Foundations of Phonology* (first published in 1939), which took the idea of opposition from Saussure's work, and made it more concrete. While Saussure emphasized the fact of distinctiveness per se and downplayed the role of phonetic content (to the point of insisting that it be ignored), Trubetzkoy shifted the emphasis to the phonetic properties on which the distinction or opposition was based.

Trubetzkoy notes that while "sound movements" constitute a "continuous flow," the system of language imposes a set of "phonic norms" that are "limited" and "enumerable" (1969: 3). These norms constitute the system by which sound is linked to meaning by the creation of contrast. All properties of a sound are important to the phonetician, but the properties that are important to the phonologist are only those that serve the function of discrimination. The properties of a phoneme that do not distinguish it from other phonemes are irrelevant to the phonologist: phonetics is only a "point of departure" for the real work of phonology (p. 14).

According to Trubetzkoy, the "phonic norms" that define a phoneme are those properties that all the variants share, and that distinguish it from other phonemes. The "essential" properties must remain unchanged in all realizations. Thus the phonologist must abstract away from predictable and non-distinctive properties. Trubetzkoy uses the example of /k/ in German, which is realized as [c] before front vowels and [k] before back vowels. The phoneme /k/ is therefore neither "palatal" nor "velar," since those terms describe only some of its realizations, but "dorsal" (p. 66).

The process for discovering distinctive features is that of phonemic analysis (also assumed by Hockett 1942, 1958; Bloch 1948; Hjelmslev 1953 and other structuralist phonologists). The phonologist begins by comparing words, chunks of sound that have different meanings, in order to find the points at which they differ. Of course, some words will differ by larger chunks, but these can be analyzed into smaller successive pieces. For example, Hockett (1958: 19) advises students to notice that "the pair *pin* and *bin* differ only at the beginning; but the pair *pin* and *pan* differ in the middle, and *pin* and *pip* contrast only at the end," The units thus derived are the phonemes: a distinctive unit that "cannot be analyzed into still smaller successive distinctive units" (Trubetzkoy 1969: 35). (Trubetzkoy notes that you can divide /k/ into closure and release phases, but you never get one without the other, so the two parts are not distinctive.) Then, once you have the inventory of phonemes, you can compare *them* to each other, to discover the properties by which they differ. These are "phonologically relevant" properties, the "distinctive marks."

The details of Trubetzkoy's system of "oppositions" are described in Section 4.3.1, followed by descriptions of some of the other major theories that have been proposed in the ensuing decades. As Section 4.3 makes clear, crafting detailed proposals for systems of distinctive features has been an activity at the center of phonological theorizing. Why have phonologists been so concerned with creating and justifying lists of possible distinctive features? As Clements (2009a) argued, if we get our features right, we will define what a possible phonology is.

4.3 THEORIES OF DISTINCTIVE FEATURES

4.3.1 TRUBETZKOY'S DIMENSIONS OF CONTRAST

Trubetzkoy spends most of *Foundations of Phonology* investigating the range of "phonic properties," both acoustic and articulatory, that languages actually use to create their systems. In creating his list of distinctive parameters, Trubetzkoy works backwards from the impressive list of inventories from Aleut to Zulu known to him. With respect to the discussion in Chapter 3, it is worth noting that Trubetzkoy does list his inventories in terms of segments, and undoubtedly he uses the same IPA character for sounds that are in fact different in the different languages he discusses. He uses the symbol not for a specific sound, but as a representation of the contrastive phoneme, a collection of distinctive features.

Based on his analysis of multiple language inventories, Trubetzkoy creates a list of possible "oppositions" or "correlations": a list of the dimensions along which phonemes/words/sentences can differ. The list comprises a set of available parameters, with different languages using different subsets and ignoring others. A list of oppositions compiled from *Foundations of Phonology* is given in Table 4.1.

There is no requirement that oppositions need to be binary: some oppositions are what Trubetzkoy terms "equipollent," an opposition between two equally common but mutually exclusive values, such as "consonant" vs "vowel." Other oppositions are privative, defined by presence vs absence, such as "nasalized" vs not. Some oppositions are "multi-valued," such as the seven possible "localizations," and others are gradable, having more or less of some scalar quality, such as aperture.

The dimensions along which languages can distinguish phonemes include a basic distinction between consonants and vowels. Then, within each category, phonemes differ in localization, timbre, and degrees of aperture (for vowels) or obstruction (for consonants). Trubetzkoy is careful to point out that by "localization" he does not mean exact "point of articulation," but rather the way the language chooses to divide up the articulatory space to create contrast. He points out, for example, that almost all languages will contrast "apical" vs "palatal" localizations, basically tongue front vs tongue body, but that "palatal" could encompass anything from palatal to uvular in a given language, and if there is allophonic variation, a given localization could encompass a variety of points of articulation. Similarly, depending on the system, vowel dimensions of contrast might be based primarily on backness (with rounding as an allophonic subsidiary variation), primarily on rounding (with backness as an allophonic subsidiary variation), or on backness and rounding independently (as in Turkish).

Any opposition can, in a particular language, be "neutralizable," meaning that it holds in some positions but not others, such as the voicing opposition being neutralized in final position. There is, however, no conception of one phoneme changing into another. In Trubetzkoy's phonology, the defining features of a phoneme must be present in all its allophones.

Table 4.1 Oppositions compiled from Foundations of Phonology

Opposition	Type	Values
consonantal vs vocalic	equipollent	obstruction/no obstruction
Vowels:		
degrees of aperture	gradable	high/low; open/close
timbre	equipollent	round/unround; front/back
resonance	presence vs absence	nasalized/non-nasalized; bright/muffled
Consonants:		
localization	multilateral	labial/apical/palatal/guttural/laryngeal/ lateral/sibilant
timbre	multilateral	palatalization/emphasis/velarization/ rounding/click
occlusion (1st degree of obstruction)	gradable	occlusives/fricatives/sonants
resistance (2nd degree of obstruction)	multilateral or gradable	fortis/lenis; heavy/light; voice/aspiration/recursion(ejective)/ release(implosive)
occlusion + resistance (3rd degree of obstruction)	presence vs absence	gemination
resonance	presence vs absence	nasal/non-nasal
Prosodic:		
	multilateral or equipollent or gradable	syllabicity; length; accent; tone movement; tone register; intensity; stod/glottal closure; intonation/sentence stress

4.3.2 JAKOBSON, FANT, AND HALLE

It was Trubetzkoy's colleague and collaborator Roman Jakobson, together with Jakobson's student Morris Halle and the electrical engineer Gunnar Fant, who decided that features need to be binary. Jakobson, Fant, and Halle's 1952 publication, *Preliminaries to Speech Analysis: The Distinctive Features and their Correlates*, laid out much of distinctive feature theory as it is understood today.

Roman Jakobson (1896–1982) was one of the most influential linguists and intellectuals of the twentieth century. In 1915, while a student at Moscow University, he helped found the Moscow Linguistics Circle, a group of young linguists dedicated to studying "linguistics, poetics, metrics and folklore" (Anderson 1985a: 84). Jakobson moved to Prague in 1920 to escape the aftermath of the Russian revolution, and a few years later, helped organize the Prague Linguistics Circle. Members of the Prague Circle included Trubetzkoy (who was then a professor in Vienna), other linguists who admired the works of Saussure and Baudouin de Courtenay, and other writers, critics and philosophers. As was noted in Chapter 2, it was the members of the Prague Linguistics Circle, prominently Trubetzkoy and Jakobson, who developed the basics of structuralist phonology and the principles of phonemic contrast. In 1939, with the advent of World War II and the German invasion of Czechoslovakia, Jakobson fled across Europe, first to Denmark, then Norway, then Sweden. In 1941 he escaped Europe for the United States, where, as a Jewish refugee, he initially did not find a warm welcome (Anderson 1985a; Waugh and Monville-Burston 1990). He continued studying, writing and teaching, however, and after a few years became a faculty member at Columbia University in New York, and later a professor of Slavic languages and of Linguistics at Harvard and MIT. Jakobson published hundreds of articles throughout his long career, on all aspects of linguistics, especially phonology: his *Selected Writings* fill ten volumes, and even the bibliography of his work (Rudy 1990) runs to 187 pages. Jakobson's most influential work, however, was in the study of phonological markedness (discussed at length in Chapter 6) and in distinctive feature theory.

In 1949, Gunnar Fant, a Swedish electrical engineer and phonetician who was working on the spectral analysis of speech for the phone company Ericsson, gave a seminar at Harvard. As Fant (2000) tells the story, Jakobson was in the audience, and saw the potential of incorporating Fant's cutting-edge acoustic analysis into the system of phonological features he was then developing. The two began collaborating, and Fant ended up staying in Boston for two years, working at MIT and the associated Research Laboratory of Electronics (RLE). Like the companies Ericsson and Bell Labs, RLE was conducting research on efficient human communication using the tools of electronics, such as the spectrograph and improved telephony, that had been invented during World War II.

The third collaborator on *Preliminaries to Speech Analysis* was Jakobson's Ph.D. student, Morris Halle (1923–2018). Halle was born in Latvia and emigrated to the United States with his family in 1940. After serving in the U.S. army, Halle earned his M.A. in Linguistics from the University of Chicago in 1948, and then attended Columbia, where he met Jakobson. He followed Jakobson to Harvard in 1950, where he earned his Ph.D. in Slavic Languages in 1955. His doctoral dissertation, *The Russian Consonants, a Phonemic and Acoustical Investigation*, was supervised by Jakobson. Halle joined the MIT faculty in 1951, as an Assistant Professor of Modern Languages (he was fluent in English, German, French, Russian, Latvian, and Hebrew), and also became a member of the RLE. When the MIT Linguistics graduate program was launched in 1961 as a collaboration between the Modern Languages departments and the RLE, Halle and Noam Chomsky were among the original faculty members. Halle taught and published at MIT for decades, his (and Chomsky's) students spreading out across American universities.

Fifty years later, in a memoir of his career as a speech scientist, Fant (2000: 2) wrote of his time at MIT, that "[t]his was a truly pioneering era in speech research as an outgrowth from linguistics, electrical circuit theory, psychoacoustics, and information theory." *Preliminaries to Speech Analysis* (henceforth JFH, for Jakobson, Fant, and Halle) embodied this confluence. Structuralist linguistic theory, based on Trubetzkoy's work, emphasized the role of "distinctive marks" and invariant "phonic properties" in creating systems of phonemic contrast. Developments in electronics allowed for these "distinctive marks" to be captured and measured in waveform and spectrographic displays, and psychoacoustic experiments confirmed their importance in speech perception. Finally, the particular theory of distinctive features proposed in JFH was based on the tenets of Information Theory.

Information Theory is concerned with the efficient transfer of information, applying statistical models to maximize the likelihood that the correct "signal" will be identified in a noisy channel. Shannon (1948: 379) in his groundbreaking article "A mathematical theory of communication," wrote that "The fundamental problem of

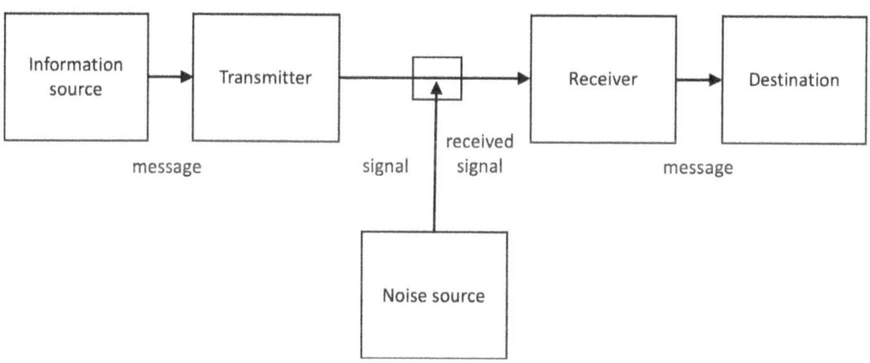

Figure 4.1 *Diagram of a "General Communication System"*
Source: Adapted from Shannon (1948: 381).

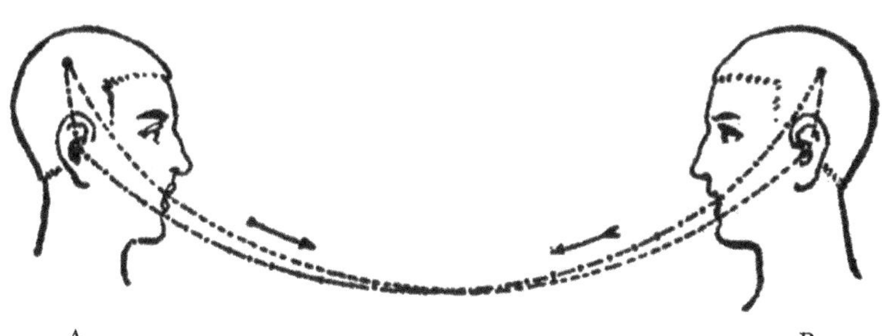

Figure 4.2 *Saussure's speech circuit*
Source: Saussure (1959: 11) (repeated from Chapter 2.)

communication is that of reproducing at one point either exactly or approximately a message selected at another point." He diagrams the problem as in Figure 4.1, with a schematic that interestingly mirrors the "speech circuit" drawn by Saussure thirty-five years earlier (Figure 4.2, repeated from Chapter 2).

Shannon's diagram includes the same five stages as Saussure's, although Shannon uses broader terms to include not only spoken language but other communication channels such as Morse code, radio, and the newly-invented television: source, transmitter, signal, receiver, destination. Where Saussure, however, was exclusively concerned with the first and final stages (the initial encoding of the source message and subsequent decoding at the destination, with transmission seen as irrelevant), Shannon focuses on the form of the signal itself. He notes that "[f]requently the messages have *meaning*; that is they refer to or are correlated according to some system with certain physical or conceptual entities" (1948: 379, emphasis original), but this pairing of concept and physical realization, the core of the Saussurean sign, is beside the point for Shannon. "These semantic aspects of communication," Shannon writes, "are irrelevant to the engineering problem," which is defined instead as the statistical problem of selecting the intended message from the set of all possible messages (1948: 379).

Crucially, JFH borrow from Information Theory (specifically from Shannon 1948 and Fano 1949) the idea of the "bit" (from "binary digit"): the insight that all more complex decisions can be broken down into a set of "elementary selections" between two equally likely choices, and thus that the most efficient way to represent information is in terms of a matrix of binary decisions. "Information Theory uses a sequence of binary selections as the most reasonable basis for the analysis of the various communication processes" (Jakobson et al. 1952: 9). JFH argue that in speech communication, "a set of binary selections is inherent in the communication process itself as a constraint imposed by the code" (p. 9). For example (p. 2), in order to recognize the word "bill," the listener has to recognize that the first sound is a stop (so it's not "fill" or "ill"), then that it's voiced (so not "pill"), then that it's labial (so not "dill" or "gill"). Since /b/ is the only voiced labial stop in English, the listener can then go on to the next sound. That is, JFH interpret concepts such as Saussure's "in language there are only differences" and Trubetzkoy's "oppositions" in the language of Information Theory as requiring the maximally efficient binary choice.

Unlike Trubetzkoy, for whom every sound was equally in opposition to every other, and for whom relations were often multilateral and multi-valued (as in Table 4.1), JFH see the "speech code" as a matrix of binary decisions. Within the matrix, phonemes are defined as "bundles" of distinctive features (p. 3). The features, then, are not just labels for referring to phonemes in useful ways, but the components of the phonemes themselves: features don't just describe phonemes, but constitute them.

As early as 1942, Jakobson was criticizing Trubetzkoy's theory of phonological oppositions on the basis of efficiency, taking the eight-vowel inventory of Turkish as a concrete example (Jakobson 1990: 236–7). If every vowel is equally in opposition to every other, then twenty-eight possible pairs of oppositions must be defined in order to fully account for the system: many more oppositions than there are phonemes. Jakobson argues, instead, that just three binary dimensions—high vs low, round vs

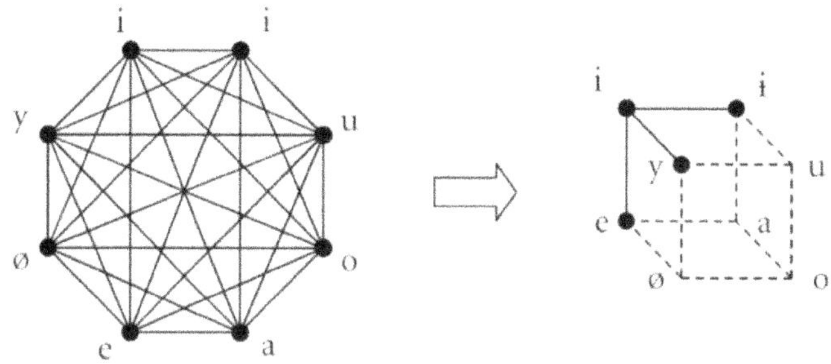

Figure 4.3 Dimensions of contrast in the Turkish vowel system
Source: Mielke (2008: 31, Figure 1.8).

unround, and back vs front—can fully define the Turkish inventory just as accurately and much more efficiently. The difference is diagrammed in Figure 4.3.

Building on this approach, JFH claim that a very small set of binary features can define the inventory of any language in the world: "[t]he inherent distinctive features which we detect in the languages of the world and which underlie their entire lexical and morphological stock amount to twelve binary oppositions" (Jakobson 1952: 40). These twelve features are listed in Table 4.2. Where possible, JFH define each feature equally in terms of production, acoustics, and perception. Their phonetic argument is supported by example spectrograph printouts, and the phonological case for each feature is made with a discussion of how the predicted classes of sounds do in fact occur in the languages of the world.

To get this economical, the feature set has to be somewhat abstract. Several features such as "mellow" and "grave" are defined in relative rather than absolute terms. For example, /θ/ is "mellow" because it is compared to /s/, even though it has more noise than a non-fricative consonant. Grave and acute are also relative. "Flattening" can be achieved by either rounding or pharyngealization. The "strength" of tense phonemes can be realized in a number of different ways: length, loudness, more extreme articulation, louder release. The set of features is general enough that the analyst can find a feature that will work to make the proper distinctions for any language.

JFH continue in the structuralist tradition of focus on contrast and inventories. Their work proceeds by examining inventories and figuring out the parameters that distinguish the different sounds. Later work, beginning with Halle's *Sound Pattern of Russian* (1959), moves toward finding a feature set that will also account for alternations.

4.3.3 HALLE AND CHOMSKY: *THE SOUND PATTERN OF RUSSIAN* AND *THE SOUND PATTERN OF ENGLISH*

Morris Halle and Noam Chomsky began working on their approach to grammar in the 1950s at MIT, drawing on the burgeoning field of Information Theory and the new field of computer science. The first publications in what came to be known as generative grammar were Chomsky's *Syntactic Structures* in 1957 and Halle's

Table 4.2 The twelve features proposed by Jakobson, Fant, and Halle

Feature	Definition and examples
Fundamental Source Features	
vocalic/ non-vocalic	Vocalic sounds are periodic and have a gradual onset, with no obstruction in the vocal tract. Vowels are specified as vocalic with no consonantal feature.
consonantal/ non-consonantal	Consonantal sounds have vocal tract obstruction that produces zeros in the spectrum. (Most) consonants are +consonantal with no vocalic feature. Liquids are +vocalic, +consonantal, and glides are -vocalic, -consonantal.
Secondary Consonantal Source Features	
interrupted/ continuant	Interrupted airflow in the vocal tract produces abrupt discontinuities in the spectrogram. Stops and tapped or trilled /r/ are interrupted; fricatives, approximants and vowels are continuant.
checked/ unchecked	Checked consonants are glottalized, with abrupt onset and sharp termination.
strident/mellow	Strident sounds have more noise and more irregular waveforms than their mellow counterparts. /s/ is strident and /θ/ is mellow. Affricates are strident stops.
voiced/voiceless	Vocal fold vibration defines voiced sounds. The voice source is evidenced by formants and/or a low-frequency "voice bar" in the spectrum.
Resonance Features	
compact/diffuse	Compact sounds have more energy in the middle of the spectrum: for vowels, a high F1, for consonants a convergence of F2 and F3. This occurs when the front resonating cavity is larger than the back resonating cavity. Compact sounds include low vowels, palatals and velars. Diffuse sounds include high vowels, labials, and alveolars.
grave/acute	Grave sounds have more low-frequency energy, especially low F2, and sound more "dark." Acute sounds have more high-frequency energy (high F2), and sound more "light." For consonants, labials are grave and diffuse, alveolars are acute and diffuse, palatals are acute and compact, velars are grave and compact. Back vowels are grave, front vowels are acute.
flat/plain	Flattening lowers the formants, either by rounding or pharyngealization.
sharp/plain	Sharpening raises the formants, especially F2, by raising the tongue body, as in palatalization.
tense/lax	Tense sounds are longer, stronger, and more extreme. The distinction applies to both vowels and consonants. Aspirated or geminate consonants may be tense.
nasal/oral	Nasal sounds have nasal resonance, oral sounds do not.

Source: Jakobson et al. (1952).

The Sound Pattern of Russian in 1959. Chomsky and Halle's co-authored work, *The Sound Pattern of English*, published in 1968, set the model for phonological analysis, and the phonology/phonetics interface, for the next three decades. In these publications, Chomsky and Halle envision grammar not as a description of the complex web of relationships among surface forms, as Sapir (for example) viewed it. Rather, the phonological grammar is built on the analogy of a computer program (Halle 1971: 12), in which input (lexical representation) and output (pronunciation) are linked by a series of rules that transform one representation into the other. Chapter 5 of this book deals at length with Chomsky and Halle's concept of a phonological derivation (an ordered series of rules), and the idea of the boundary between phonology and phonetics falling at some point in that derivation. Here, we take a closer look at how Chomsky and Halle further developed the theory of distinctive features.

They built directly on the theory of features defined in JFH a decade earlier, with a limited set of binary features: "[t]he phonetic properties in terms of which segments are characterized belong to a specific, narrowly restricted set of [articulatory and acoustic] properties called the distinctive features. All distinctive features are binary" (Halle 1959: 19). There is continued emphasis on efficiency of transmission (p. 29):

> Since we speak at a rapid rate—perhaps at a rate requiring the specification of as many as 30 segments per second—it is reasonable to assume that all languages are so designed that the number of features that must be specified in selecting individual morphemes is consistently kept at a minimum,

as long as there are enough features specified to keep all utterances distinct.

Chomsky and Halle see the set of features as part of the theory of "universal grammar," which aims not just to describe a particular language, but to delimit the set of human languages that are *possible*. The choice of "units and formal devices," including the set of features, "limit what grammars we can construct, and thus constitute a theory of what a 'linguistically significant generalization' is" (1968: 330). Chomsky and Halle argue that the set of features is innate (given by human biology) and therefore universal: "[t]he total set of features is identical with the set of phonetic properties that can in principle be controlled in speech; they represent the phonetic capabilities of man and, we would assume, are therefore the same for all languages" (pp. 294–5). This turned out to be an oversimplification. The work of Ladefoged (1980) and others showed that articulatory control does not in fact perfectly line up with the set of SPE features (see Section 4.3.4 below, as well as Chapter 5). Yet even if features do not represent every possible factor under phonetic control, one can still claim that features are innate. They may not be given by the biology of muscle groups but by a genetically-determined universal grammar that defines all the *cognitive* primitives of language structure, both phonological and syntactic. See Section 4.4 below for further discussion.

For Chomsky and Halle, the "linguistically significant generalizations" that distinctive feature theory must capture include both possible contrasts and possible rules. Possible lexical contrasts are captured in that morphemes consist of feature specifications. The phonological information in a lexical entry consists of a two-dimensional matrix of binary features, each column corresponding to a segment, and each row corresponding to a feature (Halle 1959: 32). Cells in the matrix are filled with either a plus or minus indicating the contrastive feature values for that segment, while predictable features are left blank. The set of specified features reflects the speaker's knowledge of their language, what the speaker/hearer understands to be the linguistically relevant, grammatically-determined properties (Chomsky and Halle 1968: 294). The reason for leaving predictable features unspecified was economy, again as required by Information Theory: it is more economical and efficient to state a generalization once, in the form of a rule that applies to all relevant forms, than to state the specification a separate time for each lexical entry.

Therefore, the feature matrices that specify lexical phonological representations are rather far removed from actual pronunciation. Chomsky and Halle emphasize that the purpose of features, and of phonological representation in general, is not just transcribing the physical aspects of speech (p. 293). Halle takes it as axiomatic that the binary nature of the features derives from their classificatory functions.

> The only relevant question is "does the segment possess the particular attribute?" It follows, therefore, that differences between segments can be expressed only as differences in their feature composition and that consequently segments (even in different languages) can differ from each other only in a restricted number of ways. (Halle 1959: 19)

Halle goes on to note that if the number of features just got larger and larger as we studied more and more languages, we'd have to agree that there was no limit to the ways languages could differ. But that is not the case, he argues. New languages are found to use the same old familiar features.

> Although languages have been found which possess phonetic features unknown in the Western languages, the number of such features must not be exaggerated. In examining the phonetic frameworks tested on a large number of languages . . . one cannot fail to be impressed by the small number (on the order of twenty or less) of features involved. (p. 19)

The evidence for these axiomatic conditions is "justified by the insightfulness, generality, and simplicity" of the theory that they enable (p. 19).

In addition to their classificatory function in the lexicon, Chomsky and Halle argue that features must also define the natural classes that recur in phonological rules. As they argue (Chomsky and Halle 1968: 335), all linguistic descriptions refer to "natural classes" of sounds—vowels, consonants, voiceless continuants, and so on—so linguistic theory must define what makes a class natural. For SPE, sets of segments that have features in common are natural. By requiring that rules refer to features that have phonetic content, the class of possible rules is limited, and reference to random sets of segments is ruled out. That is, at any rate, the goal. As algebraic modifications such as conjunction or disjunction are added, the rules become much less straightforward and the set of possible rules much larger (see Mielke 2008, discussed below). Such modifications are limited, however, by a criterion of simplicity: both efficiency as defined by Information Theory and naturalness as defined by feature theory favor rules for which fewer features must be specified.

It is the role of phonological rules to change or fill in phonological features depending on context. In order to complete the derivation, and get to an actual pronunciation, the abstract phonological features (which are binary in their "classificatory" function), must be mapped into "phonetic features" which consist of quantitative "physical scales" (Chomsky and Halle 1968: 297). That means that, at the end of the phonological derivation, a set of language-specific rules replace plus and minus values with integers that represent the "strength" of that feature for that language. From that specification, universal phonetic implementation takes over. For the most part, however (except for stress), these phonetic implementation rules are not spelled out in SPE, because Chomsky and Halle do not believe that they are important. "[O]ur interest in sound structure, in this book, does not extend to matters such as degree of aspiration, degree of fronting of vowels, etc." (p. 65). Much more on this phonology to phonetics translation is covered in Chapter 5.

On page 299 and following in SPE, Chomsky and Halle define their set of distinctive features, which differs from JFH in several important ways. The SPE features (pp. 299–300) are listed in Table 4.3, and the specifications of the set of English phonemes (pp. 176–7) are given in Table 4.4. As can be seen in Table 4.3, the features are divided into "Major Class" features, "Cavity" features, "Manner of Articulation" features, "Source" features, and "Prosodic" features.

Note that the feature set, while still limited, has expanded from twelve to thirty (counting the six pitch features). Chomsky and Halle are asking their features to do more than the twelve features of JFH could accomplish. Because their features do not just define inventories but also define the natural classes used in rules, and allow a straightforward mapping from feature to phonetic scale, a tighter and more transparent phonetic definition for each feature is needed. As they go through their feature list, Chomsky and Halle discuss all three aspects of each feature: the contrasts that it is meant to capture (the "classificatory function"), the rules in which it participates, and its phonetic meaning or translation (the "phonetic function"). Because the meaning of the features is generally either transparent or continued on from JFH, only a few notes are mentioned here.

For the major class features, [vocalic] and [consonantal] are preserved from JFH, but [sonorant] is added. The feature [sonorant] is defined as the set of sounds that allow "spontaneous voicing" when the vocal folds are held in a neutral position. Adding the feature [sonorant] allows for the economical description of the liquids and nasal consonants as a natural class, definitely needed to define the set of consonants that are transparent to voicing assimilation in Russian, an alternation to which Halle gave much attention in *The Sound Pattern of Russian*.

The cavity features define place of articulation. Gone are the acoustically-based "grave/acute" and "compact/diffuse" which functioned to define the contrast between labial (grave and diffuse), alveolar (acute and diffuse), palatal (acute and compact), and velar (grave and compact). Instead, the more articulatorily transparent [anterior], [coronal], [high], and [back] (along with [low] and [round], needed for vowels) are proposed. Advantages listed by Chomsky and Halle for the new feature set are greater articulatory transparency and simplicity: features for tongue body position are now the same for both vowels and consonants, for

Table 4.3 The distinctive features proposed in the Sound Pattern of English

- Major Class Features
  - sonorant
  - vocalic
  - consonantal
- Cavity Features
  - coronal
  - anterior
  - tongue body features
    - high
    - low
    - back
  - round
  - distributed
  - covered
  - glottal constrictions
  - secondary apertures
    - nasal
    - lateral
- Manner of Articulation Features
  - continuant
  - tense
  - delayed release (for affricates)
  - suction (for clicks and implosives)
  - pressure (for ejectives)
- Source Features
  - heightened subglottal pressure
  - voice
  - strident
- Prosodic Features
  - stress
  - pitch (high, low, falling, rising, elevated, concave)
  - length

Source: Chomsky and Halle (1968: 299–300).

example, allowing for a relationship between palatalization and front vowels to be expressed in the feature [-back], as well as a more straightforward description of back consonants (velars, uvulars, and pharygneals) with the features [back], [high], and [low], rather than subsidiary features such as [strident]. Subsidiary points of articulation, such as dental vs alveolar, and palatoalveolar vs retroflex, are handled with the feature [distributed], which distinguishes consonants made with a longer tongue constriction ([+distributed] laminal dentals and palatoalveolars) from those made with the apex of the tongue ([-distributed] apical alveolars and retroflexes). Less transparently, labials are [+distributed] and labiodentals are [-distributed].

Chomsky and Halle claim that these labial distinctions "fit rather naturally under the proposed distinctions" (1968: 314), though the term requires more vagueness in the definition than some others do. The feature [covered] (why that particular term was chosen is unclear) is added to account for West African vowel harmony systems, and corresponds to what later theories (see below) would call Advanced Tongue Root or Expanded Pharynx.

For the manner features, [continuant] and [tense] are used in much the same way as in JFH, though further phonetic evidence for each class is provided. The other manner features are added to account for particular classes of sounds: [delayed release] for affricates (and release distinctions among the click consonants), [suction] for clicks and implosives, and [pressure] for ejectives. For the source features, [voice] and [strident] are straightforward; [heightened subglottal pressure] is less so. It is defined as a "necessary but not sufficient condition for aspiration" (p. 326). Chomsky and Halle discuss at length how the features [voice], [tense], [heightened subglottal pressure], and [glottal constriction] account for cross-linguistic differences in stop closure voicing and voicing lag.

Note the emphasis on the universality of features: this limited set of features must be sufficient for every contrast and every rule, and each feature must be more or less straightforwardly interpretable as mapping into a single phonetic scale. The durability of the feature set over the decades and over the many languages that have been described in these terms is admirable. But the feature set is not perfect. As subsequent investigations have shown, this feature set is not entirely sufficient for all phonological contrasts and alternations, nor is the mapping to phonetic scales as simple as Chomsky and Halle make it out to be (see especially Sections 4.3.4 and 4.4 below, and Chapter 5).

Although they list the prosodic features stress, pitch, and length, Chomsky and Halle note only that "[o]ur investigations of these features have not progressed to a point where a discussion in print would be useful" (p. 329). In other words, these aspects of sound structure didn't fit easily into the segment-based, binary, feature-bundle approach they were espousing, and they weren't sure how to handle them. As it turned out, issues of stress, tone, and length, in the hands of subsequent scholars, led to new representational theories, prosodic phonology (discussed in Chapter 7), and autosegmental phonology (discussed in Section 4.3.5 below).

It is worth noting that, in SPE rules, boundaries such as those before and after a word or sentence are also considered to be feature bundles. Boundaries are defined by an additional feature [-segment], that has no phonetic correlates (1968: 364). The feature [-segment], however, seems to be just for consistency, so that all elements in the underlying representation are composed of feature bundles. Beyond making this assertion, in practice SPE never represents boundaries as feature bundles, but with symbols.

A final note on the SPE features. SPE improved phonetic (especially articulatory) transparency over the JFH system, in that the "naturalness" of natural classes was more directly represented, and "simplicity" in terms of feature counting was taken seriously. However, feature counting could not account for everything, and it was far too easy to write simple rules that were not at all natural, as Chomsky and Halle themselves point out (p. 402):

Table 4.4 Distinctive feature composition of the English segments

Vowels:	ɨ̄	ī	ū	ē	ō	ǣ	ā	œ̄	ɔ̄	i	u	e	ʌ	o	æ	ɔ	y	w	ɛ
vocalic	+	+	+	+	+	+	+	+	+	+	+	+	+	+	+	+	−	−	−
consonantal	−	−	−	−	−	−	−	−	−	−	−	−	−	−	−	−	−	−	−
high	+	+	+	−	−	−	−	−	−	+	+	−	−	−	−	−	+	+	−
back	+	−	+	−	+	−	+	−	+	−	+	−	+	+	−	+	−	+	−
low	−	−	−	−	−	+	+	+	+	−	−	−	−	−	+	+	−	−	−
anterior	−	−	−	−	−	−	−	−	−	−	−	−	−	−	−	−	−	−	−
coronal	−	−	−	−	−	−	−	−	−	−	−	−	−	−	−	−	−	−	−
round	−	−	+	−	+	−	−	+	+	−	+	−	−	+	−	+	−	+	−
tense	+	+	+	+	+	+	+	+	+	−	−	−	−	−	−	−	−	−	−
voice																			
continuant																			
nasal																			
strident																			

Consonants:

	r	l	p	b	f	v	m	t	d	θ	ð	n	s	z	c	č	ǰ	š	ž	k	g	x	ŋ	h	k^w	g^w	x^w
vocalic	+	+	−	−	−	−	−	−	−	−	−	−	−	−	−	−	−	−	−	−	−	−	−	−	−	−	−
conson.	+	+	+	+	+	+	+	+	+	+	+	+	+	+	+	+	+	+	+	+	+	+	+	−	+	+	+
high	−	−	−	−	−	−	−	−	−	−	−	−	−	−	−	+	+	+	+	+	+	+	+	−	+	+	+
back	−	−	−	−	−	−	−	−	−	−	−	−	−	−	−	−	−	−	−	+	+	+	+	−	+	+	+
low	−	−	−	−	−	−	−	−	−	−	−	−	−	−	−	−	−	−	−	−	−	−	−	+	−	−	−
anter.	−	+	+	+	+	+	+	+	+	+	+	+	+	+	+	−	−	−	−	−	−	−	−	−	−	−	−
coronal	+	+	−	−	−	−	−	+	+	+	+	+	+	+	+	+	+	+	+	−	−	−	−	−	−	−	−
round																				−	−	−			+	+	+
tense																											
voice	+	+	−	+	−	+	+	−	+	−	+	+	−	+	−	−	+	−	+	−	+	−	+	−	−	+	−
contin.	+	+	−	−	+	+	−	−	−	+	+	−	+	+	−	−	−	+	+	−	−	+	−	+	−	−	+
nasal	−	−	−	−	−	−	+	−	−	−	−	+	−	−	−	−	−	−	−	−	−	−	+	−	−	−	−
strident	−	−	−	−	+	+	−	−	−	−	−	−	+	+	+	+	+	+	+	−	−	−	−	−	−	−	−

Note: Transcriptions follow the original.
Source: Chomsky and Halle (1968: 176–7).

> [O]ur evaluation measure [feature counting] makes no distinction between a language in which all vowels are voiced and one in which all vowels are voiceless, or between a language in which obstruent clusters are redundantly voiced and a language in which they are redundantly voiceless. But surely one case is much more natural than the other.

They (famously) also note (p. 400) that it would be possible to invert the sign of every feature in every rule in SPE, and the result would still be a formally well-formed language, though not at all a natural one. The topic of naturalness of phonological inventories and alternations is taken up at length in Chapter 6.

4.3.4 PHONETICALLY-BASED FEATURE SYSTEMS: HALLE AND STEVENS, LADEFOGED

After the publication of SPE in 1968, linguists worked hard in the next two decades to produce comparable grammars of other languages. This project met with great success, and generative grammar became the dominant approach to phonology in the second half of the twentieth century. But some phonologists also argued for alternative feature theories. Two that are of particular interest to the discussion of the phonology/phonetics interface are Halle and Stevens (1971) and Ladefoged (1980). Both of these are superb examples of research at the interface, using phonetic data to solve phonological problems. They take an opposite approach to the procedure of the structuralists, JFH, and Chomsky and Halle, however. Rather than starting with the system of phonological contrasts, and then trying to determine what phonetic parameters those contrasts might correspond to, Halle and Stevens and Ladefoged start with the phonetics, specifically the physiologically-possible manipulations of the vocal tract, and then attempt to determine what phonological classes these might create.

Soon after the publication of SPE, Halle returned to the problem of glottal features with another MIT co-author, phonetician Kenneth Stevens. Halle and Stevens (1971) sought to better ground the features for consonantal source (in SPE, [glottal constriction], [voice], [tense], and [heightened subglottal pressure]) in the actual degrees of freedom in laryngeal configurations. Examining the phonetic data first, they begin with a detailed model of the aerodynamics of airflow through the glottis, and then ask the question "What are the ways in which the configuration of the vocal cords can be manipulated by the laryngeal musculature in order to produce distinctive acoustic end products that are potentially useful for the formation of phonetic categories?" (1971: 199). They determine that there are two dimensions of control—glottal stiffness and glottal opening—whose manipulation "to particular ranges of values gives rise to distinct and well-defined acoustic consequences." (The idea that continuous manipulation in one domain (e.g., glottal stiffness) can still give rise to regions of stability in another domain (type of vocal fold vibration) would be further developed in later work by Stevens on Quantal Theory (Chapter 3)). They propose four possible laryngeal actions that correspond to four different phonological features (each of which is binary).

Table 4.5 Glottal features and phonological classes according to Halle and Stevens

Features	Phonological class
+spread, -stiff, -slack	voiceless vowel
+constricted, -stiff, -slack	implosive
-spread, -constricted, -stiff, -slack	lax stop, voiced or voiceless depending on context, and vowels
+spread, +stiff	voiceless aspirated
+constricted, +stiff	ejective
-spread, -constricted, +stiff	plain voiceless
+spread, +slack	breathy voice
+constricted, +slack	creaky voice
-spread, -constricted, +slack	modal voice

Source: Halle and Stevens (1971).

1. adduction corresponds to [constricted glottis];
2. abduction corresponds to [spread glottis];
3. stiffening corresponds to [stiff vocal folds];
4. slackening corresponds to [slack vocal folds].

They then go on to consider how these features map onto the types of contrasts found in the languages of the world. Note that the feature [voice] is missing. Halle and Stevens consider voicing (actual vocal fold vibration) to be a by-product of choices made in laryngeal opening and stiffness, not a goal in itself. They argue that every possible combination of features (except for the articulatorily impossible [+spread, +constricted] and [+stiff, +slack]) is utilized contrastively by some language. They depart from Chomsky and Halle (1968) in having a feature for aspiration, but not for voicing.

Noting that [+constricted, +spread] and [+stiff, +slack] are both impossible, the combinations in Table 4.5 occur.

The set of phonological contrasts is covered, the binary features are used with near-maximum efficiency (excepting only the two impossible combinations), and the goal of a straightforward translation to phonetic scales is met. The main problem is that a set of four binary features is phonologically cumbersome. Particularly, many phonological alternations and classes do seem to refer to the feature "voiced" across different kinds oral constrictions, and in Halle and Stevens's system (by design) "voice" is not represented as a single feature. Thus, stating what should be simple rules of assimilation becomes complicated. This is the perennial challenge: the more faithful one is to phonetic detail, the more, it seems, that phonological generalization is lost.

In the end, a hybrid version of Chomsky and Halle's and Halle and Stevens's laryngeal feature set has caught on, modified by feature geometry (Section 4.3.5 below): [spread glottis], [constricted glottis], and [voice]. This makes the statement of phonological categories and alternations more straightforward because most classes can be targeted with a single feature, but the translation to phonetics more complicated,

Table 4.6 Ladefoged's articulatory parameters

List 1	
1. Front raising	10. Lip width
2. Back raising	11. Lip protrusion
3. Tip raising	12. Velic opening
4. Tip advancing	13. Larynx lowering
5. Pharynx width	14. Glottal aperture
6. Tongue bunching	15. Phonation tension
7. Tongue narrowing	16. Glottal length
8. Tongue hollowing	17. Lung volume decrement
9. Lip height	

Source: Ladefoged (1980: 486).

Table 4.7 Ladefoged's auditory parameters

List 2	
1. Voice source frequency	9. Amplitude of nasal formant
2. Voice source amplitude	10. Frequency of nasal formant
3. Frequence of formant one	11. Amplitude of aspiration
4. Frequence of formant two	12. Amplitude of fricative source
5. Frequence of formant three	13. Frequency of lower fricative pole
6. Bandwidth of formant one	14. Frequency of upper fricative pole
7. Bandwidth of formant two	15. Relative amplitude of fricative poles
8. Bandwidth of formant three	

Source: Ladefoged (1980: 493).

because [voice] is implemented in different ways depending on the supralaryngeal configuration and the surrounding context. (See further discussion of laryngeal features and their phonetic implementation in Chapter 5.)

The issue of phonetic specificity vs phonological generalization in the definition of distinctive features is raised again in Ladefoged's 1980 publication "What are linguistic sounds made of?" Drawing on decades of phonetic research, Ladefoged argues that seventeen articulatory parameters and fifteen acoustic parameters are necessary to specify the actual production and audition of the sounds of the languages of the world. His parameters are listed in Tables 4.6 and 4.7. Most of these do not correspond in any straightforward way to the phonological features proposed by JFH or SPE. The laryngeal features of Halle and Stevens fare better, in that glottal aperture and phonation tension are included in Ladefoged's list, but Ladefoged also adds larynx lowering and glottal length as parameters that must be taken into account.

For example, Ladefoged argues that two features, front raising and back raising, are necessary to adequately describe tongue shape for vowels, as opposed to a simple [+/− high] feature that only specifies the highest point of the tongue body. Further, front raising and back raising correspond directly to real muscle movements—front raising is due to the action of the genioglossus, and back raising to the action of the styloglossus—while a simplex [high] feature could correspond to various types of muscle configurations. Similar arguments are adduced for the other features.

What are we to make of this mismatch between what is required of the phonetics and what is required of the phonology? Ladefoged's first point is that Chomsky and Halle's claim that their phonological feature set "is identical with the set of phonetic properties that can in principle be controlled in speech" (1968: 294–5) is just wrong.

> Some phonological features can be readily interpreted in terms of a single acoustic parameter; thus vowel height is inversely proportional to the frequency of the first formant. But the majority of phonological features are in a many-to-many relationship with the minimal set of acoustic parameters, just as they are with the minimal set of articulatory parameters. (Ladefoged 1980: 494)

> Other aspects of linguistic descriptions, such as accounts of sound patterns within languages, are undoubtedly best stated in terms of phonological features; and if these descriptions are to be explanatory, the features must relate to articulatory or auditory (or cortical) phenomena. But phonological features are certainly not sufficient for specifying the actual sounds of a language; nor are they in a one-to-one relationship with the minimal sets of parameters that are necessary and sufficient for this purpose. (1980: 485)

> Taken as a set, [phonological features] are neither necessary nor sufficient to specify what it is that makes English sound like English rather than German. (1980: 495)

Thus, Ladefoged argues, we need two sets of features: one for phonology and one (or possibly two) for phonetics. It might be possible to use only the articulatory or only the acoustic parameters, Ladefoged argues, since one can be converted into the other, but linguists should have both available in order to have the ability to discuss the interplay of acoustic and articulatory factors. Phonologists are welcome to continue using abstract and general features for phonological descriptions, but they need to stop pretending that they are phonetically simple or transparent.

> Let me emphasize that I am not suggesting that terms like Coronal or Alveolar should be replaced in phonological descriptions by terms like Tip raising and Tip advancing. When describing the sound patterns of languages, we will want to refer to natural classes defined in terms of conventional phonological features; and these features must refer to observable phonetic phenomena. But I do advocate that, when we make a phonetic description of a language, we should not do so by trying to interpret each feature in terms of a single physical scale. (1980: 491)

If two sets of features are used, one phonological and one phonetic, then the mapping rules between them would be complicated formulas for converting feature values into physical parameters. These mapping rules would have no claim to psychological reality, just a way to relate one set of facts to another set of facts.

On balance, Ladefoged argues, allowing the phonology/phonetics mapping to be complex could be a good thing. If phonological features don't have to be simply and directly related to phonetic parameters, they can be as abstract as you want them to be, and can be evaluated not in terms of phonetic plausibility but of simplicity and elegance of the analysis. But then he goes on to question whether we need these abstract features at all. Or rather, linguists may find them useful for describing systems, but they are not necessarily part of every speaker/listener's competence. As far as producing and understanding sentences is concerned, the phonetic parameters will do the trick.

The question of the mapping of phonological features to phonetic scales is taken up again in Chapter 5, with the discussion of the derivation, and further work on the relation of phonological features to phonetic parameters. In the meantime, work on the feature set from the phonologist's point of view was moving in different directions.

4.3.5 AUTOSEGMENTAL PHONOLOGY AND FEATURE GEOMETRY

Other aspects of the SPE view of features that came under revision was the "feature bundle" and the requirement that all features be binary. As described above, in the SPE conception, each segment was characterized by a column of plus and minus feature specifications, the "feature bundle." The idea of separating feature from segment came from the study of tone languages, an area of phonology which SPE touched on only briefly.

The basic insight of the new approach (Woo 1969; Leben 1973; Goldsmith 1979) was that although tone is usually realized as a pitch contrast on vowels (high pitch vs low, or falling vs rising), tone features are independent of the segments that realize them: tones can easily jump from one vowel or syllable to the next, they can stay behind when a vowel is deleted, they can crowd onto one vowel or spread out over many. Leben (1973, 1978), for example, showed how rising tones in Etsako can be derived from a sequence of low plus high after vowel deletion. Odden (1981) described a similar process in Shona, as well as a series of rules whereby tones could shift one vowel over, or start out associated to one vowel in the root, then spread across a series of suffixes. It certainly makes sense that a rising tone would be derived from a sequence of low plus high, but the problem lies in formalizing the alternation using SPE features and conventions. There is no way to write a rule that deletes all the features of a segment except tone, to slide the tone from one vowel to the next, or to formally relate the separate features [+/–hightone], [+/– lowtone], and [+/– risingtone].

Leben, following an idea first proposed by Woo (1969), argues that [hightone] and [lowtone] should be pulled out of the feature bundle and represented as H and L "autosegments": features that are linked to the vowel, but not really part of the vowel, as in Figure 4.4.

In "autosegmental phonology," features are graphically associated to segments

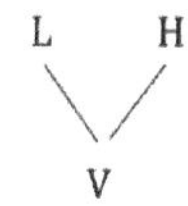

Figure 4.4 *Autosegmental representation of a rising tone*

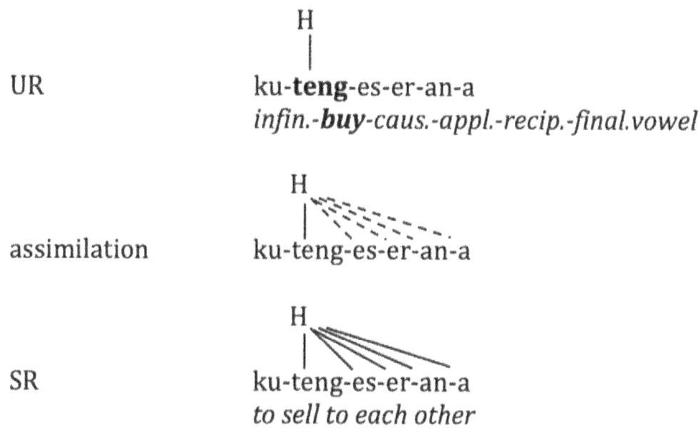

Figure 4.5 *Autosegmental derivation via feature spreading: high tone spreading in Shona. Dotted lines indicate new associations being added*

Source: Example from Myers (1990: 159).

via "association lines" which phonological rules can add or delete. Assimilation is represented as spreading of a feature from one segment to another, as in Figure 4.5.

Dissimilation or neutralization can be formalized as delinking of association lines. Association between tones and segments can be many-to-one (as in contour tones), one-to-many (as in assimilation), or zero-to-one. Segments might not be specified for tone, or morphemes can consist of tone with no underlying segmental material.

For further details on the formalism of autosegmental phonology, see McCarthy (1988), Kenstowicz (1994), Zsiga (2013a), or other introductions to phonology. While most phonologists as of this writing use autosegmental representations for tone (and intonation), see the discussion in Chapter 8 for other approaches. See also the discussion in Hyman (2014), who reminds us that autosegmental representations don't solve all the problems of tonal phonology.

From the point of view of the phonology/phonetics interface, the most important development to arise from autosegmental phonology was the generalization of the representation from tone to other features, and the development of "feature geometry."

Once autosegmental representations for tone features were proposed, phonologists began to incorporate autosegmental representations for other features as well. Just as a rising tonal contour can be represented as a sequence of two tones associated to a single vowel as in Figure 4.4, contours can be created to account for other types of segments where feature values change partway through, such as pre-nasalized stops, diphthongs, and affricates (Sagey 1986). In long distance assimilation, such as [ATR] vowel harmony in Akan (Clements 1981), backness harmony in Turkish or Finnish (Clements and Sezer 1982; Goldsmith 1985), [nasal] harmony in Sundanese (Cohn 1990; Robins 1957), or anterior harmony in Chumash (Poser 1982), a feature can begin with an association to the root vowel and then spread to suffixes, just as tone spreads in Figure 4.5.

The development of autosegmental formalism brings phonological representation closer to phonetic data. The use of association lines limits the types of phonological change that can be (straightforwardly) represented, and thus random feature switches, which would have no phonetic motivation, are predicted not to occur. Conversely, assimilation is represented not as feature copying, but as a single state extended over more than one segment, consistent with phonetic analyses showing that in harmony systems, the vocal tract configuration remains steady throughout the harmony domain (e.g., Boyce 1990; Hayes and Londe 2006; Benus and Gafos 2007). A further correspondence is developed in the theory of feature geometry (Clements 1985; Sagey 1986; McCarthy 1988).

Theories of feature geometry use association lines to represent featural groupings as graphical dependencies. While consensus has not been reached on one correct geometry (see, e.g., Padgett 1994, 2001; Clements and Hume 1995; Avery and Idsardi 2001), Figure 4.6 shows the arrangement of one commonly-used set of features.

One development is the use of unary features corresponding to vocal tract articulators: [labial], [coronal] (indicating the tongue front), [dorsal] (indicating the tongue body), [pharyngeal]. Unary features have no plus/minus specification. Either a segment entails a constriction with the tongue tip and is thus [coronal], or it does not, and thus will have no [coronal] specification. (More than one articulator can be specified in the case of complex or compound segments.) Binary features indicating exact place of articulation are dependent on the appropriate articulator feature. The feature [round], for example, entails constriction of the lips, so it is represented as dependent on [labial]. Similarly, [anterior] and [distributed], which refer to the position of the tongue tip and blade, depend on the presence of [coronal]; [high], [low], and [back], which reference tongue body position, depend on [dorsal]; and [ATR] depends on [pharyngeal]. The articulator features are grouped under a class node, "place," which allows the set of place features to be targeted as a group by a single association line, further simplifying the representation of general place assimilation. Mirroring vocal tract anatomy, in this view "place" corresponds to the supralaryngeal vocal tract. Alternative geometries have been proposed, however, to account for cases where phonological patterning is not so closely tied to vocal tract organization. Clements and Hume (Hume 1994; Clements and Hume 1995) propose separate place nodes (but identical features) for consonants and vowels, while Padgett (2002, 2011) proposes a node for "vowel color" that groups [back] and [round], prioritizing acoustic affinity over articulatory separation.

The features [spread glottis] ([SG]), constricted glottis ([CG]), and [voice] are realized with the Laryngeal articulator. Figure 4.6 follows Lombardi (1991) in representing [spread glottis] and [constricted glottis] as unary features, but continues in the

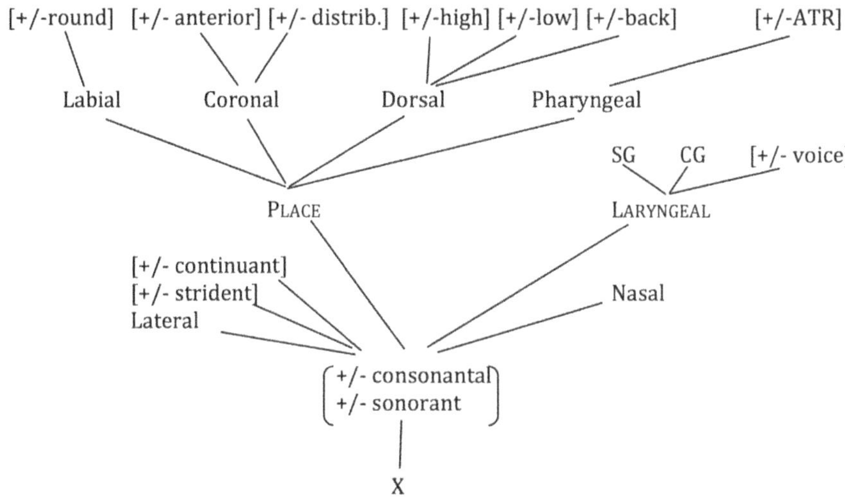

Figure 4.6 One model of feature geometry

Source: Zsiga (2013a: 293).

tradition of JFH in using [+/– voice]. If different laryngeal features are proposed, the geometry will of course be different as well (e.g., Avery and Idsardi 2001). The question of the status of the feature [voice], particularly [-voice], has been and remains contested. In addition to the references already cited, see, for example, Wetzels and Mascaró (2001), Honeybone (2005), and Zsiga (2018a).

The features [continuant], [strident], [lateral], and [nasal] are relevant across different places of articulation, and thus are not shown as being dependent on any particular articulator: stops and fricatives contrast, and nasals are found, at all different places of articulation. This reflects the consensus view, though Padgett (1994) argues for an alternative geometry where [continuant] is articulator-dependent, treating manner and place as parallel. That leaves only two features, [consonantal] and [sonorant], inside the bracketed "feature bundle." These are the two features that have *not* been shown to exhibit autosegmental behavior—they do not assimilate or dissimilate independently, and never exhibit long-distance behavior or stability when other features of the segment are deleted.

For the phonology/phonetics interface, feature geometry makes two important contributions. First, while the number of segments still can be counted through root nodes, the segment as a unit is de-emphasized while features take prominence as the basic units of phonological patterning. Features assimilate and dissimilate, through spreading and delinking, independent of the segments that sponsored them. Domains, as for vowel harmony or nasality, are defined in terms of the extent of features, not number of segments. Phonologists who see the segment as epiphenomenal in any case (e.g., Port and Leary 2005; Silverman 2006; Golston and Kehrein 2015) see this as a positive development, and cite the success of autosegmental phonology and feature geometry as evidence that segmental organization is not basic and segments aren't needed to account for phonological patterning. On the other hand, Ladd (2014) argues that treating all features as autosegmental was a wrong move, undermining a real distinction between true autosegments, such as the H and L accents of tonal and intonational representation, and segment-based assimilations such as place assimilation.

A second major influence of feature geometry is that it moved phonological theory to take a very articulatory-centric view of features, especially the articulator features [labial], [coronal], and [dorsal], which replaced earlier Jakobsonian or SPE features that were either acoustic, such as [grave] or [flat], or equally-well defined in either the acoustic or articulatory domain. The increased transparency between phonological features and articulatory states can be seen as simplifying the phonology/phonetics interface (Zsiga 1997), but it also gives rise to the question: if phonological features and articulatory states are isomorphic, why bother with the features at all? This was the argument made by Ohala (1990), who argued that autosegmental features were just the imaginary "reification" of articulatory patterns. Proponents of Articulatory Phonology (Browman and Goldstein 1986, 1988, 1989, 1990b; Goldstein et al. 2006, among others) propose that phonological features should be completely replaced by articulatory gestures as basic units (see the discussion in Chapter 9). Others argue, however, that an over-reliance on feature geometry has moved phonology too far in the articulatory direction, and room must be made for the role of acoustics and perception in phonological patterns (see the discussion in Chapter 10).

4.3.6 DEPENDENCY PHONOLOGY AND GOVERNMENT PHONOLOGY

A different set of phonological features, termed "elements" or "primes," are proposed in the theories of Dependency Phonology and Government Phonology. The two approaches are not identical, in that they propose somewhat different representations and constituents, but they share many assumptions regarding phonological features. For overviews of Dependency Phonology see Ewen (1995), van der Hulst (2016), van der Hulst and van de Weijer (2018) and references therein, and for Government Phonology see Kaye et al. (1985, 1990), Gussmann (2002), Scheer and Kula (2018) and references therein. Another similar approach is Particle Phonology (Schane 1984).

In each of these approaches, binary features in the tradition of JFH are replaced by unary elements that express basic acoustic properties and articulatory configurations. For example, the "place elements" of Dependency Phonology (van der Hulst and van de Weijer 2018: 337) are shown in Table 4.8.

The primes proposed by Government Phonology and Dependency Phonology are very similar, though not identical. These can by themselves represent segments, so that |i| is realized as [i], or they can combine to create more complex inventories. For example, the combination |i,a| is realized as [e], an [i] to which an element of additional lowness has been added. The elements apply equally to consonants and vowels: when used for a consonant, for example, |i| indicates a palatal articulation, and |l,u,a| (a lingual consonant with elements of backness and

Table 4.8 The place elements of Dependency Phonology

<table>
<tr><td>|i|</td><td>palatality, acuteness/sharpness</td></tr>
<tr><td>|u|</td><td>roundness, gravity/flatness</td></tr>
<tr><td>|a|</td><td>lowness, sonority</td></tr>
<tr><td>|ə|</td><td>centrality</td></tr>
<tr><td>|a|</td><td>advanced tongue root</td></tr>
<tr><td>|l|</td><td>linguality</td></tr>
<tr><td>|t|</td><td>apicality</td></tr>
<tr><td>|d|</td><td>dentality</td></tr>
<tr><td>|r|</td><td>retracted tongue root</td></tr>
<tr><td>|L|</td><td>laterality</td></tr>
</table>

lowness) indicates a uvular. Elements are then grouped into different hierarchical relationships ("government" or "dependency") with other elements and with higher class nodes, and these hierarchical representations determine how the elements are interpreted.

As with all other feature theories, it is argued that these featural elements better capture the inventories, alternations, and phonotactic restrictions of the languages of the world than the feature sets based on JHF. Like the features of JFH, because the elements comprise only a small set, they must be rather abstract. The |u| element of a [q] and an [u] is not realized in exactly the same way, any more than the feature [+/− back] would be.

Government Phonology proposes that the relation between phonological representation and "phonetic spell out" is essentially arbitrary (Kaye 1989; Scheer and Cyran 2018). Despite the resemblance of the phonological elements and phonetic categories in most cases, only phonological patterning can determine phonological representation, and mismatches are to be expected. An example is the case of [ɛ] in Polish: some [ɛ] vowels cause palatalization of a preceding consonant, and these are headed by an |ı| element. Other [ɛ] vowels, phonetically identical, do not cause palatalization, and these are headed by an |a| element. Thus Government Phonology is clearly in the camp of those who argue for an algebraic phonology independent of phonetics (e.g., Reiss 2018). Dependency Phonology, on the other hand, makes a clearer claim to the phonetic grounding of the phonological elements proposed, placing Dependency Phonology more in the camp of JFH and others, who argue for a more direct mapping between phonological representation and phonetic realization.

4.4 WHAT IF THERE IS NO CANON OF FEATURES?

Each of the approaches discussed above assumes that there *is* a canon of features—the same set of phonological primitives available for all languages to use to create inventories and classes—and that it is the job of the phonologist to correctly characterize what the members of that canon are. As was noted above, JFH (1952: 40) claim that twelve universal, binary features can account for all contrasts and alternations in the languages of the world, and Halle (1959: 20) states that an additional decade of linguistic data analysis has not led to any expansion of the feature set.

For Chomsky and Halle, distinctive features are universal because they are biologically given, both physically as the set of parameters that "can in principle be controlled" and cognitively as part of universal grammar (1968: 294–5). Arguments in favor of universal grammar tend to be more syntactic than phonological (see Archangeli and Pulleyblank 2015), and take us far afield from the phonology/phonetics interface. Berent (2013: 225), a strong proponent of an "algebraic" view of phonology and phonological features, reviews the phonological evidence for innate features. She summarizes studies on infant perception and production "at ages where the effect of linguistic experience is undetectable," and concludes that some of these studies have found evidence that at least some phonological primitives, including features, must be present at the "initial state." Berent (2013: 225) concludes, however:

> Such evidence hints at the possibility that several primitives and markedness reflexes are active in early development. The number of such demonstrations, however, is very small, and most findings do not dissociate grammatical and functional explanations for the results. So while these results are consistent with the possibility that core phonological knowledge is active in the initial state, the evidence necessary to fully evaluate this hypothesis is incomplete.

That is, infant studies show that at least some features *might* be innate, but the evidence is inconclusive.

Another possibility to consider is that the feature set might be universal but *not* innate. That is, features recur across languages because specific processes at the intersection of articulation, acoustics, perception, learning, and communicative needs, which do not vary from culture to culture, give rise to phonological patterns that do not vary from language to language. Such an approach is in the spirit of Lindblom's Dispersion Theory, Stevens's Quantal Theory, and Hayes' Phonetically-driven Phonology, discussed in Chapter 3. Certain intersections of phonetic parameters, and certain regions of the space of articulatory and acoustic possibility, are used again and again not because they are genetically given, but because they *work*.

But what if the features that work are different from language to language? Disagreements about feature geometry led linguists such as Padgett (1994) to propose that different languages might have different dependency relations (e.g., [lateral] and [continuant] might or might not be a dependent of [coronal]). If feature organization can be language-particular, why not the features themselves? That is, should we even be looking for a small set of universal features that work for every language? Why not let every language have its own feature set, whatever works best for that language, without worrying about whether the feature is universal?

Language-specific features are argued for in "Emergent Feature Theory" in work by Jeff Mielke (2008, 2013) and others (e.g., Hyman 2010, 2011a; Gallagher 2011; see also papers in Clements and Ridouane 2011). Emergent Feature Theory often goes hand in hand with Evolutionary Phonology (Blevins 2006; see the discussion in Chapter 6), a theory of phonology that says that synchronic inventories and processes are shaped by chains of diachronic changes that often result in patterns that are specific to a particular history and make no obvious phonetic sense. If inventories and processes are not grounded in universal phonetic principles, then the feature sets needed to describe them would not be expected to be universal either.

Mielke (2008: 48–9) begins by arguing that a feature set must correctly define all "phonologically active classes" in a language, that is, all sets of sounds that either "undergo a phonological process; trigger a phonological process; or exemplify a static distributional restriction." Emergent Feature Theory proposes that features are not universal or innate, but are "created by learners in response to phonological patterns" (p. 103). It is not that phonological patterns are the way they are because they are built on the basis of universal distinctive features. Rather, learners are presented with phonological patterns, and they infer a set of features to account for them. If a language should by chance present evidence for some unusual non-natural class, the learner is not constrained to account for the class in some predetermined way, but can infer a language-particular feature to account for it, without regard to whether this feature is useful in any other language, or whether it is phonetically motivated. That is, of the four criteria proposed by Clements et al. (2011: 6) for a successful distinctive feature set, namely phonetic motivation, recurrence across languages, formal simplicity, and comprehensiveness in accounting for the data of a particular language, Mielke proposes that only the last is really relevant.

Similar points are made by Hyman (2010), who argues that contrastive tone patterns across different languages are so different that it is a futile exercise to try to find a single small set of features (especially H and L) that will account for all of them (see further discussion in Chapter 8). Non-universal features are also assumed by Gallagher, who follows Steriade (1997) in proposing a feature [+/– ejective] (in addition to [constricted glottis]) to distinguish ejectives from glottalized sonorants (Gallagher and Coon 2009), as well as features [+/– long VOT] and [+/– loud burst] (Gallagher 2011), to deal with alternations in specific languages.

To test his prediction that features are inferred by learners on a language-particular basis, Mielke (2008) conducted a survey of the phonologies of 628 language varieties from fifty-one language families (p. 48). For each language variety, Mielke identified all of the "phonologically active classes," which, as noted above, include any set of sounds that could either "undergo a phonological process; trigger a phonological process; or exemplify a static distributional restriction" (p. 49). These are the classes that features must account for. In earlier work, these would have been known as "natural classes," but Mielke does not take their naturalness for granted.

Mielke identified 6,077 instances of phonologically active classes (2008: 147). He then attempted to characterize each class in terms of the feature sets proposed by JFH, by SPE, and by Clements and Hume. The most successful feature set was SPE, but Mielke found that even that set was only able to account for 71.0 percent of the classes with a single set of features. (Clements and Hume got 63.7 percent and JFH

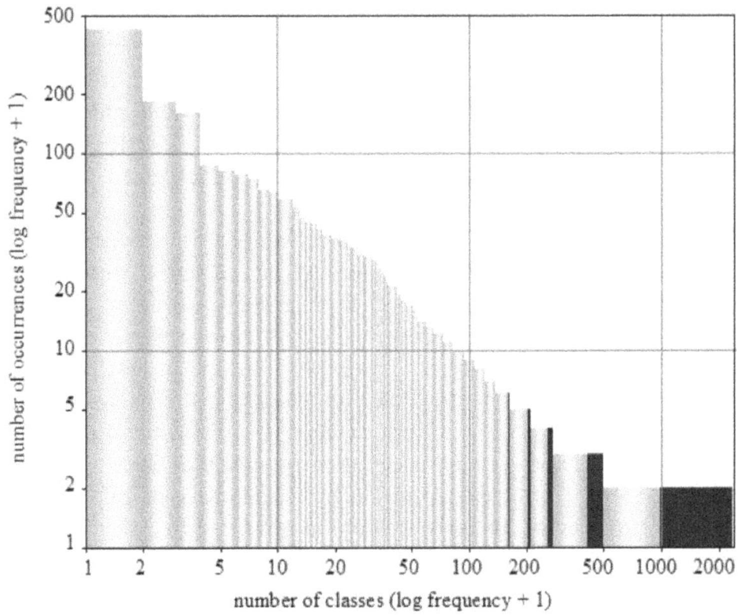

Figure 4.7 Frequency of "natural" and "unnatural" classes, based on SPE features. Light bars indicate natural classes and dark bars indicate unnatural classes

Source: Mielke (2008: 157, Figure 7.6).

got 59.0 percent). However, if you allow for one disjunction (that is, some part of the rule references two distinct classes, e.g., deletion of /p/ and /k/ but not /t/), the totals rise to 91.5 percent for SPE, 84.5 percent for Clements and Hume, and 83.7 percent for JFH. The frequency of natural and unnatural classes, using SPE features, is shown in Figure 4.7.

In Figure 4.7, the y-axis shows the number of times that a class occurred in the dataset (from 1 to almost 500), and the x-axis shows the number of classes (out of 6,077) that occurred with that frequency. Note that both scales are logarithmic, in order to show a greater range. Because the log of 0 is undefined (no number raised to any power = 0), and the log of 1 equals 0 ($x^0 = 1$), both scales start at 1, and graph the actual frequency +1. So there is one class that occurred over 400 times (Mielke tells us later (2008: 215) that it is [+syllabic]), and more than 1,500 classes that occurred only once. (The lowest bars rise to "2" on the y-axis, but because the axis graphs frequency + 1, the frequency is 1.) The light-colored bars indicate natural classes that can be defined with the set of SPE features. The dark-colored bars are unnatural classes that cannot be so defined. Thus, the graph shows that all the most frequent classes—100 percent of classes that occurred more than five times in the dataset—are successfully captured by SPE features. But the SPE feature set captures only about two-thirds of the classes that occur twice, and only about one-third (500 out of 1,500) of the classes that occur only once. Put another way, of the classes that the SPE feature set cannot capture, almost all occur cross-linguistically only once or twice.

The question is (as pointed out by Cohn 2011), does this graph illustrate success or failure? Does a 92 percent success rate argue for or against universality of features? There are definitely some "crazy" classes that the SPE features cannot account for, but crazy classes are not the norm. All of the most frequent classes are successfully characterized, and there is little overlap between "natural" and "unnatural" classes. As Cohn (2011) puts it, the SPE feature set is "approximately correct." Is that good enough?

If Mielke (2008) is right, and classes and features are language-particular, the question that the theory must answer is why so many languages choose the same distinctions and the same inventories over and over. Emergent Feature Theory answers that features recur across languages because patterns recur across languages, and patterns recur across languages because principles of sound change work pretty much the same way all the time.

In general, then, the question of which, if any, feature set is "correct" is tied up with larger questions of how phonology, and the phonology/phonetics interface, are defined and what each is accountable for. Ladefoged (2004: 6) commented on the current state of phonological feature theory: "The feature wars are clearly waning." His claim is that phonologists are worrying less about proving what the "right" universal set of features must be. Instead, phonologists seem ready to accept a near-universal set of features that is "approximately correct," with theories of why inventories and processes that are common recur so often, and some room for accounting for how languages that don't fit the general pattern might have come to be. These issues are taken up again in Chapter 6, with the discussion of the role of markedness in synchronic grammar.

Before once again taking up the issue of markedness, however, and with it markedness-based Optimality Theory, we turn in the next chapter from accounting for contrast to accounting for alternation, and the generative grammar theory of the derivation.

RECOMMENDED READING

Overviews

Zsiga, E. (2013b), "What is a possible language?" in *The Sounds of Language: An Introduction to Phonetics and Phonology*. Oxford: Wiley Blackwell, Chapter 12.

- How are distinctive features important for phonological theory?

Ladefoged, P. (2004), "Phonetics and phonology in the last 50 years," *UCLA Working Papers in Phonetics, 103*: 1–11.

- Why is it that feature theory is the place where "[p]honetics and phonology meet most obviously" (p. 1).

Jakobson, R., G. Fant, and M. Halle (1952), *Preliminaries to Speech Analysis: The Distinctive Features and Their Correlates*. (Technical Report 13). Cambridge, MA: Acoustics Laboratory, MIT.

- Read the whole thing—it's not very long. How do the authors combine phonological, articulatory, and acoustic factors in arguing for their feature set?

Exemplary research

Something old

Halle, M. and K. Stevens (1971), "A note on laryngeal features," *Quarterly Progress Report, Research Laboratory Electronics MIT 101*: 198–213.

- What data do Halle and Stevens use to argue for their distinctive features?

Something new

Avery, P. and W. J. Idsardi (2001), "Laryngeal dimensions, completion and enhancement," in T. A. Hall and U. Kleinhanz (eds), *Studies in Distinctive Feature Theory.* Berlin: de Gruyter Mouton, pp. 41–70.

- What data do Avery and Idsardi use to argue for their proposed features? How do their laryngeal features differ from those proposed by Halle and Stevens?

Gallagher, G. (2011), "Acoustic and articulatory features in phonology—The case for [long VOT]," *The Linguistic Review*, *28*(3): 281–313.

- What data does Gallagher use to argue for [long VOT]?

Opposing views

Mielke, J. (2013), "Phonologization and the typology of feature behaviour," in A. Yu (ed.), *Origins of Sound Change: Approaches to Phonologization.* Oxford: Oxford University Press, pp. 165–80.

vs

Cohn, A. C. (2011), "Features, segments, and the sources of phonological primitives," in A. Cohn, G. N. Clements, and R. Ridouane (eds), *Where Do Features Come From?* Amsterdam, The Netherlands: John Benjamins, pp. 15–41.

- How do Mielke and Cohn answer the question, "Where do features come from?"

QUESTIONS FOR FURTHER DISCUSSION

1. Discuss the ways that Jakobson et al. (1952) were influenced by Information Theory. How does Figure 4.3 represent the differences in Trubetzkoy's approach vs the approach of JFH in describing a system of vowel contrast?
2. Review Table 4.3. Why was it important for Chomsky and Halle that the SPE feature set be universally applicable? Is the set of SPE features influenced by the fact that they occur in a work on English phonology?
3. Review the four criteria proposed for feature systems by Clements et al. (2011): phonetic motivation, recurrence across languages, formal simplicity, and comprehensiveness. In what ways are these criteria potentially incompatible? Now consider how each of the feature theories discussed in Section 4.3 (and in any additional recommended readings) applied those criteria. Which criteria were most important for each theory? Are there other criteria that should be added that Clements et al. did not consider?
4. Why does Ladefoged (1980) argue that we should have three different feature sets? What are the pros and cons of this approach?

5. Does the graph in Figure 4.7 indicate success or failure? Is a feature set that is "approximately correct" good enough? Why or why not?
6. Ladefoged (2004) claimed that "the feature wars are waning." Would you agree?

CHAPTER 5

RULES AND DERIVATIONS

5.1 BACKGROUND AND ANTECEDENTS

5.1.1 WHERE IS THE DIVIDING LINE?

One approach to the phonology/phonetics interface relies on the concept of a *derivation*. In a derivational approach, phonological information is stored in the lexicon in abstract form, usually with only contrastive or unpredictable information specified. Then the details of positional variation and other types of non-contrastive information are filled in through a series of steps, one feature at a time, until a fully-specified phonetic representation is formed. In the classic model, the steps are formulated as *rules* that add or change features for particular classes of sounds in particular contexts. The process of applying the set of rules to a particular representation is that item's *derivation*.

Although it had predecessors, the serial phonological derivation became central to phonological description and analysis with the publication of *The Sound Pattern of English* (Chomsky and Halle 1968). Chomsky and Halle called their approach (to both phonology and syntax) *generative grammar*, because the rules create, or generate, output forms based on input specifications. In a generative derivational approach, the output of the syntax is the input to the phonology, and the output of the phonology is the input to phonetics. In such an approach, the main question of the phonology/phonetics interface is "Where is the dividing line?" What kinds of information are filled in by the phonology, and what kinds by the phonetics? Is there such a thing as a phonetic rule? If so, how is it different from a phonological rule?

This chapter examines derivational approaches to the phonology/phonetics interface. Section 5.1 reviews antecedents to SPE, and discusses how the theory of generative phonology arose in the context of dissatisfaction with aspects of structuralist phonology, in large part through the work of Morris Halle. Section 5.2 takes a detailed look at the SPE model, focusing on Chomsky and Halle's conception of the phonology/phonetics interface. Section 5.3 discusses subsequent phonetic findings that showed that the SPE model was too simple, and Section 5.4 discusses elaborations of the derivational model, including sub-divisions into lexical and post-lexical strata. Section 5.5 concludes with discussion of proposals for specifying the difference between phonological and phonetic rules, and thus the specifics of how phonetics may be derived from phonology.

5.1.2 ORDERED RULES IN PĀṆINI'S GRAMMAR

The first derivational phonology was the grammar of Sanskrit developed by Pāṇini (India, *c.*350 BCE). Kiparsky (1994) describes Pāṇini's grammar, the *Aṣṭādhyāyī*, as "a complete, maximally concise, and theoretically consistent analysis of Sanskrit grammatical structure" (p. 2918). As Kiparsky summarizes (p. 2918), the grammar has four components: 1) a sound inventory (the Shiva sutras discussed in Chapter 3), 2) a lexicon of 2,000 roots, 3) classes of exceptions, and 4) 4,000 grammatical rules that encompass morphology, syntax, and phonology.

Kiparsky (1994) describes Pāṇini's worldview that "human activities, even those normally carried out in an unconscious or unselfconscious way, can be analyzed by explicit rule systems, and that performing those activities in awareness of the rules that govern them brings religious merit" (2918). Another of Pāṇini's precepts was "be as simple as possible," and his rules are written so economically as to be almost in code. Classes of sounds are defined by abbreviations, then short statements (sutras) tell how different kinds of sounds are pronounced in different environments, how morphemes are put together to form words, and how words are put together to form sentences.

Each rule specifies an input, an output, and an environment for application. Further, the rules must apply in a specific order to achieve the correct output. In alternations, the form with the widest distribution is chosen as basic, to allow for the simplest statement of rules. In a principle that came to be known as Pāṇini's theorem, the most specific rule (the one applying to the smallest class) applies first, allowing subsequent rules, applying to larger classes, to be stated more simply and generally. If the more specific rule applies, the more general one does not.

Kiparsky points out multiple ways in which Pāṇini's analysis pre-figures that of twentieth-century generative grammar, implicitly arguing that such elegant principles, independently arrived at across such a distance in space and time, must be correct. The comprehensiveness of Pāṇini's grammar was also admired by the American structuralist phonologist Leonard Bloomfield, who called the *Aṣṭādhyāyī*, "one of the greatest monuments of human intelligence" (Kiparsky 1994: 2923).

5.1.3 STRUCTURALIST MORPHOPHONEMICS AND PHONEMICS

Bloomfield (1939) applies Pāṇini's ordered-rule approach to a description of allomorphy in "Menomini," published in the *Travaux du Cercle Linguistique de Prague*. Bloomfield writes:

> The process of description leads us to set up each morphological element in a theoretical basic form, and then to state the deviations from this basic form which appear when the element is combined with other elements. If one starts with the basic forms and applies our statements . . . in the order in which we give them, one will arrive finally at the forms of words as they are actually spoken. (1939: 105)

By stipulating that his statements must apply in order, Bloomfield takes advantage of Pāṇini's theorem, allowing the statements that apply later to be more simple and general. Hockett (1958: 282) describes English allomorphy (such as *wife/wives*) in similar terms: the linguist must choose a base form so that the description of "the conditions which elicit each alternate shape" can be stated most simply. To take another example, Gleason's (1955) problem book instructs the student on how to solve a problem on the form of Samoan suffixes: "The best way to describe the structure is to set up a base form for each stem. If this is done correctly, all the forms can be derived from this base form by an affix or a regular morphophonemic change" (1955: 31).

While these approaches to allomorphy are derivational in that they rely on the idea of a single basic form for each morpheme, and an ordered set of statements whose application allows one to "arrive finally at the forms of words as they are actually spoken," two important considerations distinguish this approach from the derivation as it was proposed in SPE.

First, the statements of structuralist phonology are descriptive: they describe the conditions in which each alternant appears, but they don't actually change one form into another, and thus they are not transformational in the generative grammar sense. Saussure (1959: 159) addresses this point when he writes that it is possible to speak of "laws" relating different forms of a word, but these rules simply state a regular relationship between forms. The idea of one form changing into another is rejected:

> It is possible to speak of grammatical laws of alternation, but these laws are only a fortuitous result of the underlying phonetic facts. . . . The laws of alternation, like all synchronic laws, are simple structural principles; they are not imperative. It is completely wrong to say, as people so readily do, that the *a* of *Nacht* changes to *ä* in the plural *Nächte*, for this gives the illusion that a transformation governed by an imperative principle comes between one term and the next. What we are actually dealing with is a simple opposition of forms resulting from phonetic evolution.

In this view, related forms exist side by side, but, at least for synchronic phonology, nothing changes into something else.

Second, Bloomfield, Hockett and other structuralist phonologists believed in a strict separation of "morphophonemics" and "phonemics": "A neatly-packaged description of a language can set forth its phonemics, morphophonemics, and its grammar in separate compartments" (Hockett 1958: 274). Alternations between morphemes, necessitating reference to both meaning and to grammatical structure, were not considered to be part of phonology proper. While ordered rule-like statements might be used to describe morpheme alternations that substituted one phoneme for another, as in the examples from Menomini and Samoan above, the analysis of phonemes and allophones proceeded differently. When Bloomfield refers to "arriv[ing] finally at the forms of words as they are actually spoken" he is referring to accurately specifying the phonemes that constitute the morphemes, not to deriving the allophones of the phonemes by rule.

As described in Chapter 3, structuralist phonemic analysis proceeded by careful comparison of surface forms to one another, in order to determine the predictable vs unpredictable distribution of elements. That distribution was then used to determine systems of contrastive classes and contextual realization. This methodology is described in detail by Bloomfield (1926) and Hockett (1942, 1958); see also Anderson (1985a). Crucially, the mapping between phoneme and allophone is unambiguous and bi-directional. Every sequence of phones maps to one and only one sequence of phonemes, and every sequence of phonemes maps to one and only one sequence of phones, rendering the phonemic and allophonic levels co-present and interconvertible. This bi-directionality of the phoneme–allophone mapping (Bloch 1941) was encoded in the term "bi-uniqueness" (Harris 1944). According to bi-uniqueness, the phonemic representation can be unambiguously read off the surface pronunciation, without any influence of morphology or syntax.

The grouping of phones into phonemes could make no reference to meaning or to grammatical structure. "There must be no circularity; phonological analysis is assumed for grammatical analysis, and so must not assume any part of the latter. The line of demarcation between the two must be sharp" (Hockett 1942: 21). Further, Hockett declares, there must be "no mentalism" (p. 20). Contra Sapir, who argued in favor of "psychological reality" (1949), Hockett and other structuralists, prominently Bloomfield (1926, 1933), rejected the idea that there was anything to be gained from investigating a speaker's knowledge or intuition. All analysis was based on the corpus of actually-occurring surface forms.

In the 1950s, however, phonologists such as Zellig Harris and Morris Halle began arguing that the core principles of structuralist phonological analysis were at best cumbersome and unenlightening, and at worst completely unworkable. Harris is a figure who bridges structuralist and generative linguistics. He insisted that linguistic descriptions must be rigorous and explicit, but rejected the surface-based analysis that Hockett and Bloomfield embraced for the analysis of phonemes and allophones, and explored the mathematical concept of the "transformation" to describe sets of related items. Although Harris wrote a book entitled *Methods in Structural Linguistics* (1951), the traditional structuralist Fred Householder writes in a review (1952: 260) that "the word 'structural' in the title . . . evidently reflects some sort of change of heart during the composition of the book," since much of the methods described are not structuralist at all. To make the point clear, Householder repeatedly refers to all the non-structuralist parts, such as the rejection of bi-uniqueness, the positing of abstract representations, and any proposals that would suggest a universal logical structure underlying the syntax of all languages, as "hocus-pocus" (p. 260). Harris was a mentor of Noam Chomsky, and heavily influenced Chomsky's (1957) *Syntactic Structures*.

5.1.4 THE BEGINNINGS OF GENERATIVE GRAMMAR

The first true generative phonologist was Morris Halle, student of Roman Jakobson at Columbia and Harvard and colleague of Chomsky's at MIT. Chomsky and Halle met at Harvard while Halle was a graduate student and Chomsky a junior fellow, and they had many discussions about what they saw as the shortcomings of structuralist linguistics and about possible new directions. Halle's 1959 monograph *The Sound Pattern of Russian* lays out the principles of generative phonology, and Halle's reasoning for abandoning structuralist analysis. There are two main arguments.

First, Halle was aware of recent scientific advances that were outside the field of linguistics per se, and he wanted to apply them to the description of language. As was described in Chapter 4 on the development of distinctive feature theory, Halle and colleagues at the MIT Research Laboratory of Electronics wanted to make use of recently-developed acoustical techniques such as spectography (Kopp and Green 1946; Joos 1948) and apply those techniques to create a better understanding of language contrast. That is, he wanted phonology to be based on phonetics. The MIT researchers were also very interested in the new field of Information Theory (e.g., Shannon 1948), and in applying the ideas of maximally efficient communication to language. These ideas, as previously discussed, led Jakobson, Fant, and Halle to propose that all phonological analysis should be encoded in terms of a small set of binary features. Such a representation, Halle discovered, did not fit easily into a structuralist framework.

Second, Halle was becoming frustrated with "the traditional methods of descriptive linguistics" (1959: 11), particularly bi-uniqueness, strict separation between morpho-phonemic and allophonic alternations, and the insistence that all analysis had to be based solely on the surface distribution of sounds in the dataset, with no reference to meaning or grammar. Such insistence, he wrote, led to "labored," "counter-intuitive," and "not particularly enlightening" analyses, among which he numbered the "long lists of phoneme sequences" in his own 1955 Harvard dissertation on the Russian consonants (1959: 11). His particular example was Russian voicing assimilation (an example that continued to be important to researchers in the decades following).

Halle notes (1959: 21), that /m^{j}ok bi/ ("were he getting wet") and /m^{j}og bi/ ("could he") are pronounced identically (both [m^{j}ogbi]), and that from the utterance itself it is impossible to determine which was intended. Trying to do so, he argued, is just an exercise in frustration that requires either 1) trying to find minor phonetic differences that don't exist, or 2) a complicated system of underlying contrasts that treat alternating and non-alternating consonants differently, both of which undermine the idea of a small set of universal, binary, phonetically-based features. Rather than abandon a maximally efficient and symmetrical analysis to preserve bi-uniqueness (as Bloch 1941 does), Halle abandons bi-uniqueness. Instead of arguing that the phonemic and allophonic forms can be unambiguously read off each other, Halle argues that the allophonic form is *derived from* the phonemic representation. Because the transformation goes in one direction only, it can happen that two distinct phonemic representations end up with the same pronunciation. Halle also cites studies of stress (such as Trager and Smith 1951) that showed that phonology required access to grammatical information, further arguing for a derivation in the direction of morpho-phonemes to phonemes to allophones.

As discussed above, in a structuralist analysis (e.g., Hockett 1958: 137), generalizations were also specific to one level: generalizations that dealt with contrastive phonemic substitutions (the morpho-phonology) were separate from generalizations that dealt with non-contrastive/allophonic realizations (the phonology proper). Halle argued that this strict distinction was unworkable. He points out (1959: 22–3) that certain phonetic dimensions might be crucial contrastive information for some segments but non-contrastive detail for other segments, even within a single language. In Russian, he notes, /k/ and /g/ contrast in voicing, but /ʧ/ has no phonemic voiced counterpart. Yet both /k/ and /ʧ/ undergo voicing assimilation before voiced obstruents:

(1) Russian voicing assimilation (Halle 1959)

[mʲoklʲi]	was he getting wet?	[mʲogbi]	were he getting wet
[zeʧlʲi]	should one burn?	[zeʤbi]	were one to burn

Because of the difference in phonemic status, statements about possible and impossible sequences of obstruents in Russian would have to be made twice: once as a distributional constraint on phonemes to cover /k/and /g/, and again as a case of allophonic realization to cover [ʧ] and [ʤ]. Treating voicing in the two cases as completely different, Halle argues, misses a major generalization about Russian phonology: *all* obstruents undergo voicing assimilation. The conclusion is that the focus on phonemes and allophones, on completely specifying distributions in those terms, was misguided. The focus of phonology, Halle argued, should be on the generalizations themselves—the rules—and on making them as simple and comprehensive as possible, without undue concern about whether the generalization was phonemic or allophonic. Anderson (1985a: 321) says of *The Sound Pattern of Russian* that its most important innovation "lay in the emphasis it put on the centrality of *rules* in a phonological description" (emphasis original).

In the introduction to *The Sound Pattern of Russian*, Halle notes that he and his colleague Noam Chomsky had been thinking about a different way of doing linguistics. Their new approach took as the starting point the morphemes/phonemes intended by the speaker, and then determined the rules that would produce the correct pronunciation. It is these rules, which create the well-formed sequences of the language, that should be the focus of the phonologist. Halle puts it this way (1959: 12):

> I have assumed that an adequate description of a language can take the form of a set of rules—analogous perhaps to a program of an electronic computing machine—which when provided with further special instructions, could in principle produce all and only well-formed (grammatical) utterances in the language in question. This set of rules, which we shall call the grammar of the language and of which phonology forms a separate chapter, embodies what one must know in order to communicate in the given language.

This is the essential tenet of a derivational theory: that the grammar consists of a set of rules that take some linguistic input, and output a well-formed utterance. The derivation starts with the most abstract mental representation: the underlying representation. The underlying representation consists of morphemes composed of sequences of segments that are in turn composed of bundles of binary distinctive features. Then transformational rules add features, change features, or permute segments, step by step, until a structure corresponding to the actual pronunciation, the surface structure, is obtained.

There was much back and forth at conferences and in journals in the late 1950s and early 1960s between structuralist and the generativist approaches, which the generativists slowly won, as younger scholars adopted the new theory. After 1965 (according to Anderson 1985a) they pretty much stopped arguing and just got on with the new theory.

5.2 THE SPE MODEL

The tenets of generative grammar are most closely associated with Noam Chomsky. Chomsky earned his Ph.D. in Linguistics in 1955 at the University of Pennsylvania, under the mentorship of Zellig Harris. He taught first at Harvard, and then in 1957 joined the faculty at MIT, where he remained until his retirement from teaching (though not writing) in 2005. Chomsky is best known for his contributions to syntax (and politics); his contributions to phonology came mostly through collaboration with Morris Halle.

Chomsky's 1957 book, *Syntactic Structures*, sets up the paradigm for transformational generative grammar. While this first book focuses on syntax, it proposes the basics of generative theory, and defines the different components of the grammar and their relations, including syntax, semantics, phonology, and phonetics. *Syntactic Structures* was followed in 1959 by Halle's *Sound Pattern of Russian*, which Halle notes was strongly influenced by his conversations with Chomsky, and then a decade later (1968) by their collaborative work, *The Sound Pattern of English*.

Generative grammar, as proposed by Chomsky, differed in numerous important ways from structuralist linguistics (see Anderson 1985a, Chapter 12 for extensive discussion). Chomsky argued that linguistic theory should not be just explicit and logically rigorous, but should make predictions about what can and cannot occur in Language (in general), rather than simply describing and providing a taxonomy for the set of particular languages that the linguistics community has come into contact with. Structuralist linguistics, he argued, was too concerned with coming up with a mechanical procedure that would work to analyze any corpus of data.

As the quote above from *The Sound Pattern of Russian* makes clear, generative grammar is explicitly "mentalist," concerned with what the speaker must "know," very much against the prevailing idea that "psychological reality" was "hocus-pocus," as Householder claimed (1952: 260). As Chomsky and Halle argue in SPE (1968: 3), generative grammar is a theory of *competence*, "what the ideal speaker/hearer knows about his grammar," rather than *performance*, which is what any given speaker actually produces at any given time. Language is not purely learned behavior, a habitual verbal response to an external stimulus (as argued by Bloomfield (1933) and Skinner (1957)), but is the output of a highly constrained mental system.

Generative grammar is also very much concerned with universals—aspects of grammar that all languages have in common, and which Chomsky argues are innate properties of the human mind. Linguistic theory should generate "a system of hypotheses concerning the essential properties of any human language . . . properties [that] determine the class of possible natural languages" (Chomsky and Halle 1968: 4). This "universal grammar" contains both "formal universals" that define the form of grammars, basically permissible structures and types of rules, and "substantive universals" that define the sets of elements, such as morphological categories and phonetic features.

In transformational generative grammar, all components of the grammar consist of rules. The different levels of the grammar are defined by the representations that the rules manipulate. The derivation begins with the syntax generating a tree structure that elaborates the constituent structure of the intended utterance: the subject, the verb phrase, the objects of the verb, and so on. Generative syntactic theory has undergone profound changes in the decades since *Syntactic Structures* and SPE, well beyond the scope of what can be even mentioned here, so there is little point in elaborating on the tree structure of the syntactic component. For our purposes, we can note that in the original derivational model, syntactic "phrase structure rules" reorder or delete portions of the tree that constitute the sentence. The result of the application of phrase structure rules, the output of the syntax, is the "surface structure" (Chomsky and Halle 1968: 9) which consists of "terminal elements" that include lexical items, suffixes, and so on. The syntactic surface structure example from SPE, for the sentence "We established telegraphic communication," is shown in Figure 5.1.

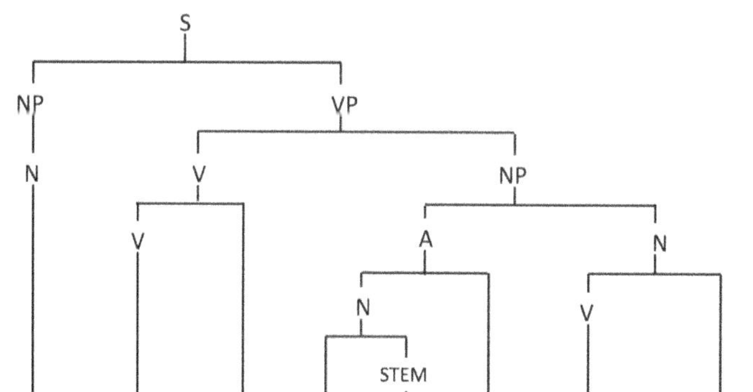

A. +we+ +establish+ +past+ +tele+ +graph+ +ic+ +communicate+ +ion+

B. $[\,[\,[\text{we}]_N\,]_{NP}\,[\,[\,[\text{establish}]_V\ \text{past}]_V\,[\,[\,[\text{tele+graph}]_N\ \text{ic}]_A\ [\,[\text{communicate}]_V\ \text{ion}]_N\,]_{NP}\,]_{VP}\,]_S$

C. # # # we # # # # # establish # past # # # # tele+graph # ic # # # communicate # ion # # # #

Figure 5.1 Syntactic surface structure for the utterance "We established telegraphic communication"

Source: Adapted from Chomsky and Halle (1968: 9–10).

As shown in the figure, the tree representation can be linearized to a bracketed string. The linearized syntactic output is then submitted both to the phonological component, to be transformed into a pronounceable string, and to the semantic component, to be transformed into an interpretable string. Again, we must leave discussion of the interface between syntax and semantics for other texts.

At the syntax/phonology interface are certain necessary "readjustment rules" (p. 9ff.) that render the syntactic structure suitable for phonological interpretation. Morphemes are spelled out: *sing+past* becomes *sung*, and *mend+past* becomes *mend-d*, and so on. The readjustment rules may also alter the syntactic bracketing: for example, long syntactic elements may be broken down into smaller phrases, or some brackets may be eliminated (see further discussion of phrasing in Chapter 7). Finally, the readjustment rules spell out lexical items as strings of segments, and syntactic brackets as boundary symbols (in SPE, #, as in Figure 5.1). The boundary symbols show depth of embedding but provide no information on syntactic or morphological categories, to which the phonology is assumed to have no direct access. Although syntax and morphology feed into phonology, the phonological rules themselves have access to, and are able to manipulate, only segments and boundaries.

The "surface structure" of the syntax is the "underlying representation" of the phonology. It is the job of the phonology, then, to "take the transformed terminal strings, which consist entirely of special kinds of segments and of boundaries, and complete the assignment of phonetic features to these symbols" (p. 26). The first step of the phonological derivation is to represent segments as bundles of distinctive features. While SPE uses the symbols of the phonetic alphabet, these are always to be understood as abbreviations for bundles of features (p. 5).

As was discussed in Chapter 4, SPE spends considerable space on feature theory, revising the set of features proposed by JFH to be more workable to express alternations in addition to contrasts. Importantly, in the underlying representation, only a minimum number of contrastive features is specified. Following the tenets of Information Theory, economy and efficiency is prized (Halle 1959: 29). Only enough information is specified to keep all utterances distinct. Features predictable from other features or from surrounding context are left out of the underlying representation. It is the job of the rules to fill them in.

An example of underlying featural representation from SPE (Chomsky and Halle 1968: 166) is shown in Table 5.1. Zeroes represent unspecified features.

While the underlying representation of generative phonology often seems to correspond to the phonemic level of structural analysis, SPE itself does not use the term "phoneme." Chomsky and Halle explicitly state (p. 11) that they will not use the term "phoneme" because they argue that no particular level corresponds to the "phonemic" level as it had been defined in structuralist linguistics. Further, there is no requirement that an underlying segment contain an "invariant core" of features that define it at every level of analysis, as Jakobson envisioned for the definition of the phoneme. Rather, features are specified underlyingly as needed to keep lexical items distinct, and are changed subject to whatever is stated in the set of phonological rules.

Table 5.1 Underlying featural representation of the words "inn" and "algebra"

	inn		algebra						
	i	n	æ	l	g	e	b	r	æ
consonantal	–	+	–	+	+	–	+	+	–
vocalic	0	0	0	+	–	0	–	+	0
nasal	0	+	0	0	–	0	–	0	0
tense	–	0	–	0	0	–	0	0	–
stress	0	0	0	0	0	0	0	0	0
voice	0	0	0	0	+	0	+	0	0
continuant	0	0	0	0	–	0	–	0	0

Source: Chomsky and Halle (1968: 166).

Important to the discussion of the phonology/phonetics interface, recall that Chomsky and Halle argue that the distinctive feature set is universal, set by human biology. "The total set of features is identical with the set of phonetic properties that can in principle be controlled in speech; they represent the phonetic capabilities of man and, we would assume, are therefore the same for all languages" (pp. 294–5). They argue that a universal phonetic component establishes the set of features, as well as universal constraints on what features can be simultaneous (e.g., [-consonantal] and [+strident] are universally incompatible). Some sequences (e.g., long strings of consonants) are also prohibited by universal phonetic principles. Note that this echoes the view of Sapir, who describes non-contrastive differences as "purely mechanical phonetic variations" (1949: 47) and refers to "absolutely mechanical phonetic laws" (p. 49).

By requiring that rules refer to features that have phonetic content, the class of possible rules is limited (Chomsky and Halle 1968: 170). Rules apply to natural classes, and the natural classes are defined by the intersection of their phonetic features. (It turned out, as Chomsky and Halle themselves recognized, that this concession to "naturalness" did not go nearly far enough in constraining possible phonological rules. We return to this topic at length in Chapter 6.)

In the course of a derivation, the underlying form is subject to the set of the phonological rules of the grammar, which change feature values according to the context, or insert predictable feature values ("lexical redundancy rules"). Chomsky and Halle (p. 332) argue that all phonological rules are of the form A → B / X __ Y, "A is changed to B in the context of X and Y," where A, B, X, and Y can be feature matrices, boundary symbols, or null. Some examples of rules (both simple and complex) are shown in Examples (2–4).

(2) Non-nasal consonants agree in voicing (SPE rule #11)

$$\begin{bmatrix} +\text{cons} \\ -\text{voc} \\ -\text{nasal} \end{bmatrix} \rightarrow [\alpha\ \text{voice}] \ / \ \_\_\_ \begin{bmatrix} +\text{cons} \\ -\text{voc} \\ -\text{nasal} \\ \alpha\ \text{voice} \end{bmatrix}$$

(3) Dipthongization (SPE rule #21)

$$\emptyset \rightarrow \begin{bmatrix} -\text{voc} \\ -\text{cons} \\ +\text{high} \\ \alpha\ \text{back} \\ \alpha\ \text{round} \end{bmatrix} \quad / \quad \begin{bmatrix} +\text{tense} \\ \alpha\ \text{back} \end{bmatrix} \ \underline{\qquad}$$

(4) Glides are deleted after palato-alveolars (SPE rule #122)

$$\begin{bmatrix} -\text{cons} \\ -\text{voc} \end{bmatrix} \rightarrow \emptyset \ / \ \begin{bmatrix} +\text{cor} \\ -\text{ant} \\ -\text{sonor} \end{bmatrix} \ \underline{\qquad}$$

The whole set of rules apply in sequence, such that the output of rule *i* becomes the input to rule *i+1*.

In SPE, Chomsky and Halle treat, in order, the stress system, the vowel system, and then the consonant system of English. The rules can be quite complex and rely on relatively abstract lexical representations, and much of SPE is thus closer to morphology than to phonetics.

For the stress system in particular, Chomsky and Halle argue (e.g., p. 60) for the "transformational cycle" to account for the fact that morphology, particularly the addition of affixes, can change stress patterns. Cyclic application means that the set of rules applies more than once. Recall that the string of segments contains boundary symbols indicating morphological and syntactic constituency. The set of rules applies first on the inmost constituent. Then the innermost brackets are erased and the rules apply again, iterating until all internal brackets within the phrase have been removed. Allowing cyclic application makes the statement of stress rules much simpler than trying to do everything based on a single surface string, and allows for sensitivity to morphological structure.

In addition, the derivation allows for greater abstraction and for "opacity"—the statement of rules that are not "surface true." The structural description of a rule may be met in underlying representation, or at some intermediate stage, but after the rule applies another rule might alter the context, making the rule application "opaque." McCarthy (2007), arguing for the necessity of a derivation even in constraint-based phonologies, points out that "classical structuralist phonology was based on the premise that all authentic phonological generalizations are categorically true statements about the distribution of allophones" (p. 2), but "the idea that phonological knowledge is reducible to surface-true generalizations has turned out to be an intellectual dead end" (p. 3). The derivation creates intermediate stages, "temporary truths" (p. 3). Phonological opacity remains a challenge for non-derivational phonological theories. It is a strength of the derivational model that it handles opacity straightforwardly.

At the end of the last cycle of rule application, every segment will be fully specified with the appropriate feature values. Phonology having completed its work, the representation is sent to the phonetics.

In the SPE model, features are both "classificatory," defining inventories and classes needed for rule application, and "phonetic," referring to phonetic content (Chomsky and Halle 1968: 297). The phonology/phonetics interface consists of filling in numeric feature values, changing the classificatory binary features into numeric phonetic scales representing the "degree of intensity" with which that feature is realized in a particular language. For example, a [+voice] stop in Russian might be [8 voice] and a [+voice] stop in English might be [2 voice]. The example is made up, because while SPE assumes rules of phonetic implementation, examples are not given, because "our interest in sound structure, in this book, does not extend to matters such as degree of aspiration, degree of fronting of vowels, etc." (p. 65, see also p. 164). Because the features are defined as phonetic parameters that represent all the controllable aspects of pronunciation, it is assumed that these scalar features can be interpreted as a set of instructions to the articulatory system, or as a code for the perceptual system.

This phonetic interpretation is argued to be universal and non-linguistic. Since the features represent "the phonetic capabilities of man," fully specifying the features gives a complete set of instructions for articulatory implementation and acoustic interpretation. Phonetic "interpretation" is the same for all languages. Language-specific differences (such as amount of voicing, aspiration, length, etc.) are encoded only in different numerical specifications of the distinctive features.

Chomsky and Halle discuss in their introduction why they place so much emphasis on morpho-phonology, such as levels of stress, and so little emphasis on differences such as gradations of aspiration: because their interest is in linguistic theory and "the nature of mental processes." They state (p. vii):

> It seems to us that the gradations of stress in English can be explained on the basis of very deep-seated and non-trivial assumptions about universal grammar, and that this conclusion is highly suggestive for psychology, in many ways that we will sketch. On the other hand, gradations of aspiration seem to shed no light on these questions, and we therefore devote no attention to them.

If the phonology of stress is deep-seated, non-trivial, and capable of providing insight into universal grammar, then by comparison, phonetics is shallow, trivial, and of no psychological or linguistic interest. While SPE was anti-structuralist in many ways, it did keep the structuralist assumption that phonetics was outside the purview of grammar.

5.3 NON-UNIVERSAL PHONETICS

It didn't take long, however, for phoneticians to counter that the SPE model of the phonology/phonetics interface was too simplistic. (Other shortcomings of the SPE model are discussed in Chapter 6.) An early paper that makes this point is Keating (1985). Keating points out that because in the SPE model phonetics is assumed to be universal and automatic, a function only of biology, anything language-specific or planned must be part of the phonology. She then goes on to show three examples of small, non-contrastive details of phonetic implementation that must be planned. It is known, for example, that low vowels are in general longer than non-low vowels (Lindblom 1967). This *might* be a biomechanical effect: since the jaw has to travel farther, it just takes longer. But Keating found that speakers use greater initial force (as measured by electromyography (EMG)) for low vowels. Recruiting greater energy before the movement begins shows that the greater duration must have been planned. Keating then points out that in many languages, vowels are shorter before voiceless consonants and longer before voiced: a non-contrastive sub-featural phonetic detail according to SPE. But not every language shows this effect, and different languages have different degrees of lengthening. Further, in Russian and German, vowels are longer before *underlyingly* voiced consonants, even those that have undergone word-final devoicing. If vowel lengthening were a result of automatic phonetic implementation, lengthening would be consistent across languages, and could be sensitive only to the amount of consonant voicing found on the surface. The third example is word-final devoicing itself. While there are good aerodynamic reasons to prefer devoicing in final position and voicing in medial position, word-final devoicing and word-medial voicing are far from automatic. Keating's conclusion (1985: 123) is that "[e]ach language must specify its own phonetic facts by rule." If everything rule-based is "grammar," then phonetics is part of the grammar as much as phonology is.

> I am suggesting here that we consider all phonetic processes, even the most low-level, to be phonologized (or grammaticized) in the sense that they are cognitively represented, under explicit control by the speaker, and once-removed from (that is, not automatic consequences of) the physical speaking machine. (p. 128)

Phonetics is rule-governed, and also part of the grammar.

Other examples of "low-level" phonetic rules that are shown by phonetic research to be language-specific, and thus necessarily learned and controlled rather than automatic, include velar fronting (Keating and Cohn 1988; Keating and Lahiri 1993), vowel nasalization (Huffman 1989; Cohn 1990), degrees of voicing and aspiration (Kingston 1985), and vowel coarticulation (Whalen 1990). In one of Whalen's experiments, talkers were instructed that they would be producing one of four bisyllables: [abi], [abu], [api], or [apu]. In half the trials, they saw a computer screen with "AB__" or "AP__," but the second vowel did not appear until the talker began to pronounce the token. In the other half of trials, the speaker saw "A __ U" or "A __ I" and the consonant appeared as soon as the talker began to speak. In the latter condition, when the second vowel was known, there was a significant influence on the formant structure of the first vowel: talkers adjusted the first vowel to be more like the second. In the former condition, where the second vowel could not be planned for, there was no influence. Whalen argues that if coarticulation were automatic it would be seen in both conditions. The conclusion, he argues, is that contra SPE, "coarticulation is largely planned" (p. 3).

The complexity, planning, and language-specificity of phonetic implementation is further explored in Kingston and Diehl's (1994) discussion, "Phonetic knowledge." Kingston and Diehl argue that speech implementation only seems/feels automatic because it is "so thoroughly overlearned" (p. 420). Nonetheless, they argue that speakers have control over very small details of phonetic implementation, and use them to enhance phonological contrast. There are certainly physiological constraints and influences on speech—for example, tongue raising pulls on the hyoid bone, which pulls on thyroid cartilage, which stretches the vocal folds and raises pitch. But speakers can control this influence, and either accentuate it or compensate for it, to optimize their phonetic behavior, either to increase clarity or decrease effort. They use the analogy of driving a car: while experienced drivers don't directly control each wheel bearing, they know how the car is going to respond to every slight turn of the wheel. Speakers aren't directly conscious of raising the hyoid bone, but are fully capable of controlling the strap muscles of the neck to get the required effect.

Following up on Kingston's (1985) work, Kingston and Diehl discuss the feature [voice] as an example of speaker-controlled phonetics. First, they note that the feature [voice] is realized in many different ways, with different degrees of vocal fold tensing or other vocal tract accommodations such as pharyngeal expansion required to maintain the contrast in different contexts. Thus the feature [voice] is not a one-dimensional command to a single parameter setting, but a goal that requires complex and varied implementation involving multiple parts of the vocal tract. They go on to note, however, that regardless of how the [voice] contrast is realized, F0 is uniformly lowered adjacent to [+voice] stops. They argue that this change would not be automatic across all contexts, but instead is an active accommodation that speakers make in order to enhance the difference between voiced and voiceless stops. In general, different "sub-properties" of voiced stops (duration, low-frequency energy, F1 lowering and voice onset time, in addition to F0 lowering) can be emphasized to enhance a contrast. The speaker knows what the auditory/perceptual effect of different changes will be, and controls them. This is a far more complex picture of "phonetic knowledge" than the difference between [8 voice] and [2 voice].

Keating (1990a, 1990b, 1996) proposes an explicit model of the translation of phonological features to phonetic parameters, taking into account the complex relationship between features and articulators, and addressing the need for language-specific phonetic implementation. She calls this the "targets and interpolation" model (1990a: 331), and in later work the further elaborated "window model of coarticulation" (1990b, 1996). In this derivational model of the phonology/phonetics interface, the output of the phonology and input to the phonetics consists of features organized into segments, consistent with the SPE view of the interface. The phonological features are then translated into continuous, quantitative articulatory and acoustic "parameters" such as vocal fold tension or velum opening (or on the acoustic side, closure duration and formant values). The mapping between features and parameters, however, is many-to-many rather than one-to-one (1990a: 331). The articulatory parameter of "vocal fold opening," for example, is influenced by the features [voice], [spread glottis], and [constricted glottis]. At the same time, the feature [voice] may also recruit parameters of larynx lowering and pharyngeal expansion. The different parameters are recruited, as was also argued by Kingston

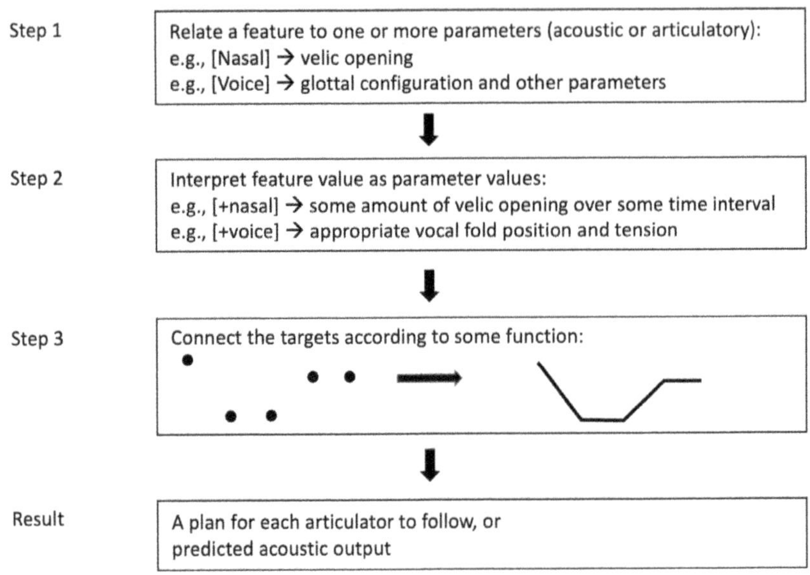

Figure 5.2 Keating's model of targets and interpolation

Source: Adapted from Keating (1996: 264).

and Diehl, "so as to guarantee a robust and salient instance of the defining property of the feature value" (p. 332). Each mapping specifies a language-specific target value for each parameter. Parameter targets have extent in time, and may be a range of values (a "window") rather than an exact specification. Once the targets are specified, interpolation functions connect the targets to create a smooth trajectory. For the listener, there is a corresponding mapping from perceived acoustic parameters back to features. The target–interpolation model is illustrated in Figure 5.2, adapted from Keating (1996).

In the "window model of coarticulation" (Keating 1990b, 1996), the size of the window of allowable values is related to the presence or absence of contrast. If there is contrast, the window of acceptable values is small, because contrast must be maintained. If there is no contrast, the window is large (or no target is specified) so contextual effects are greater. In careful articulation, windows may narrow; in more casual or fast speech, they may widen.

Keating's model thus maintains the central SPE concepts of a derivational relation between phonology and phonetics, and of binary distinctive features that are instantiated in phonetic parameters, but allows for a more complex mapping that is language- and context-specific. Keating notes that in this derivational approach, phonology and phonetics cannot be thought of as two different ways of looking at the same phenomena. Rather, it must be the case that "phonology and phonetics are accounts of different phenomena—different levels in the grammar which carry out different computations" (1996: 276). This has come a long way from Saussure, who argued that the true linguist must ignore phonetics.

5.4 LEXICAL AND POST-LEXICAL PHONOLOGY

At the same time that phoneticians were developing theories of phonetics as modules of the grammar, phonology itself was becoming more modularized. Phonologists including Anderson (1974, 1975), Mascaró (1976), Rubach (1981), Kiparsky (1982, 1985), Mohanan (1982), and Kaisse and Shaw (1985), among others, argued for separating out the phonological rules that interface with morphology ("lexical" phonology) from those that are more clearly linked to phonetics ("post-lexical" phonology), and then for separate levels within each of these. While Chomsky and Halle argued for the cyclic application of the whole block of phonological rules in a language, these later phonologists argued for assigning different rules to different components. The theory was called "Lexical Phonology" (Kiparsky 1982, 1985; Mohanan 1982) because it sought to explain the interaction of phonological rules with morphology and the lexicon. The key insights are summed up in Kenstowicz (1994: 195–243).

Kenstowicz notes that there are two kinds of rules in phonology, which display different clusters of properties. The lexical rules:

- can access morphological information, such as root vs affix or part of speech;
- may have exceptions;
- may not have clear phonetic motivation;
- apply cyclically, and only in "derived environments" (that is, simplifying somewhat, they require the addition of an affix or some new morphological element to trigger their application); and
- are "structure preserving" (that is, they may change one contrastive segment into another, but will not alter the language's basic underlying inventory by creating new non-contrastive segments, or filling in non-contrastive features).

Kenstowicz's example of a lexical rule is the English rule of tri-syllabic laxing, proposed by Chomsky and Halle, which relates the tense vowels in certain morphemes to a lax counterpart when certain suffixes are added, as shown in Table 5.2. The "tri-syllabic" part of the name references the fact that the lax vowel is most often third from the end in the derived form.

Each of the tense and lax vowels is part of the underlying contrastive inventory of English, thus the rule is structure preserving. It references morphological information in that it is only triggered by certain suffixes: the rule applies to "gradual" but not "gradation," for example, and does not apply in non-derived environments, such as mono-morphemic "ivory." Even with the right suffix, however, certain morphemes, such as "obesity," are unpredictably exempt. Further, there is no clear phonetic motivation: why should [aɪ] change to [ɪ] in the third from last syllable of a suffixed word? The answer lies in the history of English, when the suffixed and unsuffixed forms had short and long versions of the same vowel, but the contemporary learner of English has no access to that historical information in the process of acquisition, so for her the rule must be learned as unmotivated.

In contrast to lexical rules, the post-lexical cluster of rules has the following properties:

Table 5.2 Tri-syllabic laxing

Alternation	Examples
aɪ ~ ɪ	div**i**ne / div**i**nity subl**i**me / subl**i**minal der**i**ve / der**i**vative
i ~ ɛ	ser**e**ne / ser**e**nity pl**ea**se / pl**ea**sure imp**e**de / imp**e**diment
e ~ æ	gr**a**de / gr**a**dual gr**a**ve / gr**a**vity prof**a**ne / prof**a**nity
aʊ ~ ʌ	prof**ou**nd/prof**u**ndity ab**ou**nd / ab**u**ndance pron**ou**nce / pron**u**nciation
u ~ a	g**oo**se / g**o**sling f**oo**d / f**o**dder sch**oo**l / sch**o**lar
o ~ a	m**o**de / m**o**dify s**o**le / s**o**litude prov**o**ke / prov**o**cative

Source: Examples from Zsiga (2013a: 421).

- don't apply cyclically;
- apply "across the board," (that is, whenever the phonological environment for application is present, with no exceptions and no reference to morphological or syntactic category);
- have clear phonetic motivation;
- where rule-ordering can be shown, apply after the lexical rules.

The classic example of a post-lexical rule is English tapping: /t/and /d/ become taps between vowels, the second of which is unstressed, as in "kitty, kiddie, electricity," and so on. There are no exceptions to the rule, which applies in nouns, verbs, derived and underived forms, and even across word boundaries, when the structural description is met: "atom, better, fatter, add 'em up." A tap is not a contrastive phoneme of English, but a contextual allophone of /t/ or /d/. The theory of lexical phonology argues that these clusters of properties always occur together: for example, a rule that references morphological information will never create a non-underlying segment. All phonological rules, it is argued, can be classified into lexical or post-lexical types, depending on the cluster of properties displayed.

Further, within the domain of lexical phonology, certain (blocks of) affixes are found to trigger certain (blocks of) rules. So, for example, the "level 1" affixes of English such as "in-," "-al" and "-ity" trigger assimilation ("imbalance"), "velar softening" ("electric/electricity"), stress changes ("parent/parental"), and tri-syllabic laxing ("divine/divinity"), while the "level 2" affixes, such as "un-," "-hood" and "-cy" do not trigger any of these ("unbalanced, parenthood, piracy").

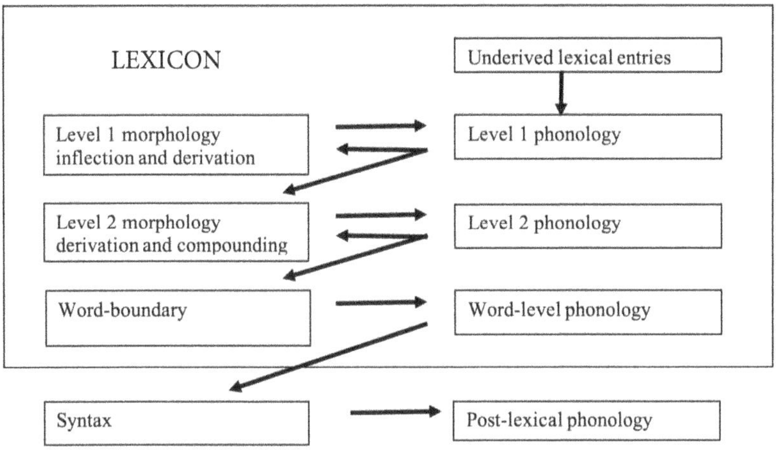

Figure 5.3 Level-ordering in lexical phonology
Source: Adapted from Kiparsky (1982: 5).

In order to make level-ordering work, phonology and morphology must be interleaved, as in Figure 5.3 (based on Kiparsky 1982: 5). The lexicon is structured as a system of levels, each of which is associated with a set of phonological rules, which apply each time an affix from that level is added. The output of each cycle of rules is then sent back to the lexicon to pick up a new morpheme, and then will undergo the rules associated with that morpheme.

The model becomes quite complex, with levels multiplying in order to account for different morphological effects: Kiparsky (1985) argues for three levels in English, Mohanan (1982) for seven in Malayalam. Even so, assigning rules to blocks also turned out to be an oversimplification, as affix ordering, rule application, and level ordering did not always perfectly align. More recent theories continue to investigate the interaction of phonology and morphology, and propose new models of that interface. Specific models of the phonology/morphology interface are beyond the scope of what can be covered here, but references include Kiparsky (2000), Itô and Mester (2003), Pater (2010), Scheer (2011), and Inkelas (2014).

For the discussion of the phonology/phonetics interface, the crucial component is the block of post-lexical rules, which apply across the board, and which have clear phonetic motivation. The problem is, if the phonetic component, or phonetic implementation, also consists of a set of learned, language-specific rules, what is the difference between post-lexical phonology and phonetics? Kaisse (1990) argued for two types of post-lexical rules, which she terms P1 and P2 rules, mirroring to some extent the modularization of the lexical component of lexical phonology. The P1 rules are phonetically-motivated and apply across the board, but are still categorical and apply consistently at all rates of speech. The P2 rules (such as some cases of devoicing) are still considered by Kaisse to be part of the post-lexical phonology, but are gradient and variable, and depend on the style of speech. They may also be called the "fast speech rules." However, there seems to be no distinction between what Kaisse would call a P2 rule and what Keating would call a phonetic rule. How can one tell the difference between a phonetically-motivated phonological rule, and a rule of the phonetics per se? Where does the boundary between phonology and phonetics lie?

5.5 PHONOLOGICAL AND PHONETIC RULES

Underlying the discussion of lexical phonology is the assumption that if the processes are assigned to the right stratum—lexical, post-lexical, or phonetic—the overall system is simpler and better understood. Properties of the alternations come from the characteristics of the stratum in which they operate, not from special properties of the rules themselves, or the segments on which they operate. Thus, a number of studies began to argue that if rules that were thought to be phonological could be reassigned to the phonetics, the phonology would be the better for it.

An early argument along these lines was Hayes (1984), "The phonetics and phonology of Russian voicing assimilation." Hayes addresses the anomalous behavior of Russian /v/, which sometimes behaves like a sonorant and sometimes like an obstruent. In Russian, obstruents assimilate in voicing to the rightmost obstruent in a cluster, as was shown above (Halle 1959):

(5) Russian voicing assimilation (Halle 1959)

[mʲoklʲi]	was he getting wet?	[mʲogbi]	were he getting wet
[zeʧlʲi]	should one burn?	[zeʤbi]	were one to burn

Sonorant consonants neither trigger nor undergo voicing assimilation. The segment /v/, however, has an intermediate status. Like a sonorant, it fails to trigger assimilation (/s-vami/ "with you" is pronounced [svami] not *[zvami]); but like an obstruent, it undergoes devoicing (/korov-ka/, "little cow" becomes [korofka]). If all the facts about the pronunciation of Russian /v/ are assumed to be phonological, then the solution to the behavior of /v/ requires a set of otherwise unmotivated ordered rules. A purely phonological account, according to Hayes, would argue that /v/ is underlyingly /w/, and as such, along with all other sonorants, it fails to trigger voicing assimilation in an adjacent obstruent. Then, according to this account, *all* segments, obstruents, and sonorants alike are subject to devoicing in a cluster with a voiceless obstruent. After devoicing has occurred, [w] is "strengthened" to [v] (and [w̥] to [f]). Once that one sonorant is changed to an obstruent, then all remaining sonorants are *revoiced*, leaving /v/ as the only underlying sonorant that can be voiceless on the surface.

This works; but the problem, of course, is that there is no independent evidence that the other sonorants have undergone the devoicing and revoicing process—they look as though they were simply voiced all along. Such "Duke of York" rule ordering schema (Pullum 1976), where a subsequent rule reverses the result of a previous one (as the Duke of York, in the nursery rhyme, "marched them up to the top of the hill and then marched them down again"), are problematic for phonological theory. If a rule is going to apply, there ought to be some independent evidence for it in the output. What would prevent a grammar from containing multiple pairs of rules in which A makes a change that B then reverses, leaving no trace of either rule application? Hayes solves the Duke of York problem in Russian by proposing that the "revoicing rule" is not phonological, but phonetic. He argues that all sonorants do indeed undergo a rule assigning [-voice] if they occur in a cluster that ends with a voiceless obstruent. But he then argues that the feature [+/− voice] references an articulatory state, not necessarily actual vibration, and he defines [-voice] as a laryngeal configuration that results in lack of vibration for obstruents, but not necessarily for sonorants (due to the higher airflow in sonorant configurations). On this account, sonorants can undergo phonological devoicing, defined as assignment of the feature [-voice], but still surface as phonetically voiced. No phonological rule of "revoicing" is needed.

Thus, Hayes argues, if certain rules (such as Russian devoicing) are recognized as phonological feature assignment, and other rules (such as Russian "revoicing") are recognized as phonetic implementation, the overall system is simplified. There is no need for special conventions, exceptions, or otherwise unmotivated rules to account for what looks like special cases (such as Russian [v]). The overall grammar is made simpler and more general when phonological and phonetic rules are distinguished and correctly assigned.

Liberman and Pierrehumbert (1984) make the same point. In their paper, they argue for a simplified representation of intonation that relies on only two primitives, H (high tone) and L (low tone). In order to allow for this simplified phonology, however, they require a more sophisticated model of phonetic implementation (involving quantitative parameters such as a speaker's overall pitch range and degree of final lowering) to derive actual pitch trajectories from abstract sequences of tones. "Better understanding of phonetic implementation," they write, "removes from phonology the burden of representing those sound patterns for which its natural descriptive mechanisms are inappropriate. As a result, the theory of phonological features and rules should be clarified and simplified" (p. 229). Much more discussion of intonation, and the Liberman and Pierrehumbert system, is found in Chapter 8. The point here is that Liberman and Pierrehumbert argue that by separating out the complicated and speaker-specific phonetic rules, a simpler, more general phonology is revealed.

A more recent example comes from McCarthy (2009, 2011). McCarthy argues for a phonological account of vowel harmony that relies on alignment between features and domains, and thus cannot countenance "directional" spreading, where a language might have two different harmony processes, one that applies from left to right and the other from right to left. He is confronted with the counterexample of Akan ATR vowel harmony, as described by Clements (1981). According to Clements, Akan has two rules of ATR harmony: one that applies in the rightward direction (from root to suffix), the other that applies in the leftward direction (from root to prefix). Crucially, however, the leftward spreading is gradient: Clements describes Akan leftward-spreading ATR vowel harmony as producing "increasingly raised variants in a gradual 'crescendo' as the conditioning syllable is approached" (Clements 1981: 157). McCarthy argues that if this leftward spreading is dismissed as phonetic coarticulation rather than phonological rule, the counterexample is eliminated and the alignment theory is unscathed.

If it's important to assign rules to the correct module, there must be criteria (other than the need to sweep theoretically inconvenient rules onto the other side of the fence) for deciding what's phonological and what's phonetic. Liberman and Pierrehumbert ask: "Once we have granted that phonetic implementation is non-trivial, how can its appropriate domain of application be determined?" (1984: 230). Their argument was made from intonation, but applies to any phonological process. Given a particular process, be it devoicing, vowel harmony, epenthetic stop insertion, aspiration, nasalization, or vowel lengthening, how is one to tell whether it belongs to the phonology or to the phonetics?

For SPE, the criterion was language-specificity—anything non-automatic had to be phonology—but that criterion was shown by researchers such as Keating to be incorrect. Any rule that met the criteria for a "lexical" rule (reference to morphology, necessarily cyclic application, structure preservation, lack of phonetic motivation) was clearly phonological, but that criterion would reassign all of the post-lexical phonology to the phonetics. This is in fact the position taken by Ohala (1990), who, as discussed in Chapter 2, argues that the "phonology/phonetics interface" is a distinction without a difference. All phonetically-motivated rules, he argues, are simply phonetics, and there is nothing to be gained by stating them in the form of abstract algebraic transformations. Morphologically conditioned "rules," on the other hand, are simply a matter of selection of different allomorphs from the lexicon. Thus, Ohala argues that there is no need for formal phonology at all. As mentioned in Chapter 2, he claims that formal phonologists are guilty of "reification" and "myopia" (among other failings), their insistence on dealing in abstractions causing them to miss the really interesting generalizations that are found in phonetics.

Less radically, Kingston and Diehl (1994) argue that the dividing line is not morphology but structure preservation. Following a line of reasoning that goes back to the definition of the domain of phonology advocated by Saussure and Trubetzkoy, they argue that phonology deals only with contrasts, phonetics with non-contrastive detail. Kingston and Diehl differ from the structuralists, of course, in their insistence on the complexity and language-specificity of that non-contrastive knowledge. Nonetheless, allowing phonology to deal only with contrast still leaves all allophonic realization, and again much if not all of the post-lexical phonology, to the phonetics.

Most commonly, however, the dividing line was taken to be categorical vs gradient application. Phonological rules apply categorically, changing a segment from one featural category to another, and phonetic rules apply gradiently, resulting in more or less of given property. This division is argued explicitly by Keating's (1990a, 1990b, 1996) articles on the definition of the phonology/phonetics interface: "The phonetic component of the grammar is concerned with quantitative operations, and . . . the phonological component of the grammar is concerned with operations on symbols" (1990a: 333). Phonological features indicate labels that apply to segments as a whole, while phonetics deals with "numbers on continuous dimensions" (1996: 263).

An extended example of distinguishing categorical and gradient processes is Cohn's (1990, 1993) examination of rules of nasalization in French, Sundanese, and English. Using nasal airflow data, Cohn shows that rules of nasalization in some cases produce a steady plateau of nasality across an entire segment, and in other cases a "cline" or

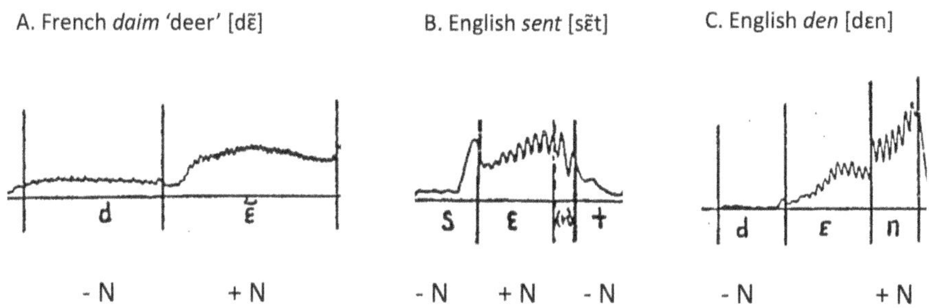

Figure 5.4 Nasal airflow showing categorical and gradient nasalization
Source: Data from Cohn (1990), adapted from Cohn (2006: 27).

gradually increasing nasal airflow. The former rules, because they produce a segment of the category [+nasal], are phonological. The latter rules are phonetic. In the case of phonetic nasalization, the affected segment is not changed to [+nasal], but receives some degree of nasal coarticulation. Specifically, in terms of Keating's window model, the nasalized segment simply demonstrates a transition between a [-nasal] specification on one side and a [+nasal] specification on the other. Example airflow traces are shown in Figure 5.4. Nasalization in French "daim" *deer*, pronounced [dɛ̃], is categorical. Some cases of English nasalization, as in "sent" pronounced [sɛ̃t], are pronounced with similarly steady, categorical nasalization, and are thus argued to be the result of phonological rule. In other cases, as in "den," the gradually changing "cline" of nasalization diagnoses a product of phonetic coarticulation.

Cohn (2006) revisits the results of her earlier work, asking the question "Is there gradient phonology?" The 2006 article confirms the earlier view that the kind of partial assimilation seen in Figure 5.4C is still indicative of phonetics rather than phonology, but that gradient acceptability judgements and variable application are phenomena that phonology itself has to deal with. See Section 5.6 below, and further discussion in Chapter 10.

The position that categorical vs gradient application marks the boundary between phonology and phonetics is defended from a somewhat different theoretical perspective by Zsiga (1993, 1995, 1997). These papers argue for a derivation in which phonological features are interpreted by the phonetics as articulatory gestures, rather than as parameters that are both acoustic and articulatory as in Keating's window model. This view follows the theory of Articulatory Phonology (Browman and Goldstein 1986, 1990c) in arguing that a number of processes that had been described as phonological are better captured by specifying the target position, duration, and coordination in time of articulatory gestures (see Chapter 9). Zsiga's account differs from standard Articulatory Phonology in arguing that not all phonology is gestural. Rather, the argument is that categorical phonology should be expressed in terms of categorical features, which are then mapped into gestures at the phonology/phonetics interface. Following Keating and Cohn, Zsiga adopts the position that categorical application diagnoses a phonological (featural) rule, and gradient application a phonetic (gestural) process.

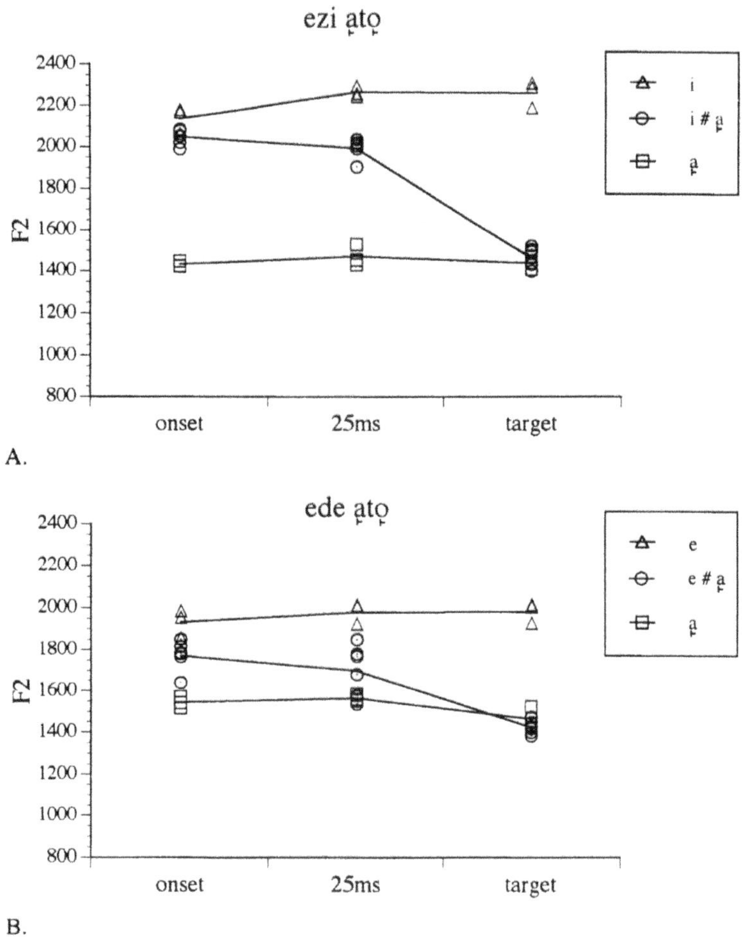

Figure 5.5 Igbo vowel assimilation at word boundaries

Source: Zsiga (1997: 245).

One example is Igbo vowel assimilation. (Another case, that of English palatalization, is discussed in Chapter 9.) In Igbo, when two vowels become adjacent at a word boundary, the first assimilates to the second. The result, however, is again gradient and partial. Figure 5.5 (from Zsiga 1997: 245) shows the result of assimilation in /i#a/ and /e#a/ sequences. The value of the second formant at the beginning of the sequence, and at 25ms into the first vowel, is intermediate between the values expected for the first vowel and the values expected for the second.

This partial and intermediate assimilation is impossible to accurately characterize as a feature-changing phonological rule: the [-back] or [-high] feature can't spread halfway from one vowel to the previous. The gradient application can be explained if vowel assimilation is understood as (phonetic) gestural overlap rather than (phonological) featural assimilation.

Other processes (cited by Keating 1996) that were in previous accounts written as feature-changing phonological rules, but are now argued to be attributed to phonetics, either gestural overlap or interpolation between targets, include English final consonant deletion (Browman and Goldstein 1986), Korean lenis stop voicing (Jun 1995), vowel allophony in Marshallese (Choi 1992), and velar fronting in a number of languages (Keating and Lahiri 1993).

These publications are all examples of the "border wars" referenced by Myers (2000). Phonology and phonetics are defined as separate domains, with different primitives and different operations, and researchers argue over whether a particular process belongs on one side of the fence or the other. Myers concludes (2000: 260) with a list of properties that define the two domains. Only phonology can reference morphological categories or morpheme boundaries: "There are no gradient phonetic patterns that distinguish among morpho-syntactic categories." Only phonology allows lexical exceptions. Only phonetics can be sensitive to rate, style, carefulness or other aspects of the "speaking situation." Only phonetics is gradient. And neither phonology nor phonetics is automatic, universal, or purely biologically given, as was originally argued by SPE.

5.6 A POROUS BORDER?

The whole idea behind modularization, whether lexical vs post-lexical phonology, or post-lexical phonology vs phonetics, is that each module has well-defined properties that are mutually exclusive of those of other modules. Each module is simplest and works best when it does not have the "burden of representing those sound patterns for which its natural descriptive mechanisms are inappropriate" (Liberman and Pierrehumbert 1984: 229). In a derivational modular system, as has been described in this chapter, the output of one module is the input to the next: lexical phonology feeds into post-lexical phonology feeds into phonetics.

Both modularity and the derivation have been questioned. Pierrehumbert (1990b) began to question the derivational relationship between phonology and phonetics. In this article she confirms the claim that phonology is categorical and phonetics is gradient, but argues that attempts to actually and accurately map features into acoustics or articulation via a "syntactic" transformation have not been successful. Rather, she argues that the relation between phonology and phonetics is "semantic," not syntactic. "[P]honetics describes what phonological entities mean in terms of events in the world" (p. 392). A semantic interpretation is not "derivational" in the sense of SPE at all.

While many researchers continue to argue for separate modules for phonology and phonetics (e.g., Reiss 2007), others have argued against modularization, and these non-modular approaches are discussed in detail in the following chapters. Pierrehumbert et al. (2000) argue that the cutting edge of the field has "moved beyond dualism." Cohn (2006: 43) offers the schematic in Figure 5.6.

The modular approach is illustrated in Figure 5.6A, with non-overlapping circles. Phonology and phonetics are separate domains. Another possibility is Figure 5.6B: there is no difference between what is phonological and what is phonetic. The more

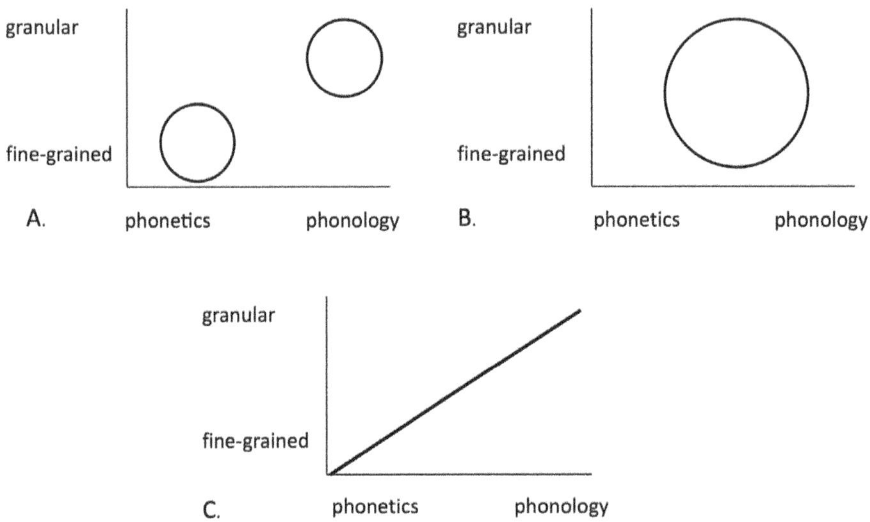

Figure 5.6 Three different views of the phonology/phonetics interface
Source: Adapted from Cohn (2006: 43).

nuanced possibility is Figure 5.6C: a continuum between "fine-grained" phonetic knowledge and "granular" phonological knowledge. Speakers can be sensitive to both, and different kinds of knowledge and processes can be explored without the need to draw a definitive line separating the phonology from the phonetics.

In current theories, there are two arguments against rule-based derivational phonology. The first is that phonology is not distinct from phonetics, and thus can be absorbed into it (Ohala 1990; Pierrehumbert et al. 2000). The other is that abstract phonological rules are not phonetic enough. Chomsky and Halle pointed out the difficulty themselves (1968: 400):

> The entire discussion of phonology in this book suffers from a fundamental theoretical inadequacy. Although we do not know how to remedy it fully, we feel that the outlines of a solution can be sketched, at least in part. The problem is that our approach to features, to rules, and to evaluation has been overly formal. Suppose, for example, that we were systematically to interchange features or to replace [αF] by [-αF] (where α = +, and F is a feature) throughout our description of English structure. There is nothing in our account of linguistic theory to indicate that the result would be the description of a system that violates certain principles governing human languages. To the extent that this is true, we have failed to formulate the principles of linguistic theory, of universal grammar, in a satisfactory manner. In particular, we have not made any use of the fact that the features have intrinsic content. By taking this intrinsic content into account, we can, so it appears, achieve a deeper and more satisfying solution to some of the problems of lexical redundancy as well as to many other problems that we have skirted in the exposition.

On page 402, they extend the naturalness problem to inventories:

> To take another example, our evaluation measure makes no distinction between a language in which all vowels are voiced and one in which all vowels are voiceless, or between a language in which obstruent clusters are redundantly voiced and a language in which they are redundantly voiceless. But surely one case is much more natural than the other.

Chomsky and Halle argue here that rule schema and distinctive features can be made to say anything, even rules and inventories that are the exact opposite of what actually happens in human language. It's not just the form of the rules and features, but their content, that must be addressed. It was the idea of "naturalness," the idea that phonological rules must have phonetic content, that led many phonologists to abandon derivational theories, and open new ways of thinking about the phonology/phonetics interface. This is how Prince and Smolensky phrase it (2004: 4) in their opening description of why a different approach to phonology, one without rules, is needed:

> The standard phonological rule aims to encode grammatical generalizations in this format:
> A → B / C _ D
> The rule scans potential inputs for structures CAD and performs the change on them that is explicitly spelled out in the rule: the unit denoted by A takes on property B. For this format to be worth pursuing, there must be an interesting theory which defines the class of possible predicates CAD (Structural Descriptions) and another theory with defines the class of possible operations A → B (Structural Changes). If these theories are loose and uninformative, as indeed they have proved to be in reality, we must entertain one of two conclusions:
> (i) phonology itself simply doesn't have much content, is mostly "periphery" rather than "core", is just a technique for data-compression, with aspirations to depth subverted by the inevitable idiosyncrasies of history and the lexicon; or
> (ii) the locus of explanatory action is elsewhere.
> We suspect the latter.

It is to the challenge of naturalness that Chapter 6 now turns.

RECOMMENDED READING

Overviews

Anderson, S. (1985b), "Generative phonology and its origins," in *Phonology in the Twentieth Century: Theories of Rules and Theories of Representations.* Chicago: University of Chicago Press, Chapter 12.

- Discuss this quote: "The conception of a language as a system of rules (rather than a set of representations) . . . lies at the heart of Halle's sound pattern of Russian" (p. 323).

Keating, P. A. (1990b), "The window model of coarticulation: articulatory evidence," in J. Kingston and M. Beckman (eds), *Papers in Laboratory Phonology I.* Cambridge, UK: Cambridge University Press, pp. 451–70.

- What is Keating's view of the phonology/phonetics interface in this article? What aspects of the SPE model does she retain and what aspects does she reject?

Exemplary research

Something (really) old

Kiparsky, P. (1994), "Pāṇinian linguistics," in R. E. Asher (ed.), *Encyclopedia of Language and Linguistics.* Oxford and New York: Pergamon Press, pp. 2918–23.

- What does Pāṇini's grammar have in common with twentieth-century generative grammar?

Something new(er)

Keating, P. A. (1985), "Universal phonetics and the organization of grammars," in V. Fromkin (ed.), *Phonetic Linguistics: Essays in Honor of Peter Ladefoged.* Orlando: Academic Press, pp. 115–32.

Kingston, J. and R. L. Diehl (1994), "Phonetic knowledge," *Language, 70*: 419–54.

- How do the findings of Keating and of Kingston and Diehl undercut SPE arguments for universal and mechanistic phonetics?

Opposing views

Prince, A. and Smolensky, P. (2004), *Optimality Theory: Constraint Interaction in Generative Grammar.* Oxford: Blackwell.

vs

Halle, M. (1959), *The Sound Pattern of Russian.* The Hague, The Netherlands: Mouton.

- Read the first chapter of both Halle (1959) and Prince and Smolensky (2004). What are Halle's arguments in favor of rule-based grammar? What are Prince and Smolensky's arguments against it?

QUESTIONS FOR FURTHER DISCUSSION

1. What aspects of structuralist phonological analysis did Halle argue against in *The Sound Pattern of Russian*? What is the significance of the forms below?

		[mʲogbi]	could he
[mʲoklʲi]	was he getting wet?	[mʲogbi]	were he getting wet
[zetʃlʲi]	should one burn?	[zedʒbi]	were one to burn

2. Discuss how this quote (Chomsky and Halle 1968: vii) encapsulates the SPE conception of the phonology/phonetics interface:

> It seems to us that the gradations of stress in English can be explained on the basis of very deep-seated and non-trivial assumptions about universal grammar, and that this conclusion is highly suggestive for psychology, in many ways that we will sketch. On the other hand, gradations of aspiration seem to shed no light on these questions, and we therefore devote no attention to them.

How was the concept of a universal feature set an important part of this view?

3. Where did different derivational approaches to the phonology/phonetics interface (Chomsky and Halle 1968; Kaisse 1990; Keating 1990a; Kingston and Diehl 1994; Zsiga 1995) place the dividing line between phonetics and phonology? How do these relate to the position of Trubetzkoy in *Foundations of Phonology*? To the position of Ohala (1990)?
4. How is the concept of modularity central to a derivational approach to the phonology/phonetics interface? How do the approaches diagrammed in Figures 5.6A and 5.6B reject modularity in different ways?
5. What did Chomsky and Halle mean by the terms "competence" and "performance"? What do the terms mean for their approach to the phonology/phonetics interface? In what way is a focus on competence a rejection of structuralist principles? Is competence the same as Kingston and Diehl's "phonetic knowledge"?
6. Are Prince and Smolensky (2004) advocating a return to a more structuralist approach, where all generalizations are based on surface forms?

CHAPTER 6

MARKEDNESS, NATURALNESS, AND ABSTRACTION

6.1 WHAT IS "MARKEDNESS"?

Markedness theory asks two related questions. The first is "What is universal, or at least common, across languages?" De Lacy (2006: 1) defines markedness this way: "Certain structures are often avoided while others are generated; the avoided structures are called 'marked' while the generated ones are 'unmarked.'" Thus, the first part of markedness theory is to study what is and isn't avoided. The second question is "Why?" Can we define markedness in a way that allows us to *predict* what will be avoided? As noted by Hayes and Steriade (2004: 5), research in both phonetics and phonology

> suggests that a connection can be found between constraints governing the production and perception of speech and markedness patterns. Certain processes (cluster simplification, place assimilation, lenition, vowel reduction, tonal neutralisation) appear to be triggered by demands of articulatory simplification, while the specific contexts targeted by simplification (e.g., the direction of place assimilation, the segment types it tends to target) are frequently attributable to perceptual factors.

To the extent that we can identify patterns that are common and "natural," do they arise because of properties of the language code itself (a "formal" approach), properties of language use in the world (a "functionalist" approach), or a combination? If a combination, then how do formalist and functionalist pressures interact? How is language use in the world reflected in the language code?

The question of markedness thus causes us to look at the phonology/phonetics interface from a different perspective. A derivational approach to phonology and phonetics as described in Chapter 5 considers the two disciplines as separate, and the main questions concern where the dividing line between them lies and what properties separate one from the other. In markedness theory, the phonetic constraints of articulation and perception define a set of properties: those that are "natural." The main questions then focus on the extent to which these properties are reflected in the phonology per se. Is all phonology necessarily natural? If not, how abstract and "unnatural" can phonology be? What is the role of phonetics in determining what is a possible rule? As noted in the discussion of lexical and post-lexical phonology, it is easy, for many rules, to find a "phonetic grounding" in that the output seems to be obviously easier to articulate or to hear. But other rules don't seem to have any phonetic grounding at all. Which is the default case? Are phonological rules, in general, phonetically based, with occasional special consideration for non-phonetic cases, or is phonology essentially independent of phonetics, with phonetic conditioning accidental, historical, and/or beside the point?

More thorough examinations of the role of markedness in phonological theory include Hume (2003), de Lacy (2006), Haspelmath (2006), and Rice (2007). Here, we consider more specifically what studies in markedness tell us about the phonology/phonetics interface. In this chapter, we begin in Section 6.1 by discussing different definitions of markedness that have been offered. Section 6.2 looks at the broad question of the roles of naturalness and abstraction in phonology, concluding that there must be some room for both. Section 6.3 then surveys the main points of Optimality Theory (Prince and Smolensky 2004), a currently broadly-accepted theory that takes a specific view of the role of markedness constraints in phonological theory. Section 6.4 considers how Optimality Theory makes it possible for detailed phonetic information to be directly accessed by the phonological component, eliminating the modularity imposed by a rule-based derivational theory. Section 6.5 considers the theory of Evolutionary Phonology (Blevins 2006), which argues that markedness plays no role in synchronic phonology, and that language universals can be explained via diachronic processes that are parallel to biological evolution. Learning biases, which allow for some middle ground between purely formal and purely functional approaches, are discussed in Section 6.6, and Section 6.7 closes the chapter with a summary of different views of the role of markedness in phonology.

When Trubetzkoy defined his poles of opposition in phonology, he argued that one pole would always be the "marked" option (1969: 75ff.). That is, for every phonemic contrast, one member of the pair will have a default value, termed "unmarked," and the other member will be "marked" as having a different quality. Trubetzkoy argued that the values can be determined by examining cases of neutralization: neutralization of contrast always consists of "removing a mark," so that the result of neutralization will always be the unmarked value. Thus, voicelessness is the unmarked option for obstruents, because final voicing neutralization involves removing the mark of voicing. Trubetzkoy further notes that the unmarked option will be more frequent (not independent of the facts of neutralization): in the absence of a contrast, either in a given position or across the whole inventory, only the unmarked value will be present.

Cross-linguistically, then, markedness could be defined as an implication: the presence of the marked value in a language implies the presence of the unmarked. If a language has voiced stops, it will also have voiceless; if a language has affricates, it will also have fricatives, and so on. Thus, any sound that is universally present across languages is unmarked.

Jakobson developed the idea further, adding other criteria for defining markedness. In his *Child Language, Aphasia and Phonological Universals* (1968), he argued that the segment with the marked value would be learned later by children, and lost sooner in aphasia. The order of segment acquisition, he argued, was the same for all languages and all children, and the order of segment loss was universally its inverse. Jakobson posited the markedness ordering shown in (1), where "<" means both "is less marked" and "is acquired earlier and lost later."

(1) Markedness scales from Jakobson (1968)
consonants: p, t, m, n < b, d < k, g, ŋ < fricatives < affricates < l, r
vowels: i, a < u < e, o

Identifying absolute universals, however, has proved to be elusive. Jakobson's predicted order of acquisition turned out not to be the case. Dunbar and Idsardi (2016) point out that even deciding when a contrast is officially "acquired" can be tricky—does the child have to just produce the sound, or produce it consistently in all contexts in which an adult does? Dunbar and Idsardi also find that the proposed sequence does not hold universally, citing, among other counterexamples, Macken (1978) who found that Spanish children acquire fricatives before voiced stops, and Pye et al. (1987) who found that Quiché children acquire [t͡ʃ], [ʃ] and [l] very early.

Universals of cross-linguistic occurrence are also difficult to substantiate. Greenberg (1963) and Comrie (1981) propose lists of universals, almost all syntactic, usually having to do with word order (e.g., if the basic word order of a language is verb–subject–object, then the language will use prepositions rather than postpositions (Greenberg 1963: 78)). Linguists at the University of Konstanz have compiled a "Universals Archive" that aims to catalog every claim of universality in the linguistics literature, including typology, phonology, syntax, and semantics (https: // typo.uni-konstanz.de/archive/). The website, as of this writing, lists 2,029 proposed universals or universal tendencies, 543 pertaining to phonology. Of these, 422 are implicational (e.g., if a language has [t], [d] and [dʰ], it will also have [tʰ] (Jakobson 1968)). Some "universals" are better substantiated than others, but almost all have documented counterexamples (e.g., Blust (1973) claims that the Austronesian language Kelabit has [t], [d] and [dʰ] but not [tʰ]).

There *are* absolute, unconditional universals in phonology—true all the time of every language—but they are few, and very simple (as pointed out by Maddieson (1984) and by Hyman (2008)). Based on available databases, we can safely say that:

- All languages have consonants and vowels.
- All languages contrast sonorants and obstruents.
- All languages have stops, and for all languages, stops contrast in place of articulation.

But for phonological absolutes, that's about it. There are some very-near universals, but these also have been argued to have counterexamples. Almost all languages have /t/, except Hawai'ian and Samoan. All languages have CV syllables, except maybe Arrernte.

One can, of course, also posit formal universals, referencing abstract structures, such as "all languages have syllables" or "all languages respect the prosodic hierarchy." But as Hyman (2008: 113) points out,

> Anyone determined to maintain [formal universals such as syllable structure] can continue to do so, the worst consequence being an indeterminate or more awkward analysis. One can establish syllables in Gokana, assign extrasyllabic consonants to syllables in Bella Coola and Piro, and reanalyze Arrernte in terms of CV syllables. Architectural universals have this property: It all depends on your model and on what complications you are willing to live with.

If you are convinced that a language must be analyzed in terms of a specific abstract structure, it is probably possible to do so.

Nonetheless, even if absolute universals can't be found, there are real asymmetries, or "statistical universals." Almost all languages have /t/ (perhaps three or four out of 7,000 don't). Almost all languages have CV syllables (maybe one doesn't). Even if the absolute universal is not true, the asymmetries still have to be explained. It is these asymmetries and implications that the theory of markedness aims to address.

Asymmetries, if not perfect universals, abound. But what determines which value is marked and which is unmarked in a given opposition? *Why* is one value more common than the other? Phonologists may refer to that one value as being "simpler" or "more natural" or "easier" to either produce or perceive. Rice (2007: 80) lists the properties of segments (Table 6.1) that linguists have used to describe marked and unmarked sounds.

Some of the characteristics listed by Rice are purely phonetic, such as "harder to articulate"; others, such as "unlikely to be epenthetic" are purely phonological. From

Table 6.1 Properties of marked and unmarked sounds

Marked	Unmarked
Less natural	More natural
More complex	Simpler
More specific	More general
Less common	More common
Unexpected	Expected
Not basic	Basic
Less stable	Stable
Appear in few grammars	Appear in more grammars
Later in language acquisition	Earlier in acquisition
Early loss in language deficit	Late loss in language deficit
Implies unmarked feature	Implied by marked feature
Harder to articulate	Easier to articulate
Perceptually more salient	Perceptually less salient
Smaller phonetic space	Larger phonetic space
Subject to neutralization	Neutralization targets
Unlikely to be epenthetic	Likely to be epenthetic
Trigger of assimilation	Target of assimilation
Remains in coalescence	Lost in coalescence
Retained in deletion	Lost in deletion

Source: Rice (2007: 80).

the outset, Trubetzkoy insisted on phonological criteria only, and resisted defining markedness in any phonetic way. In fact, he argues against a claim by Zipf (1935) that the cross-linguistic frequency of a segment is determined by how complicated the sound is, with less complicated sounds being more frequent. Trubetzkoy counters that "complicated" is hard to quantify, and argues instead that markedness and therefore frequency should be defined solely in phonological terms based on the facts of neutralization. He rejects trying to "explain phonological facts by means of biological, extra-linguistic causes" (1969: 262).

As pointed out by Ohala (1990), there is a grave danger of circularity in trying to *explain* "markedness" in terms of "naturalness." Why do we say a sound is unmarked? Because it's more common. Why is it more common? Because it is more natural. How do we know it's more natural? Because it's unmarked. Adjectives such as "simple" and "natural" do not have any definition that's independent of the distribution of sounds that they are called on to explain. Further, some of the characteristics that make a sound unmarked are probably mutually exclusive: sounds that are easier to perceive are generally harder to produce, and vice versa. If one can call on either ease of perception or ease of production as a justification, some phonetic reason for any proposed universal or universal implication can usually be found. As Odden (1996: 154) states the problem, "[A] substantive independent theory of phonetic motivation seems unattainable at present. Natural phonological correlations are generally arrived at by observation of databases; then, based on the assumption that such correlations have a phonetic basis, a plausible phonetic explanation is constructed." An example is the argument, discussed below, over whether an [mb] or [mp] cluster should be considered more marked.

Haspelmath (2006) surveys the many different ways the term "markedness" is used in linguistics, including phonological, morphological, syntactic, and semantic studies. He concludes that the term has become too vague to be useful, and that the different uses, even in phonology, are not really a single phenomenon that should have a single label.

> An abstract markedness notion, whether intended as a claim about the cognitive code, as a convenient metagrammatical term, or as an explanatory concept in an autonomous theory of markedness . . . only makes it harder to understand the cross-linguistic and language-particular patterns. (2006: 58)

Linguists would do better, he argues, to just refer directly to cross-linguistic frequency or ease of articulation, if that's what they mean. Haspelmath does not take a stand, however, on exactly how these more "substantive factors" (2006: 27) should be incorporated into phonological theory.

Regardless of exactly how the term "markedness" is defined, asymmetries and implications remain, and linguists seek to explain these patterns. It is true that infants (in general!) say [mama] and [papa] before [t͡ʃet͡ʃe] and [ruru], that more languages have [t] than have [q], that there is something hard about getting your tongue in the right configuration for English [ɹ] or for a pharyngeal fricative, and that [f] and [θ] are easily confused. Cross-linguistically, neutralizations do tend to go in a certain direction: post-nasal position favors voicing, and word-final position favors voicelessness. Phonetic tendencies are documented: opening the velum increases airflow and thus promotes voicing (Hayes and Stivers 1995), consonants are easier to identify when they have an audible release (Repp and Lin 1989). The question isn't whether such tendencies exist, but how they influence phonological patterns. Some alternations seem very clearly phonetically motivated. Others don't. To the extent that we know what sounds are "marked" and "unmarked," what difference does it make?

6.2 NATURAL AND UNNATURAL PHONOLOGY

As discussed in Chapter 2, it was axiomatic in the early definitions of the field that phonology should not take phonetic content into account. It was probably Hjelmslev that took this position to the extreme, arguing, that "language is a form and that outside that form . . . is present a non-linguistic stuff, the so-called substance" (1953: 49), and that linguistics should be "an algebra of language, operating with . . . arbitrarily named entities without natural designation" (p. 50). It was impossible to ignore, however, the fact that the "entities" out of which languages built inventories and classes were *not* in fact arbitrary, and did overwhelmingly correspond to "natural" designations of articulation and perception (vocal fold vibration, velum opening, fricative noise, etc.), as codified for example in the feature definitions of Jakobson et al. (1952) or Chomsky and Halle (1968).

Nonetheless, it was recognized that the fit between phonetic description and phonological patterning was not always perfect. Fudge (1967) argued against the strongly phonetic basis of the distinctive feature set proposed by JFH, arguing that the features needed to describe phonological alternations would not always correspond to a single phonetic parameter. Fudge wrote that there is a

> lack of fit of phonological classes with phonetic ones over the totality of the world's languages. . . . The logical conclusion of this is that phonologists (above all generative phonologists) ought to burn their phonetic boats and turn to a genuinely abstract framework. (1967: 26)

While SPE worked with phonetically-based *features*, the phonological *rules*, with their binary designations and logical operators, were completely algebraic. Chomsky and Halle themselves saw this as a problem, as was discussed earlier. They write (1968: 400), as quoted at the end of Chapter 5, that SPE had "a fundamental theoretical inadequacy," in being "overly formal." They go on to suggest (1968: 402) that the problem might be addressed at least in part by returning to some of the ideas of Trubetzkoy and Jakobson, incorporating "the Praguian notions of marked and unmarked values of features into our account in some systematic way." This task, however, was left for others.

As many phonologists saw it, the problem was that the algebraic formalism of SPE was too powerful. Feature-changing rules could clearly and concisely represent alternations, but there was no theory of why some alternations were common, and others, that were equally easy to represent, were vanishingly rare or non-existent. That is, there was no theory of what was and wasn't "natural." Push-back against the "overly formal" nature of SPE thus included proposals to change the structure of the grammar to privilege natural changes over unnatural ones. As Anderson (1981: 528) put it, if SPE is "too abstract" we must "build into phonological theory some set of constraints on the kind of relation that can exist between phonological and phonetic representations."

An early alternative to the overly-formal nature of SPE was the theory of Natural Phonology (Stampe 1979; Dressler 1984; Donegan and Stampe 1979, 2009 and references therein). (Articulatory Phonology (Browman and Goldstein 1986, 1989, 1992a; Goldstein et al. 2006), which has some close affinities to Natural Phonology, is discussed in Chapter 9.) Natural Phonology argues that there is a set of universal natural "processes" that either enhance perception ("fortitions") or ease production ("lenitions"). These arise from the nature of the human speech perception and production apparatus, they do not need to be learned, and "apply unwilled and unconsciously" (Donegan and Stampe 2009: 2). What needs to be learned are the language-specific "inhibitions" that override the processes and result in language-specific variation.

An example (from Donegan and Stampe 2009) is the interaction of denasalization and nasalization. Denasalization is a fortition that requires all continuants to be oral, thus maximizing formant clarity. Nasalization is a lenition that spreads nasality from a nasal stop to neighboring vowels. Inhibition of denasalization results in the emergence of contrastive oral and nasal vowels, and inhibition of nasalization stops the creation of allophonic nasal vowels, as shown in Table 6.2.

In English, denasalization applies, so there are no contrastive nasal vowels, and nasalization also applies, so nasal vowels occur only as allophones adjacent to nasal stops. In Hawai'ian, denasalization applies and nasalization is inhibited, so all vowels are oral in all contexts. In French, both are inhibited, so nasal and oral vowels are contrastive in all positions. The fourth possible combination is illustrated by Hindi: denasalization is inhibited, so nasal and oral vowels contrast in oral contexts, and nasalization applies, so nasal vowels are also derived in nasal contexts. The point is that both patterns of contrast (inventories) and allophonic variation are driven by principles fully grounded in articulation and perception. Unnatural morphophonological alternations can also be learned, but these are not really phonology. "Our system," writes Dressler (1984: 40), "does not exclude [morpho-phonological rules], but assigns to them, deductively and correctly, a marginal rather than a central role." Because natural, universal processes are at the center of phonology, phonology as a whole will tend to be natural.

Table 6.2 A Natural Phonology account of nasalization

Process:	If applied, results in:	If inhibited, results in:
nasalization (lenition)	allophonic nasality on vowels	no allophonic nasalization
denasalization (fortition)	no contrast: all vowels are underlying oral	contrastive nasality on vowels

Source: Donegan and Stampe (2009).

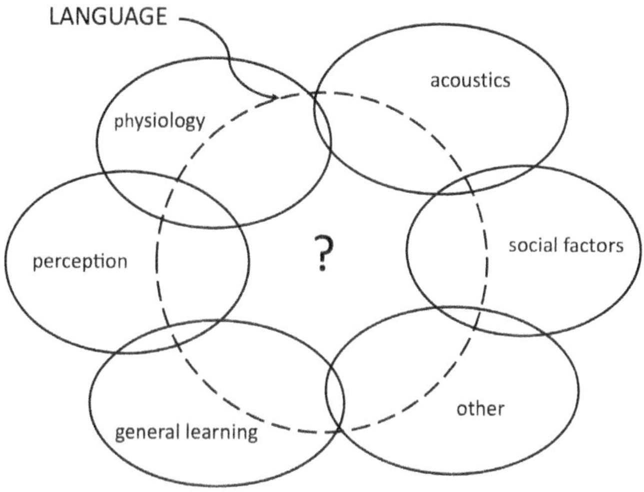

Figure 6.1 *Linguistics should study what's left of language patterns when all the other explanations have been factored out*

Source: Adapted from Anderson (1981: 494).

This thesis of the central role of phonetic grounding in phonology is explicitly argued against by Anderson (1981), in the pointedly-titled "Why phonology isn't 'natural'." Anderson argues that it is exactly that part of phonology that *can't* be explained by general properties of perception and production that linguists ought to be concerned with, and offers the diagram in Figure 6.1. Some parts of language can be explained by "functional" factors such as phonetics, learning mechanisms, and social pressures, but not all, and not, in his view, the most interesting. In making this claim, he follows Saussure, and is himself followed by, for example, Hale and Reiss (2008) and Reiss (2018).

There isn't any particular reason, Anderson argues, that we should expect that phonology as a module of linguistics should be grounded in phonetics. Requiring that phonology be based in phonetics requires "limiting the scope of inquiry in linguistics in arbitrary and undesirable ways" (1981: 496). "The language faculty is an intrinsically autonomous aspect of human mental organization" (p. 535), influenced, but not determined, by other considerations. Anderson argues that if we are to have an adequate phonology, we cannot a priori constrain how abstract underlying representations can be, or the kinds of rules that phonologies might require.

For phonological features, Anderson argues, the exact phonetic realization cannot be read directly off the features, and the phonological features cannot be read directly off the phonetic realization. Anderson follows Ladefoged (1980) and others (see Chapter 4) in arguing that the mapping from phonological contrast to phonetic implementation is complex and many-to-many, and the set of phonologically-adequate features cannot be directly translated into phonetic scales. For example, what counts as a [+high, +back] will be very different across languages. Conversely, two sounds that might be identical in articulation and acoustics could have different featural representations. A [w] in one language might be revealed to be a velarized labial because it alternates with [b], yet be pronounced identically to [w] in another language that is revealed to be a labialized velar because it alternates with [g]. Such cases do not argue against phonetic grounding per se, but they do argue that the relation between phonology and phonetics is not always so simple as the more straightforward cases would seem to imply.

Anderson further argues that even phonological rules that seem natural cannot be completely explained by "mechanical" principles, because there is always some degree of language-specific arbitrariness. We may be "tantalizingly close" to complete phonetic explanation, but phonology is not under the "rigid deterministic control" of phonetics, and the rule cannot be completely "identified with" its phonetic underpinning (1981: 509). For example, because vowel nasalization is not the same in all languages, the "process" cannot be universal, as Donegan and Stampe claim it to be. For their part, Donegan and Stampe (2009) and Dressler (1984) argue that Anderson is attacking a straw man here: processes can be cross-linguistically universal without being completely mechanistic.

More difficult to address are Anderson's examples of "crazy rules," a term popularized by Bach and Harms (1972). As both Anderson and Bach and Harms point out, phonological rules tend over time to become detached from their phonetic grounding. Rules may arise out of mechanical phonetic tendencies, but once they are phonologized (that is, take on a controlled, systematic role in the language) their status changes, and they are unmoored from their phonetic conditioning. (See also the discussion of Evolutionary Phonology below.) Phonetics may determine how phonological rules get started, but once phonologized, rules are not constrained by phonetics. Examples cited by Anderson include velar fronting in Icelandic (once all front vowels caused velar fronting, but now only some do), compensatory lengthening (only some consonant deletions cause vowel lengthening), and vowel allophony in Breton (there is no phonetically-transparent way to state which vowels alternate in which contexts).

The counterargument is, if markedness has nothing to do with synchronic phonology, why aren't seemingly-random patterns more common? "Crazy" rules exist, but some functional pressure keeps them from predominating. One account of the interaction of formal and functional pressures in phonology is offered by de Lacy and Kingston (2013). They propose a Venn diagram that is similar to Anderson's (Figure 6.2).

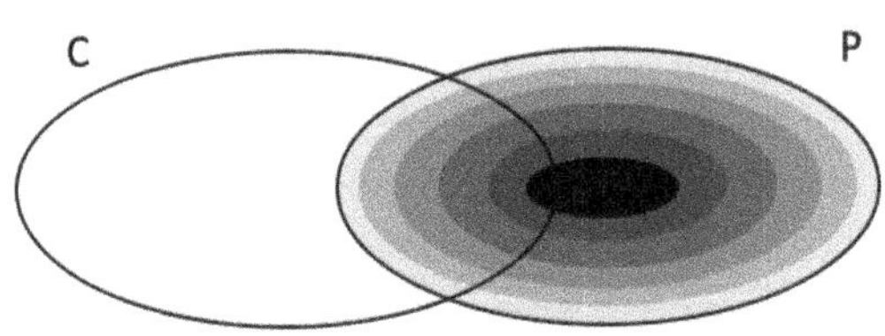

Figure 6.2 The intersection of possible "formal" grammars ("C"), and possible "functional" grammars ("P")

Source: Adapted from de Lacy and Kingston (2013: 343).

In the diagram, "C" represents all grammars that phonological formalism allows. (The "C" stands for "competence" as in formal phonological knowledge.) These may be natural and unnatural, including the English of SPE, and English', with all feature values reversed. "P" (for "performance," as in extra-linguistic factors) represents all the grammars that could arise due to functional pressures. De Lacy and Kingston argue that many of the grammars in P also don't exist—for example, grammars that allow epenthesis of dorsal consonants, that could easily arise due to sound change. C and P constrain each other—grammars that could arise in P are ruled out by the principles of C, and vice versa. Assuming that de Lacy and Kingston are right, and that no grammar is either all formal or all functional, how does that interaction play out? What are the mechanisms by which the formal grammar accesses phonetic information?

One theory that attempted to constrain the application of phonological rules is "Grounded Phonology" (Archangeli and Pulleyblank 1994). In this theory, the physical requirements of articulation and perception create a system of "grounding conditions" that are stated as implicational relations, such as "if nasal, then voiced" (based on aerodynamics) or "if [+high] then [+ATR]" (based on physical constraints on tongue musculature). Rules that create representations that accord with these phonetic grounding conditions are favored. Phonological rules are further restricted by considerations of simplicity and recoverability, as well as by the constraints imposed by feature geometry (see Chapter 4) resulting in a much more restricted set of possible rules. The rule-based Grounded Phonology was important in making the case that phonology *should be* grounded in phonetics, and arguing for an explicit method for doing so within a formal theory, but it was eclipsed by fully constraint-based Optimality Theory.

6.3 OPTIMALITY THEORY

The most successful, widely-accepted recent theory of the role of markedness in phonology is Optimality Theory (OT). Much current work in phonology, and at the phonology/phonetics interface, is now couched in OT terms, so it is worth taking some time to explore the mechanics of the theory. (For a more complete introduction, see Kager 1999; McCarthy 2008; Gussenhoven and Jacobs 2011; Zsiga 2013a and other basic phonology texts.) What is now called "classic" OT was developed as a collaboration between scholars who brought differing perspectives to the problems of phonology: Alan Prince, a student of Chomsky's and professor of linguistics, and Paul Smolensky, a cognitive scientist and professor of computer science. Their original manuscript began to circulate in 1993, and was published in book form in 2004.

Prince and Smolensky (2004: 4) begin by critiquing rule-based approaches to generative grammar. As was quoted at the end of Chapter 5, they claim that rule-based theories have not provided any "interesting" or informative theory of possible structural descriptions or structural changes, and that thus "the locus of explanatory action is elsewhere." They develop the argument that while linguists have good theories of what is natural and common ("Praguian markedness" in Chomsky and Halle's terms), that knowledge has not been incorporated into rule formalisms—there is no adequate theory of possible rule-based predicates or operations. Prince and Smolensky acknowledge the progress made by theories such as Natural Phonology and Grounded Phonology, though they argue that the solutions proposed by these theories are incomplete.

Rules make changes (A → B), and in most cases (except for the "crazy rules" discussed above), these changes are seen to bring about an output that is more natural, easier to produce or easier to perceive, that is, less marked. In most cases, some phonetic grounding is usually evident. The issue is how to get the requirement that rules be phonetically grounded into the theory of rule formalism itself. As the problem is stated by Hayes and Steriade (2004: 2) in their paper on the role of phonetic grounding in phonology, "the *what* is phonetically difficult is not the same as the *how to fix it*." Prince and Smolensky argue that we shouldn't try to incorporate markedness into rule formalisms (as in Grounded Phonology), but get rid of rules altogether.

In Optimality Theory, Prince and Smolensky thus abandon the idea, proposed by Halle (1959) and Chomsky and Halle (1968) and accepted by virtually all phonologists in 1993, that the grammar consists of a set of rules. Instead, they argue that the grammar consists basically of a set of *constraints*. OT is still a generative grammar: the grammar takes an underlying representation (input) and generates a unique surface representation (output) from it, but the input/output relation is mediated by constraints rather than rules. How does that work?

An OT grammar in fact has three parts: GEN (the "generator"), EVAL (the "evaluator") and CON (the constraint set). For every input lexical item, GEN creates a set of possible output pronunciations. It is the job of EVAL to decide which input/output pair is the "optimal" solution, and thus the actual pronunciation. EVAL does its job based on the set of ranked constraints in CON. In "classic" OT, constraints are argued to be universal, and come in two types: markedness and faithfulness constraints.

The markedness constraints directly encode states of affairs that are either allowed or prohibited, based on grounding in articulation, perception, or processing. Unlike rule-based grammars, where the phonetic basis of the rules is inferred or implicit, markedness constraints make direct statements of phonetic principles. Simple examples include those in (2):

(2) Syllables must have onsets.
    Syllables must not have codas.
    Obstruents are voiceless.
    Sonorants are voiced.
    Vowels are oral (not nasal).
    Vowels are nasal adjacent to a nasal consonant.

The markedness constraints directly state the principles proposed by Trubetzkoy, Jakobson, Greenberg, and other researchers who have investigated what is cross-linguistically common, easy to articulate, or easy to perceive. Constraints may not reference random states of affairs—there is no "Syllables must not have onsets" or "Syllables must have codas." Thus, phonetic grounding is directly encoded in the phonological grammar.

Though they are stated as absolutes, and are argued (by some) to be present universally, the markedness constraints are not, of course, universally "surface true." Voiced obstruents, codas, and nasal vowels exist. Crucial to OT is the tenet that constraints are violable. Constraints are ranked, and violation of one constraint is allowed only in case the violation allows the form to obey a higher-ranked constraint. Ranking is language-specific, allowing for cross-linguistic variation.

Markedness constraints may sometimes conflict with each other. In many cases "general" markedness constraints such as "vowels are oral" and "stops are voiceless" may conflict with "contextual" markedness constraints such as "vowels are nasal adjacent to a nasal consonant" and "intersonorant segments are voiced" (see Kager 1999 for exposition). There may also be conflicting dimensions of markedness (see Gouskova 2002; de Lacy 2006). For example, syllable structure constraints require an intervocalic consonant to be an onset, but this might conflict with a constraint in the stress system that requires stressed syllables to be heavy (have codas). Perceptual constraints requiring strong cues may conflict with articulatory constraints favoring conservation of energy.

More commonly, however, markedness constraints conflict with faithfulness constraints. Faithfulness constraints require that inputs and outputs be identical. Thus, if an input has a marked structure, perhaps a consonant cluster, then in the pronunciation, either the input structure can be maintained, violating the markedness constraint, or a change must take place to make it less marked (the output might break up the cluster with an extra vowel) thus violating faithfulness. Something has to give, depending on language-particular ranking. The lower-ranking constraints will be violated in order to obey constraints that are higher-ranked.

To illustrate, the following paragraphs work through an OT analysis of the nasalization patterns from Donegan and Stampe (2009). Example languages are 1) Hawai'ian, where all vowels are oral, 2) French, where oral and nasal vowels contrast in every position, 3) English, where oral and nasal vowels are allophones, and 4) Hindi, where oral and nasal vowels are contrastive after oral consonants but only nasal after nasal consonants. That is, the cases represent 1) unmarked only, 2) contrast, 3) allophony, and 4) positional neutralization. (Kager 1999 gives a parallel example with a more thorough explication of constraint ranking and of formal devices.)

First, we need a set of constraints: a general markedness constraint, a contextual markedness constraint, and a faithfulness constraint, as in (3).

(3) Constraints pertaining to oral and nasal vowels
General markedness: Vowels must be oral.
Contextual markedness: Vowels adjacent to a nasal consonant must be nasal.
Faithfulness: Nasality in input and output must match.

All languages have these three constraints, but differ in how they are ranked. In Hawai'ian, general markedness outranks the other two constraints, and thus all vowels are oral, regardless of the context, and regardless of how they might have been specified underlyingly. In French, faithfulness is highest ranked. Vowels are marked as nasal or oral in the input, and surface that way in all positions, so contrast is maintained. In the case of allophony (English), faithfulness is lowest ranked and contextual markedness is highest ranked, with general markedness in the middle. Nasal and oral vowels are completely contextually determined, and are in complementary distribution. No matter how the vowel is specified underlyingly, it will surface as nasal adjacent to a nasal consonant (by contextual markedness) and oral elsewhere (by general markedness). Finally, in Hindi, contextual markedness is ranked over faithfulness which is ranked over general markedness. In the context of a nasal consonant, contextual markedness applies and the vowel becomes nasal, but otherwise, the vowel is faithful to its underlying specification, and contrast is maintained in other positions.

Languages have the option of choosing faithfulness over markedness, and allowing contrast. Or they can choose markedness over faithfulness, allowing neutralization and alternation. All of the OT literature since 1993 has been the working out of this premise in particular situations. What an OT analysis *can't* do is reverse markedness, and allow change from an unmarked to a marked state. To the extent, then, that phonological alternations create unmarked outputs, as originally proposed by Jakobson, then OT is more successful than rule-based theories because it's more constrained, and provides a theory of a possible structural change. To the extent, however, that alternations cannot be shown to fix markedness violations, OT has a problem.

There are two kinds of counterexamples. The first is "opacity," which was mentioned in Chapter 5 in the context of ordered rules. In an opaque interaction, an alternation is plausibly triggered by markedness constraints, but the triggering context is obscured by some other alternation. For example, vowel nasalization may be triggered by a nasal consonant that is present in the input but is prohibited from surfacing in the output by some other high-ranked constraint. The problem of opacity has been addressed in various ways that elaborate the architecture of the OT grammar and allow reference to intermediate levels. These elaborations include reintroducing some additional derivational structure into the grammar (with "stratal OT" (Kiparsky 2000) or "candidate chains" (McCarthy 2007)), allowing faithfulness constraints to reference other output forms ("paradigm uniformity" (Benua 1997; Burzio 2005; McCarthy 2005)), or allowing markedness constraints to reference more abstract representations ("turbidity theory" (Goldrick 2001), "two-level markedness" (Ettlinger 2008)). While these may make OT more complicated and therefore less attractive, they have more impact on the phonology/morphology interface than on the phonology/phonetics interface.

More problematic are the markedness reversals, which get at the very idea of OT. If constraints are *not* necessarily grounded, but can prohibit or enforce unmotivated configurations, then OT also has no theory of possible structural changes and structural descriptions, and is no better than its competitor rule-based formalizations. One example of this dilemma is the constraint *ND proposed by Hyman (2001).

The debate begins with Pater (1999), who argues for a markedness constraint, *NT, that disallows voiceless stops after nasals, supporting it on both typological and phonetic grounds. Pater cites phonetic sources showing that aerodynamic factors favor voiced consonants after nasals, and argues that these factors promote a cross-linguistic preference for [mb] over [mp], driving changes such as post-nasal voicing in Puyo Pungo Quechua: /kam-pa / → [kamba], *yours* (p. 21; Orr 1962). If [mb] is less marked than [mp], as Pater argues, then OT predicts that a language may prioritize *NT and allow only voiced stops after nasals (as in Quechua and the other languages Pater cites), or a language may prioritize faithfulness to underlying representations and allow both [mb] and [mp], as English does in "camper" and "amber." But no language should allow the marked option only, or prefer [mp] to [mb].

Hyman (2001) argues that Setswana does exactly that, turning underlyingly voiced stops into voiceless stops after nasals: [bóná] "see!" [m-póná] "see me." What are we to make of the Setswana alternation? Hyman argues that post-nasal devoicing in Setswana and other languages demonstrates a "limit to phonetic determinism." He argues that post-nasal devoicing arose in Proto-Tswana as a phonetically-grounded fortition of continuant sonorants to obstruent stops in post-nasal position, at a point in time when all obstruents in the language were voiceless. Subsequent sound changes obscured the original phonetic motivation, leaving only the apparently-unnatural devoicing in the synchronic phonology. If OT is to account for this devoicing, another constraint is needed, which Hyman names *ND: a constraint that rules out *voiced* stops after nasals, the exact opposite of *NT.

Hyman allows that one could argue that there is *another dimension* of markedness where [mp] is in fact preferable to [mb], so that the alternation is phonetically grounded after all. This is the tack taken for this data by Gouskova et al. (2011), Hamann and Downing (2017), and Zsiga (2018a), each arguing, from somewhat different theoretical perspectives, that [mp] might be preferable to [mb] because [p] (or [p']) is a "stronger" consonant with a more salient release, making [mp] less marked than [mb] on the basis of perception rather than articulation. A concerted search for *some* aspect of markedness that supports an alternation must be undertaken carefully, however, to avoid danger of circularity and post-hoc reasoning. As was suggested above, the savvy analyst can usually think of *some* reason a configuration might be preferred. Hyman acknowledges that this sort of reanalysis is always possible, but he argues that a more straightforward solution is just to recognize that unnatural alternations are part of phonology, even if it means OT loses predictive power.

These challenges for OT continue to be addressed in the literature. Many, but not all, phonologists working in formal theory as of this writing accept OT as the best way of incorporating markedness into phonological theory.

6.4 PHONETICS IN PHONOLOGY

The advent of a constraint-based theory opens new ways of thinking about the phonology/phonetics interface, and the role of phonetics in phonology. One line of research within Optimality Theory has sought to explain the *origin* of markedness constraints. This line continues to hold to the traditional view that phonology and phonetics should not be conflated, and that phonological constraints, though

Table 6.3 Relative difficulty of producing a voiced stop based on place of articulation and environment. Values in shaded cells are > 25 (repeated from Table 3.1)

Environment	Place of articulation		
	b	d	g
After an obstruent	43	50	52
Word-initial	23	27	35
After a non-nasal sonorant	10	20	30
After a nasal	0	0	0

Source: Adapted from Hayes (1999: 251).

phonetically grounded, abstract away from their phonetic basis. The question then is how that abstraction takes place. Another line of research takes advantage of the non-serial nature of OT to blur or obliterate the modular division between phonetics and phonology. If phonological rules do not follow one another sequentially, is it necessary to assume that the phonetic module follows the phonological? Why not let phonological constraints have direct access to phonetic information, bypassing the process of formalization? For example, why not posit a phonological constraint that evaluates "F2 > 750 Hz" instead of the feature [+low], or "audible release" instead of Onset? These two views of the role of phonetics in phonology are discussed in the next two sections.

6.4.1 WHERE DO MARKEDNESS CONSTRAINTS COME FROM?

The theory of "Phonetically-driven Phonology" (Hayes 1999; see also Hayes and Steriade 2004) was introduced in Chapter 3, in the context of discussing the interacting pressures of phonetic ease and phonological symmetry in inventory selection. Here, we return to the theory in the context of specific markedness constraints in OT. Table 6.3 repeats the "landscape of difficulty" for voiced stops in different positions that was proposed by Hayes (1999).

As was discussed in Chapter 3, Hayes argues that languages will prefer simpler over more difficult configurations, for example preferring voiced stops after nasals and voiceless stops after obstruents, but will implement these preferences in ways that are formally simple and symmetrical, making reference to classes defined by features rather than directly encoding phonetic difficulty. That is, phonological constraints are based on phonetic difficulty, but are not the same as the direct expression of phonetic forces.

Steriade (1994, 2001) approaches the problem from the perceptual side. Parallel to Hayes' map of articulatory difficulty, she proposes a "P-map" of perceptual similarity. Neutralization or alternation is likely to take place in contexts where perceiving difference is difficult, while contrast is expected where perceptual differences are clearer. Thus, nasals assimilate in place of articulation to a following stop because place cues within nasals are muddied by nasal formants, whereas the cues to stop place at release are very clear. Thus [np] is not perceptually distinct from [mp], and the difference is likely to neutralize. Conversely, the perceptual cues to retroflex stops are clearest in the formant transitions leading into the consonant closure rather than at release, so retroflex consonants are more likely to occur in syllable codas rather than onsets (reversing the usual preference for onset position). Like Hayes, Steriade proposes that the P-map grounds the phonological constraints, but the constraints themselves abstract away from it.

Hayes and Steriade (2004) argue that speakers have "phonetic knowledge" in the sense of Kingston and Diehl (1994): they know (implicitly if not explicitly) what is easy and hard to say and to hear. Language learners turn this phonetic knowledge into phonological constraints through a process of induction. They argue that principles of markedness are not genetically encoded in a long list of pre-existing (innate) constraints that are part of universal grammar, but can be induced from language experience. What universal grammar provides is the ability and propensity to make these phonological generalizations. Infants gain phonetic knowledge through the use of their own articulators in babbling, and their perception of the acoustic effects of these articulator movements along with observation of the adults around them.

The learner constructs a "profuse" set of constraints, referencing many different aspects of phonetic difficulty. The process of learning phonology, then, consists of evaluating the constraints for "effectiveness" by using the cognitive tools provided by universal grammar. The learner is able to figure out the set of constraints, and the constraint ranking, that account for the ambient language in the most efficient and symmetrical way. "To generalize," Hayes writes (1999: 252),

> I believe that constraints are typically natural, in that the set of cases that they ban is phonetically harder than the complement set. But the "boundary lines" that divide the prohibited cases from the legal ones are characteristically statable in rather simple terms, with a small logical conjunction of feature predicates. In other words, phonological constraints tend to ban phonetic difficulty in simple, formally symmetrical ways.

Using the tools of OT formalism, Hayes and Wilson (2008) show that the problem of formulating constraints from the data and then ranking them to account correctly for the ambient language is a computationally-solvable problem.

Pierrehumbert (2000), discussing the "phonetic grounding of phonology," also argues that phonologies are shaped by the interaction of physical and cognitive constraints. In her analysis of vowel systems, Pierrehumbert cites the example of the vowel [æ], which belongs to the class of phonologically front vowels (as evidenced for example by vowel harmony alternations) but does not share the raising and fronting articulatory gestures of /i/ and /e/. The class of front vowels is based in articulation and acoustics, but is also shaped by symmetry and cognitive classification. Pierrehumbert does not agree that there is a strict dividing line between phonetics and phonology, stating that "there is no particular point on the continuum from the external world to cognitive representations at which it is sensible to say that phonetics stops and phonology begins" (p. 17). Nonetheless, she argues that the language system, like any other complex system, can be analyzed at different levels of granularity, with multiple levels of representation, and she agrees with Hayes in arguing against reductionism (what Hayes (1999: 275) terms "greedy reductionism"). No one level of description or generalization gives us all the answers. Languages "represent nature's solution to multiple constraints, including physical, biological, and cognitive constraints" (Pierrehumbert 2000: 18). Learners are able to generalize over different levels of abstraction, and the formalism of continuous mathematics and the formalism of logic can both provide important insights. On the one hand, phoneticians need to recognize the successes of "algebraic" phonology:

> To a phonetician whose work requires Laplace transforms or signal detection theory, treatments of phonology using the mathematically restricted resources of logic alone surely have a surprising degree of success, in terms of the coverage of the phenomena they address. This suggests that the cognitive system makes a simplified or impoverished use of the wealth and complexity of the phonetic domain—a very interesting fact about language which we have by no means gotten to the bottom of. (p. 9)

On the other hand, phonologists need to recognize that "the large scale analyses made possible by recent technology, however, show that even core areas of phonology are not as categorical as more intuitive methods of data collection once suggested" (p. 9). In short, phonological formalism and phonetic substance are not inconsistent. The question remains, however: can they be co-present?

6.4.2 DIRECT ACCESS OF PHONOLOGY TO PHONETICS

In a serial derivational model, the syntax is input to the phonology and the phonology is input to the phonetics. But in a non-derivational model, all information is co-present. If OT markedness constraints must be grounded in phonetic principles, why not make direct reference to phonetic principles instead? A number of researchers have suggested this approach.

On the perceptual side, Flemming (2002, 2004, 2017) unites ideas from Dispersion Theory (Liljencrants and Lindblom 1972; see Chapter 3) with Optimality Theory. Flemming notes that while markedness constraints on articulation may assess the difficulty of individual sounds, auditory constraints must assess the difficulty of maintaining a particular contrast *between* sounds. For example, there is nothing particularly difficult about articulating a high central vowel [ɨ], and some languages that have a small "vertical" vowel inventory (such as Kabardian and Marshallese; Choi 1991, 1992) do without a front/back contrast and use only the central space: [ɨ ɜ a]. The only problem with [ɨ] is that, because of its intermediate F2 value, it is hard to maintain a contrast between [ɨ] and both [u] and [i]. Flemming formalizes the requirement to maintain contrast with "minimal distance" (MinDist) constraints. These constraints evaluate inventories (or parts of inventories) by referencing perceptually-relevant parameters (such as F1, F2, duration, nasalization, voicing, etc.), assigning violation marks for inventories that fail to maintain the specified minimal distance. (Distance is schematized as steps on a scale: MinDist > 2, MinDist > 3, etc.) Minimal distance constraints are ranked with respect to other constraints that require a minimum number of contrasts (thus setting the size of the inventory within limits required for a communication system to function) and with other (usually articulatory) constraints on specific articulations or configurations (such as *implosive or *pre-nasalized stop) that determine language-specific properties.

Flemming's MinDist constraints introduce several important innovations. They differ from (most) previous markedness constraints introduced in OT in referencing perception rather than articulation. They evaluate inventories rather than individual sounds or lexical items. Further, they make direct reference to acoustic parameters such as F2 rather than mediating that reference through more abstract phonological features.

On the articulatory side, Kirchner (1998, 2004) suggests a set of "Lazy" constraints that penalize effort above a certain level. Kirchner doesn't quantify exactly how effort should be measured, but relies on an understanding that tighter constrictions of longer duration require more effort than those that are more open and shorter.

The interaction of Lazy constraints with faithfulness constraints drive processes of lenition, flapping, spirantization, and debuccalization. For example Lazy(voiced stop) can be read as "do not expend the effort needed to create a voiced stop" and Lazy(voiced approximant) can be read as "do not expend the effort needed to articulate a voiced approximant." Because voiced stops, with complete closure, require more energy than more open voiced approximants, Lazy(voiced stop) is universally ranked above Lazy(voiced approximant). If a faithfulness constraint, such as Preserve(continuant), is ranked above Lazy(voiced stop), the faithfulness constraint wins and the stop is articulated without lenition. If, however, Lazy(voiced stop) is ranked higher, then the "easier" lenited sonorant is produced instead. By referring directly to quantitative scales of effort, Kirchner's analysis unifies processes of lenition that cannot easily be captured in terms of feature changes, such as the Florentine Italian process in which [k] is realized as [x], and [x] is realized as [h].

Another dimension of quantitative information in phonological constraints is proposed by Gafos (2002). Gafos argues that phonological constraints can make direct reference to timing relations between articulatory gestures, accounting for patterns of epenthesis in certain dialects of Arabic. Because Gafos's approach is couched in the theory of Articulatory Phonology (Browman and Goldstein 1986, 1990a, 1992a), it is discussed as part of that theory in Chapter 9.

Boersma's theory of Functional Phonology (Boersma 1998, 2007, 2008, 2009; Boersma and Hamann 2008, 2009; Boersma and van Leussen 2017) utilizes direct reference to both perception and production. Rather than referencing a schematized scale of perceptual distinctiveness, Boersma's "cue constraints" reference specific auditory cues and articulatory implementations. They "express the speaker-listener's knowledge of the relations between continuous auditory cues and discrete phonological surface elements" (Boersma 2008: 3). Examples include "don't perceive [F2 = 2000 Hz] as [a]," "don't perceive spectral noise of 26 Erb as [s]" and "don't produce a lengthened vowel before a voiceless stop." In Boersma's model, these constraints are in addition to the markedness and faithfulness constraints that mediate between

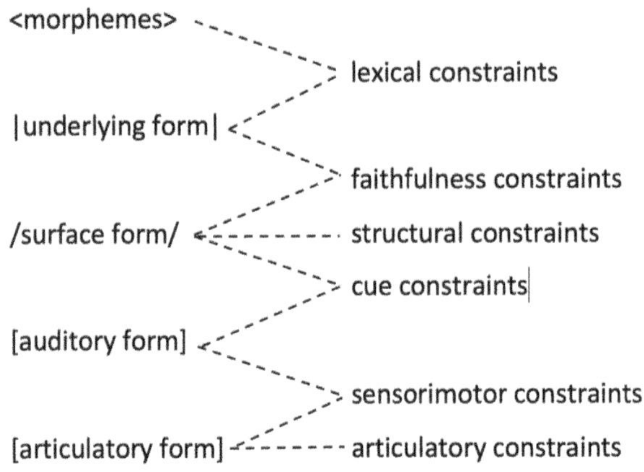

Figure 6.3 Representations and constraints in Functional Phonology
Source: Adapted from Boersma (2009: 60).

underlying and surface phonological representations, but they are evaluated in parallel in a single ranking. Figure 6.3 (Boersma 2009: 60) illustrates his model of the phonology/phonetics interface.

The phonological form and phonetic form are kept separate, but constraints that regulate the mapping from surface phonology to phonetic implementation are interleaved with constraints that regulate the mapping from underlying to surface phonological representation. Because these constraints are interleaved, pressures to minimize effort and maximize contrast are directly expressed by ranking quantitative constraints on both perception and articulation with other markedness and faithfulness constraints. The overall model gets quite complex, with the profusion of quantitative constraints, but Boersma argues that all of these mappings and constraints are needed independently. By evaluating phonological mapping and phonetic implementation in parallel rather than sequentially, he argues, one gains the ability to model interactions more straightforwardly.

But is all of this phonetic information *necessary* to do phonology? The phonetically-explicit constraints proposed by Flemming, Steriade, Kirchner, Gafos, and Boersma account for the data they were designed for. But do we need all of this phonetic information directly encoded in our phonology? Other researchers answer no.

6.5 EVOLUTIONARY PHONOLOGY

On the other side of the argument for "grounded" phonology is *Evolutionary Phonology*. Grounded Phonology, Functional Phonology, Natural Phonology and Articulatory Phonology, as well as most implementations of Optimality Theory, all take as axiomatic the premise that markedness principles based in articulation and acoustics drive synchronic phonology, whether phonetic principles are expressed quantitatively or more abstractly. Evolutionary Phonology (Blevins 2004, 2006; Yu 2011, 2013), on the other hand, argues that markedness plays no role in synchronic phonology, and that typological markedness (what is and isn't common cross-linguistically) is entirely due to diachronic pressures.

Evolutionary Phonology argues that universals and near-universals come about because of sound change, in the same way that biological characteristics of organisms emerge because of gene mutations over time. If there are many diachronic paths to a particular sound pattern, then that pattern will be common. If there are few paths, rare. If no paths, impossible. Paths that induce sound change include "change," "chance," and "choice" (Blevins 2006: 126). A canonical "change" takes place when a listener misperceives an utterance for something similar, such as [ampa] for intended /anpa/, and incorporates the incorrect form into her grammar. This is the path explored by Ohala (1981) in "The listener as a source of sound change." Phonological change can also take place if the utterance is perceived correctly, but the cues are misattributed by "chance." For example, the speaker says [ʔa̰ʔ] for /aʔ/, but the listener thinks /ʔa/ was intended, resulting in apparent metathesis. Finally, if there is variation, the listener may choose the wrong form as underlying. For example, a speaker may vary between [CVCiCVC] and [CVCCVC]. For the speaker, the tri-syllabic form may be underlying, but the listener might mistakenly choose the bisyllabic form instead.

All of these changes take place in transmission and are explained by forces (auditory, aerodynamic, articulatory) that are outside the grammar per se. Evolutionary Phonology argues that if external pressures that result in sound change can account for the occurrence of a pattern, then synchronic phonology does not need to account for it: "[t]he premise is that principled *extra-phonological* explanations for sound patterns have priority over competing *phonological* explanations unless independent evidence demonstrates that a purely phonological account is warranted" (Blevins 2006: 124, emphasis original). As with biological evolution, sound change is not teleological: there is no pressure for the resulting system to conform to principles of "naturalness," "ease of articulation," or the like. Over time, the effect of multiple sound changes may create generalizations that are quite "unnatural," such as post-nasal devoicing or word-final voicing, but the synchronic grammar should account for them with equal ease.

Thus, in the synchronic grammar, constraints will not necessarily be grounded in phonetics. "Naturalness plays no role in constraining synchronic systems" (Blevins 2004: 70). Learners will induce features and constraints to account for the particular language data they encounter, without regard to phonetic grounding. For example, OT might argue that word-final voicing is impossible and word-final devoicing is common because of a universal markedness constraint against word-final voiced stops. An OT phonologist might argue that this constraint is given by universal grammar, or (following Hayes 1999) that it is induced from practice with phonetic difficulty, but in either case would argue that it is the presence of this constraint in the grammar, and its ranking with respect to a faithfulness constraint on voicing, that gives rise to the pattern of final devoicing. Since no constraint preferring final voicing is given by universal grammar (or could be induced from phonetic difficulty), final voicing is predicted not to occur. The evolutionary phonologist, on the other hand, would point to Lezgian (Yu 2004; Blevins 2006: 150) and alternations such as [rapár]/[rab] *needle pl/sg*, or [mukár]/[mug] *nest pl/sg*, to show that that final voicing is indeed possible. Final devoicing is common, Blevins argues, because it is easy to produce phrase-final stops with variable voicing, and easy to perceive this variability as intended devoicing, and then to generalize the pattern to all word-final stops. Because it is easy, it happens often. Yu (2004) argues that the Lezgian pattern, on the other hand, arose due to a series of sound changes including gemination, devoicing, degemination, and stress shift, such that an originally voiced stop ends up voiceless in onset position. Because this path is complex and specific, it's not surprising that few languages instantiate it. But once the learner is confronted with the pattern, she has no problem making final voicing part of her phonology: she is not constrained to only allow phonetically-natural alternations.

One could preserve the "no final voicing" analysis by recapitulating the sound change in the synchronic grammar (e.g., Kiparsky 2008), but Yu argues that this is an unnecessary complication. The "unnatural" final voicing analysis is simpler, and there is no need to create otherwise unmotivated abstraction and complication in order to preserve synchronic naturalness, when all is perfectly well explained diachronically. Evolutionary Phonology is committed to the influence of phonetic factors on phonological patterns. The influence, however, is all diachronic: there is no preference for markedness such as that encoded in the synchronic markedness constraints of OT.

De Lacy and Kingston (2013, mentioned briefly above), building on work by Kiparsky (2006, 2008), push back against a purely diachronic account of typological patterns, and argue that there are indeed "non-trivial cognitive restrictions" (that is, markedness constraints) active in the synchronic phonological grammar (De Lacy and Kingston 2013: 288). De Lacy and Kingston note, for example, that there are processes, such as a coda neutralization of /t/ to /k/, that are found diachronically, but that do not survive as synchronic alternations. They attribute this typological fact to the activity of a synchronic markedness constraint disfavoring dorsal consonants (motivated by de Lacy 2002, 2006).

In addition to typological evidence, de Lacy and Kingston also adduce evidence for the synchronic activity of markedness constraints in production and perception. They argue that some sound changes, such as vowel monophthongization or raising, can be "optimizing," in the sense that speakers make these changes to actively enhance perceptual distinctiveness. Perception is also argued to be influenced by the activity of markedness constraints: English speakers perceive an ambiguous stimulus as legal [dw] rather than illegal [dl] (Moreton 2002), and Japanese speakers hear illegal [ebzo] as legal [ebuzo] (Dupoux et al. 1999). If, as Evolutionary Phonology argues, language users process whatever the data presents them with, without the influence of synchronic markedness constraints, listeners ought to be able to accurately hear the stimuli they are presented with. Instead, listeners encode incoming stimuli as forms that are phonotactically licit, according to the markedness constraints that are high-ranked in their language. Infants prefer sequences that obey markedness constraints (Jusczyk et al. 2002), and adult English speakers prefer clusters that obey sonority sequencing (Berent et al. 2007, 2009), even for clusters such as [ml] vs [md] that have zero probability of occurring as onsets in the ambient language.

In summary, while Blevins argues that diachronic explanation should always replace synchronic mechanisms (2004: 5), de Lacy and Kingston argue that the two must coexist, and sometimes converge. The fact that diachronic processes prefer some unmarked state does not mean that synchronic markedness constraints therefore must play no role.

6.6 LEARNING BIASES

The typological studies summarized thus far clearly show that "crazy" non-phonetically-grounded patterns exist, but are rarer than those that are phonetically grounded. Evolutionary Phonology argues that the reason for the asymmetry is "channel bias": certain sound changes are more likely over time because of the ways sounds are produced and perceived. The acoustic channel makes [anba] confusable with [amba], so the change from /anba/ to /amba/ is likely. Inaccuracies in the ways sounds are produced and perceived result in changes to the system. Another possible bias, however, is "inductive bias." Certain types of patterns may persist over time because of cognitive biases that make these patterns easier to learn. Two types of inductive bias that have been argued for in recent literature are "substantive bias" and "structural" or "analytic bias."

According to proponents of inductive bias, anti-markedness patterns *can* be learned, but just not as easily.

Hypotheses about inductive bias are tested through artificial learning paradigms. In these experiments, adults in a laboratory setting are exposed to a set of words that exemplify a particular pattern or process. After exposure, they are tested to see whether they can identify new examples that fit the pattern, or generalize the pattern to new contexts. They might be asked "Have you heard this word before?" or "Is this word likely to be part of the language you just heard?"

Wilson (2006), for example, taught his participants two "mini-grammars" of velar palatalization. In the first condition, listeners heard examples of /k/ becoming /t͡ʃ/ before /i/, and in the second they heard examples of /k/ becoming / t͡ʃ / before /e/. In natural language typology, palatalization of /k/ is argued to be grounded in the raising and fronting of the tongue body, and in the perceptual similarity of [ki] and [t͡ʃi], so that palatalization before /i/ is more common than palatalization before /e/, and any actually-occurring language that has palatalization of /ke/ also has palatalization of /ki/. Wilson's English-speaking participants had no knowledge of the typological facts, and do not have a rule of palatalization in their native language grammar. Nonetheless, Wilson wanted to find out if the underlying markedness relations (/ki/ is more marked than /ke/, and palatalization of /ke/ implies palatalization of /ki/) would influence their learning. He found that it did. Listeners trained on the /e/ condition were likely to generalize to /i/ as well, but listeners trained on /i/ were significantly less likely to generalize to /e/. Wilson concludes that language learning is subject to a substantive bias, such that patterns that reflect universal markedness tendencies are easier to learn than patterns that do not.

Finley (2017) finds a similar result for learners exposed to a mini-grammar of metathesis. She finds that learners prefer patterns where metathesis improves syllable structure as opposed to patterns where it does not. She concludes (pp. 143–4) that "language users are biased toward phonological patterns that are grounded in phonetic or structural improvements."

Not all artificial learning experiments have been as successful, however. Moreton and Pater (2012) survey the results of artificial language learning experiments to that date, and conclude that results for artificial learning experiments that attempt to teach natural vs unnatural patterns "have been mixed at best" (Moreton and Pater 2012: 708). In many of the studies cited, natural and unnatural patterns are learned equally well. Instead of a "substantive" bias, Moreton and Pater argue that the successful studies can be accounted for in terms of a "structural bias" that favors the learning of patterns that can be stated more simply, whether or not they are phonetically grounded. The point that linguistic patterns are more easily learned if they are formally simple and symmetrical has a long history in generative linguistics (e.g., Bach and Harms 1972). With Hayes (1999), Moreton and Pater argue that linguistic typology is shaped both by phonetic grounding and by criteria of simplicity and symmetry. Ongoing research is testing the ways these two biases interact.

Another possible interacting variable is the role of frequency, also mentioned by Moreton and Pater, and explored in the context of a markedness bias by Zsiga and Boyer (2017). (The role of frequency in general in phonology is discussed more thoroughly in Chapter 10.) In general, frequency of occurrence and markedness go together: structures that are less marked are also more common. But sometimes they can be decoupled. Zsiga and Boyer describe a case of "a natural experiment in learning an unnatural language," in the contact between the languages Setswana and Sebirwa. Setswana, as was described in Section 6.3 above (Hyman 2001 and others), is well known for its "unnatural" process of post-nasal devoicing. Setswana is the national language of Botswana, spoken by the majority of the population and ubiquitous (along with English and Afrikaans) in media and education. Sebirwa is a related Bantu language, spoken by only a few thousand people, and not widely used outside the home. (It is in fact moribund, not being much used at all by adults of child-bearing age, and thus not being acquired by children.) Sebirwa's historical evolution from Proto-Tswana diverged from Setswana's, and unlike Setswana it has a fully symmetrical set of voiced and voiceless obstruents. As late as several decades ago, Sebirwa, unlike Setswana, had no post-nasal devoicing (Chebanne 2000). But in recent decades, Sebirwa has borrowed heavily from its culturally- and numerically-dominant cousin. This language contact sets up in a real-life situation the scenario that artificial learning paradigms emulate in the lab: Sebirwa speakers are exposed to the "unnatural" pattern of post-nasal devoicing.

Making the situation even more marked, because of Setswana's skewed consonant inventory (no [g] in native words, and [d] occurring only as an allophone of /l/ before high vowels), the majority of Setswana words in which Sebirwa speakers will hear evidence of post-nasal devoicing involve labials. As Hayes (1999) argued (see Table 6.3 above), labial place is the one that most *favors* voicing. Nonetheless, Zsiga and Boyer (2017) found that Sebirwa speakers reliably produced post-nasal devoicing for labial, but not coronal or velar stops. It goes against markedness that they would borrow post-nasal devoicing at all, and it is even more surprising that they would borrow it only in the context in which is it phonetically *least* favored. Zsiga and Boyer conclude that frequent exposure to evidence of an alternation of [b] changing into [mp] overcame any markedness bias against it.

Zsiga and Boyer conclude that overall, the data described does not support a role for synchronic markedness. "Still," they write (p. 363):

> one would like to follow the situation into the future, to see what further historical developments might ensue. Would children acquire the labial-only devoicing, or would they regularize the pattern? If so, in what direction—to devoicing of all voiced stops, or away from any devoicing at all? Sadly, it appears that we will not have the chance to find out. The inundation of Sebirwa by Setswana set up the situation for borrowing in the first place, but barring immediate revitalization efforts or the discovery of a truly more vigorous community of Sebirwa speakers . . . Sebirwa will not survive.

Independent of the fate of Sebirwa, work still continues on whether the "unnatural" post-nasal devoicing found in Setswana might be grounded after all, perhaps in the necessity to "strengthen" a syllable onset, as argued by Gouskova et al. (2011), and by Zsiga (2018a), or in the need to enhance perceptual contrast, as suggested by Hyman (2001). The situation is complex and multifaceted. There are multiple dimensions of markedness, and no phonological system is shaped by one factor alone.

6.7 THE ROLE OF MARKEDNESS IN PHONOLOGY

In conclusion, markedness and phonetic grounding do play a role in phonology, but researchers disagree over how important and direct that role is. Constraint-based theories of phonology, specifically Optimality Theory, have allowed for a direct statement of markedness principles (and possibly detailed phonetic information) in the content of phonological constraints. Phonetically-grounded markedness constraints allow a more straightforward representation of phonetic influence in phonology than did the solely feed-forward rule-based paradigm. In rule-based phonology, phonetics has to follow phonology and thus has no direct influence on what phonological rules can do. Some researchers (such as Boersma and Kirchner) take full advantage of the non-serial nature of OT to argue that phonological constraints must make direct reference to detailed phonetic information. Others (such as Hayes, Moreton, and Pater) argue that phonological constraints should be phonetically grounded, but that the grounding is mediated by pressures toward formal simplicity and more abstract encoding. Still others (Hale and Reiss, Blevins, Mielke) argue that it is a good thing for phonetic pressures to be completely separate from phonological encoding, as in rule-based systems, with all of the phonetic shaping coming through diachronic forces. These debates are far from decided, and research into all of these questions is ongoing.

RECOMMENDED READING

Overviews

Rice, K. (2007), "Markedness in phonology," in P. de Lacy (ed.), *The Cambridge Handbook of Phonology*. Cambridge, UK: Cambridge University Press, pp. 79–97.

- Do the definitions of markedness offered by Rice escape Ohala's charge of circularity?

Pierrehumbert, J. (2000), "The phonetic grounding of phonology," *Bulletin de la Communication Parlée, 5*: 7–23.

- Discuss this quote (p. 9): "the cognitive system makes a simplified or impoverished use of the wealth and complexity of the phonetic domain—a very interesting fact about language which we have by no means gotten to the bottom of."

Hayes, B. and D. Steriade (2004), "Introduction: the phonetic basis of phonological markedness," in B. Hayes, R. Kirchner, and D. Steriade (eds), *Phonetically based Phonology*. Cambridge, MA: Cambridge University Press, pp. 1–32.

- How do Hayes and Steriade propose to resolve "the tension between formal symmetry and phonetic effectiveness"?

Exemplary research

Something old

Anderson, S. R. (1981), "Why phonology isn't 'natural'," *Linguistic Inquiry, 12*: 493–539.

- Summarize Anderson's arguments.

Something new

Moreton, E. and J. Pater (2012), "Structure and substance in artificial-phonology learning. Part I, Structure. Part II, Substance," *Language and Linguistics Compass* 6(11): 686–701, 702–18.

- What can artificial learning experiments tell us about the role of markedness in phonology?

Zsiga, E. C. and O. T. Boyer (2017), "A natural experiment in learning an unnatural alternation: Sebirwa in contact with Setswana," in J. Kandybowicz and H. Torrence (eds), *Africa's Endangered Languages: Documentary and Theoretical Approaches*. Oxford: Oxford University Press, pp. 343–66.

- What was the result of this "natural experiment"?

Opposing views

de Lacy, P. and J. Kingston (2013), "Synchronic explanation," *Natural Language and Linguistic Theory, 31*(2): 287–355.

vs

Blevins, J. (2006), "A theoretical synopsis of Evolutionary Phonology," *Theoretical Linguistics, 32*: 117–65.

- Invite Blevins and de Lacy or Kingston to debate the statement "Markedness plays no role in synchronic phonology." What arguments would be presented for and against?

Hyman, L. (2001), "On the limits of phonetic determinism in phonology: *NC revisited," in E. Hume and K. Johnson (eds), *The Role of Speech Perception Phenomena in Phonology*. San Diego: Academic Press, pp. 141–85.

vs

Gouskova, M., E. Zsiga, and O. Tlale (2011), "Grounded constraints and the consonants of Setswana," *Lingua, 121*: 2120–52.

vs

Hamann, S. and L. J. Downing (2017), "*NT revisited again: an approach to postnasal laryngeal alternations with perceptual cue constraints," *Journal of Linguistics, 53*(1): 85–112.

- Do we need a *ND constraint? How are different dimensions of markedness relevant to deciding?

QUESTIONS FOR FURTHER DISCUSSION

1. How did Trubetzkoy define markedness? Was it a phonological or phonetic property?
2. What are the main claims of Natural Phonology? Describe the different roles of lenitions, fortitions, and inhibitions. How does Natural Phonology handle "crazy" rules?
3. What does the question mark refer to in Figure 6.1 (from Anderson 1981)? What do *C* and *P* stand for in Figure 6.2 (from de Lacy and Kingston 2013). How do the two diagrams illustrate different approaches to the role of markedness in phonology?
4. What is the role of markedness constraints in OT? Why do markedness reversals "get at the very idea of OT"? Conversely, why doesn't every utterance come out as maximally unmarked [papa] (or something similar)?
5. According to Hayes (1999), what is "greedy reductionism" and how can we avoid it?
6. How do MinDist and Lazy constraints incorporate phonetic knowledge directly into phonological analyses? In your opinion, is this a good idea? Why or why not?
7. Discuss Boersma's Functional Phonology from the point of view of modularity. As shown in Figure 6.3, Boersma proposes to take on all aspects of phonology and phonetics simultaneously, with a complicated system of interleaved constraints. How might Boersma answer a critic (such as Myers 2000) who might suggest that the overall grammar would be simpler if each part was computed separately?

CHAPTER 7

SUPRASEGMENTALS: SYLLABLES, STRESS, AND PHRASING

7.1 "SEGMENTAL" VS "SUPRASEGMENTAL"

Most of the examples of phonological alternations and phonetic realizations in the previous six chapters have been about vowels and consonants, how inventories are structured, and how strings of segments assimilate, dissimilate, lenite, and so on. This chapter and the next look at larger units: syllables, stress feet, phonological and intonational phrases, as well as the representation and realization of lexical tone, which is also often argued to be associated to domains larger than the segment. (Intonation and tone are the topics of Chapter 8.) Important work at the phonology/phonetics interface has often examined these hierarchical units, and work on suprasegmentals has been vital in moving the discussion of the interface forward.

Suprasegmentals, while often discussed in the phonological and phonetic literature, are sometimes treated as secondary to segmental aspects of speech, phenomena "around the edges of language" (Bolinger 1964). Bolinger later (1986: 3) refers to intonation as "the Cinderella of the communication complex," dismissed as "an accompaniment to the message rather than an inseparable part of it." Ladd (2014: 63) suggests that the modern use of the term "prosody" to refer to aspects of speech as diverse as stress, tone, and intonation is an "artefact of alphabetic literacy." The alphabet, with its string of segments, is considered first, and then "prosody" lumps together all aspects of pronunciation that are left over.

Early twentieth-century phonologists and phoneticians alike focused on speech as a linear "chain" of phonemes, though they also discussed syllables, stress, and intonation. Saussure, for example, states both that "the main characteristic of the sound chain is that it is linear" (1959: 103), and that "the ear perceives syllabic division in every spoken chain" (p. 58). In *The Pronunciation of English*, Jones emphasizes production of the necessarily linear string of IPA symbols. (Although this work was first published in 1909, there were numerous new editions over the decades, and citations in this text are from the 4th edition, published 1966). In addition to the focus on segments, however, Jones does discuss the importance of the "carrying power" and "force" of stress (1966: 141), the musical patterns of intonational tunes (p. 149), and phrasing into "sense groups" and "breath groups" (p. 148). Similarly, Trubetzkoy's oppositions (see the discussion in Chapter 3) are mostly localized in vowels and consonants, but he does include a set of "prosodic" oppositions, including syllabicity, length, accent, tone movement, tone register, and intensity. He includes a detailed discussion of how languages can differ in the choice of "prosodeme" (the smallest prosodic unit), and use pitch, length, and stress to create different kinds of word-level contrast (e.g., 1969: 170–82). He further notes that the same properties that create word-level oppositions can also create sentence-level oppositions (sentence stress and intonation), but he complains that at the sentence level it is difficult to separate emotional and grammatical influences. "At the present stage of investigation," he writes, "it is impossible to treat sentence phonology with the same detail and certainty as word phonology. Far too little material is available, and what is available is mostly unreliable" (1969: 202).

One structuralist phonologist, J. R. Firth, went against the trend of taking segments as basic (see further discussion in Anderson 1985a). Firth, Anderson argues (1985a: 178), had "grave reservations" about the validity of segments and phonemic analysis, and did not take the segment to be the basis of phonological structure. For Firth, phonological properties (including stress and tone, but also other features such as palatalization or labialization) are not "placed" or anchored to particular segments, but are distributed over domains of different sizes. Intonation is a property of the sentence, stress a property of the word, tone a property of the syllable. Every feature has a "prosody" which specifies the domains in which it is realized. Anderson's example (p. 186), is that English might be analyzed as having a prosody of aspiration that is localized to syllable-initial position. Because properties are localized to positions, variants of the "same" sound (allophones, as they would be described by Jones, Sapir, Trubetzkoy, or Jakobson) are not the same at all for Firth.

Firth's ideas were not widely accepted at the time. In fact, Anderson notes that Firth seemed to enjoy being hard to understand. The clearest statement of his theory is probably "Sounds and prosodies" (Firth 1948), but even that is "full of obscure and allusive references and completely unclear on essential points" (Anderson 1985a: 179). Firth's work does preview, however, current representations of processes like vowel harmony, in which autosegmental features including rounding, fronting, ATR, nasalization, and others, are associated to domains larger than the segment.

When Jakobson et al. (1952) defined their set of universal distinctive features, they focused only on segmental contrasts (again, see the discussion in Chapter 4). While they acknowledge that properties of syllabicity, stress, tone, and length are potentially contrastive, these are classed together as "prosodic" features that are "superposed upon" the "inherent" features of segments (1952: 13ff.). Inherent features are defined solely by their own phonetic properties (voicing, nasality, fricative noise, etc.), while the prosodic features are necessarily relative, requiring comparison to other parts of the string to determine relative loudness, pitch, or length. Jakobson et al. further note that prosodic features also often serve a "configurational" or "expressive" role, marking boundaries or signaling attitude or emphasis. Having briefly mentioned the prosodic features, however, they set them aside to focus exclusively on segments.

Chomsky and Halle (1968), with their emphasis on linear strings of feature bundles, also focused almost exclusively on segmental alternations. Though stress placement plays an important role in the discussion of English phonology, and forms the basis of the argument for cyclic rule application, stress is represented solely as a feature of vowel segments, determined by the place of the vowel in the segmental string. For example, the "alternating stress rule" (p. 84), is written as in (1):

(1) Chomsky and Halle's alternating stress rule
$V \rightarrow [1\ \text{stress}]\ /\ \_\_\_\_\ C_0\ V\ C_0\ V^1\ C_0\ ]_{NAV}$

This rule has the effect of assigning secondary stress to alternating syllables preceding the main stress, without any reference to the concept of syllable or foot. The rule for assigning main stress in the first place (V^1 in the rule) is considerably more complicated (see SPE p. 84), and later rules will reduce the multiple main stresses produced by the alternating stress rule to secondary or tertiary status.

Because SPE is linear and segment-based, there is no clear way to incorporate units larger than the segment into any phonological description in SPE formalism. As was discussed in Chapter 5, in these early instantiations of generative grammar, syntax was understood to be hierarchical and the translation of syntactic structure to phonological structure was described as a process of linearization or "flattening" (p. 9). The richly embedded phrases of syntax are reduced to a linear string of pronounceable segments, separated by strings of boundary symbols (#) to indicate groupings. The number of symbols indicates the strength of the boundary, as in (2), repeated from Figure 5.1 (Chomsky and Halle 1968: 10).

(2) Boundary symbols in SPE
a. Nested syntactic constituents
$[\ [\ [\text{we}]_N\ ]_{NP}\ [\ [\ [\text{establish}]_V\ \text{past}]_V\ [\ [\ [\text{tele+graph}]_N\ \text{ic}]_A\ [\ [\text{communicate}]_V\ \text{ion}]_N\ ]_{NP}\ ]_{VP}\ ]_S$

b. Strings of boundary symbols
###we#####establish#past####tele+graph#ic###communicate#ion####

The edge of every hierarchical syntactic constituent corresponds to one # mark, so the boundary between noun and adjectival suffix in "telegraphic" gets one #, while the main sentence division between subject and verb phrase gets five. If you count, there is one # in (2b) for every left or right bracket in (2a).

Phonological rules can then make reference to the boundary symbols in the same way as to segmental feature bundles. While phrasing is thus based on syntactic constituents, "readjustment rules" (p. 372) can adjust boundaries (and simplify syntactic structure) to make phrasing align with pronunciation rather than syntax. In the sentence "This is the cat that chased the rat that stole the cheese," phrase boundaries after "cat" and "rat" do not correspond to the major syntactic boundaries.

Lees (1957: 404) specifically criticizes generative grammar for ignoring suprasegmentals: "One currently confused issue about which this conception of grammar has as yet little to say is the question of the suprasegmental phonic elements in the sentence, the stresses, pitches, junctures, features of vowel harmony etc." Lees argues, in fact, that suprasegmentals must be part of the syntax, not the phonology, because only the non-linear representations of the syntax can cope with the "extended scope" of suprasegmental elements.

Lehiste (1970) is an important publication that moved the discussion forward, bringing the precision that characterized the relationship between phonetics and phonological features in works such as Jakobson et al. (1952) to the study of suprasegmentals. Lehiste defines "suprasegmentals" as features of length, stress, and pitch, writing that "suprasegmental features characteristically constitute patterns in time" (1970: 154). These temporal patterns are overlaid on the features inherent to segments, such as voicing. Later work in feature geometry (e.g., Clements 1985) defined length in terms of a "timing tier" that consisted of a sequence of X-slots or segment-sized root nodes (see Chapter 4), so that "length" is seldom now considered a feature, either segmental or suprasegmental. (On the other hand, as discussed below, syllable and phrase structure, and other hierarchical groupings not considered by Lehiste, *are* considered to be a part of suprasegmental phonology.)

Following Jakobson et al. (1952), Lehiste argues that suprasegmental features differ from segmental in that they are *sequentially* contrastive: high pitch follows low, stressed and unstressed syllables alternate. Lehiste notes that "prosody" has always been part of linguistic study, as described above, but she contends that the discussion has been characterized by a "certain degree of vagueness" and that prosodic features "seem more elusive than segmental features, and their incorporation into a linguistic system sometimes seems to strain the limits of an otherwise coherent framework" (1970: 1). The book discusses the phonetic details of articulation and perception of duration, length, stress, and pitch, and then discusses how languages recruit these features for linguistic contrast.

Lehiste argues that most current linguistic theories "either ignore or deny" (p. 167) hierarchical suprasegmental structure, but "any satisfactory theory" must take suprasegmentals into account. She concludes that phonetic experimentation is absolutely necessary for testing linguistic hypotheses, because "linguistics is an empirical science" and "phonetic realizations of utterances are the only aspect of language directly subject to observation" (pp. 167–8).

As Lehiste attests, phonetic studies have been crucial in arguing for the existence of these larger, hierarchical units. But phonologists and phoneticians continue to probe the extent to which they have a "life of their own" as abstract units, beyond their physical instantiations in segmental strings, durational patterns, or pitch tracks. That is, how does the hierarchical organization of speech units work at both the phonological and phonetic levels? To that end, this chapter reviews influential papers and current issues in the study of different kinds prosodic patterning. Section 7.2 introduces the prosodic hierarchy. Subsequent sections are organized by unit: 7.3 discusses syllable structure, 7.4 the stress foot, and 7.5 the phonological word and phrase. Tone and intonation are discussed in Chapter 8.

7.2 THE PROSODIC HIERARCHY

As described above, SPE phonology was strictly linear, with the phonological representation consisting of a string of feature bundles and boundary symbols. Work soon developed, however, including Selkirk (1978, 1980, 1984b, 1986), Kaisse (1985), and Nespor and Vogel (1986), that considered phonological structure to be just as

```
(____________________________________________________________________)Utterance
(_________________________________)(_________________________________)Intonational Phrase
(______________)(__________________)(_________________________________)Phonological Phrase
(______________)(______)(__________)(_____________________)(_________)Phonological Word
(___)(_________)(______)(___)(______)(_________)(___)(___)(_________)Foot
(___)(___)(___)(___)(___)(___)(___)(___)(___)(___)(___)(___)(___)(___)Syllable
```

Figure 7.1 The prosodic hierarchy, showing exhaustive parsing and strict layering of units

Source: Adapted from Selkirk (1986: 384).

hierarchical as syntactic structure. These works proposed an articulated "prosodic hierarchy," a series of nested domains that was based on, but not exactly isomorphic to, syntactic structure. Selkirk's proposed hierarchy is shown in (3) and in Figure 7.1.

(3) Prosodic hierarchy
Utterance
Intonational Phrase
Phonological Phrase
Phonological Word
Foot
Syllable

On this view, "The phonological representation consists of a hierarchy of prosodic constituents or categories which may serve as the characteristic domains for phonological rules" (Selkirk 1986: 383). Phonological rules can be specified to apply either within a domain ("domain span" in the terminology of Selkirk 1980), at the edge of a domain ("domain limit"), or across domain boundaries ("domain juncture"). For example, for the word constituent, vowel harmony might be a domain span rule, final devoicing a domain limit rule, and palatalization across word boundaries (as in "this year") a domain juncture rule. Extensive further examples are discussed in the following chapter sections.

Figure 7.1 shows a hierarchy that obeys both "exhaustive parsing" and "strict layering" (Selkirk 1986: 384). Together, these two principles require that all units at level n be grouped into a well-formed unit of level $n+1$. Words are made up of feet, phrases are made up of words, and so on. Exhaustive parsing prohibits representations with units that are not properly incorporated, such as syllables that are attached at the edges of words but are not part of a foot. Strict layering rules out recursion: words within words or phrases within phrases. Both of these principles have been called into question, however; see, for example, the papers in Kabak and Grijzenhout (2009) for discussion.

Much work on the prosodic hierarchy has focused on the relationship between syntactic and prosodic structure (Inkelas and Zec 1990, 1995; Selkirk 2011; Elfner 2018). Specifically, these works (and references therein) address the question of what syntactic information is needed to determine the boundaries of prosodic phrases. In the sections that follow, we look instead at the relationship between prosodic structure and phonetic realization.

7.3 THE SYLLABLE

7.3.1 THE SYLLABLE AS PHONOLOGICAL DOMAIN

Overviews of the role of the syllable in phonological patterning include Blevins (1995), Zec (2007), Cairns and Remy (2010), and Goldsmith (2011). Beginning soon after the publication of SPE, phonologists began to argue for the necessity of referencing syllable structure in phonological rules and descriptions (e.g., Hooper 1972; Vennemann 1972; Kahn 1976; Vogel 1977; Selkirk 1978; McCarthy 1979; Steriade 1982; Clements and Keyser 1983). These works, many of them dissertations, argued that numerous phonological rules in diverse languages, including assimilations, neutralizations, epenthesis and deletion, could be stated more simply and comprehensively if the rules applied based on syllable boundaries rather than specific segmental strings. For example, a frequently occurring environment in phonological rules involves the disjunctive environment "before a consonant or at the end of a word." To take just a few examples from the references above, in this environment English /t/ is glottalized, English /l/ is dark, the Korean liquid consonant is realized as a lateral, voiced consonants devoice in German, both voicing and aspiration are neutralized in Thai, epenthesis takes place in Japanese, and many more. On the assumption that rule complexity should reflect markedness, it was troubling that such a common environment had to be expressed as a disjunction rather than a single natural class. The property that unifies "before a consonant" and "at the end of a word" is syllable structure. In both cases, the consonant is in syllable-final position. Incorporating syllable structure in phonology allowed all of these rules to be stated more simply.

The earliest generative work on syllable structure (e.g., Hooper 1972; Kahn 1976) considered the syllable to be a linear string of segments, with the syllable boundary inserted at a particular place in the string, just as word boundaries were. Later work, however (e.g., Selkirk 1978, 1982, 1986; Kaisse 1985; Nespor and Vogel 1986; Itô 1986; Halle and Vergnaud 1987) considered the syllable to be a unit in the prosodic hierarchy. Processes such as epenthesis or deletion can then be seen as applying to bring the string into conformity with strict layering and exhaustive parsing. Epenthesis, for example, is argued to take place so that all segments can be grouped into well-formed syllables.

Most phonologists working in the decades since the publications cited above have taken the existence of syllable structure, and its relevance to phonological patterning, for granted, and have incorporated syllable structure into their phonological descriptions and analyses. There have, of course, been disputes about exactly *how* the syllable is structured: Is there an onset constituent? Should the mora be part of the hierarchy? What exactly is the status of those trailing coronals in words like "sixths"? Phonologists have also debated how syllable structure is built (templates? rules? constraints?) and what aspects of syllable structure are universal vs language-specific. Hyman (2011b) argues that there are some languages (his example is Gokana) that lack any positive phonological evidence for syllable structure at all, and thus that syllables themselves might not be universal. From the point of view of the phonology/phonetics interface, however, the most interesting question is probably "What *is* a syllable?" Syllables certainly seem to be psychologically real. If you ask an English speaker to tell you how many syllables in "Appalachicola" they will be able to answer "six" (though they will probably have to count them out on their fingers). Syllable structure has real effects in phonology, as argued above. But syllables have been notoriously difficult to define in phonetic terms.

7.3.2 DEFINING SYLLABLES 1: THE SONORITY HIERARCHY

In the easy cases, a syllable consists of a vowel and surrounding consonants, as in "cat," for example; but not all cases are easy. Sometime syllables have no vowels, such as [vlk], Czech for "wolf," or the English word "ladle," which is judged to have two syllables even when the /dl/ sequence is produced as a laterally-released stop, with no lowering of the tongue tip and thus no vowel. Sometimes two vowel qualities count as a single syllable, a diphthong. The word "pies" [pʰaɪz] is one syllable, but "pa is" [pʰɑɪz] is two; similarly, "Joy" [d͡ʒɔɪ] is one syllable and "Joey" [d͡ʒoi] is two. One can, of course, write the "offglide" of a diphthong as a glide ([pʰajz] and [d͡ʒɔj]) but that doesn't answer the underlying question of *how* the syllabicity is determined. In addition, while the majority of speakers agree on the syllable count of most words, there are ambiguous cases. How many syllables in "hour" or "Karl"? We need a phonetic definition that explains the clear cases, as well as why the problematic ones are problems.

Stetson (1928) hypothesized that one could measure the phonetic realization of the syllable in muscle contractions. He argued that each syllable is a "chest pulse" caused by a contraction of the muscles of respiration. Unfortunately, subsequent phonetic studies (Ladefoged 1967; Ohala 1977) showed that this was not the case. They were unable to find any physiological correlate of the syllable in muscle contractions, jaw movements, or breathing patterns.

Some of the most successful phonetic definitions of the syllable are based in the idea of sonority (Jespersen 1904; Saussure 1959; Hooper 1972; Selkirk 1984a; Zec 1988, 1995; Gouskova 2004; Parker and Lahiri 2012; among others). Sonority can be defined as "openness" or "loudness," which are correlated: the more open the vocal tract is, the more sound that escapes. Sounds can then be ranked, or "scaled" according to their sonority, with voiceless stops as the least sonorous (completely closed, with complete silence during the voiceless closure) and low vowels as the most sonorous (as open as possible, with the highest relative amplitude), and other segments in between. Given a sonority scale, a syllable can be defined around a peak of sonority. All sources agree that more sonorous sounds make better syllable nuclei, and less sonorous sounds make better syllable margins. Sonority rises from onset to nucleus, and falls from nucleus to coda, as in "print," where the sequence of obstruent–sonorant–vowel–sonorant–obstruent constitutes a single syllable peak. Vowels are the most open sounds, and therefore they make the best (least marked) syllable nuclei, but sonorant consonants will do if they are the most open articulation in the sequence (as in [vlk]). Diphthongs usually consist of one low vowel quality and one high, as in [aɪ], preserving a single sonority peak (but see Petersen 2018 for further discussion of diphthong typology). Cases of ambiguous syllable count often have two contiguous high-sonority segments ("hour," "Karl").

Defining the syllable in terms of sonority thus goes a long way toward defining possible and impossible syllables cross-linguistically, though it doesn't solve every problem (such as what to do about "sixths" and other words that do not exactly obey sonority sequencing). There is disagreement, however, over exactly how the "sonority scale" should be formulated. One simple sonority scale that still works for most cases (Clements 1990; Kenstowicz 1994) is shown in (4).

(4) A sonority scale, from highest to lowest
    Vowels
    Glides
    Liquids
    Nasals
    Obstruents

A more elaborated scale (5) is proposed by Gouskova (2004).

(5) A more elaborated sonority scale
    Non-high vowels
    High vowels and glides
    Rhotics
    Laterals
    Nasals
    Voiced fricatives
    Voiced and plain voiceless stops
    Voiceless fricatives
    Aspirated and ejective stops

As sonority becomes more fine-grained, the exact phonetic definition becomes more problematic. Gouskova assumes that rhotics are more sonorant than laterals; other sources (e.g., Clements 1990) treat rhotics and laterals as equally sonorous. The ranking of fricatives vs stops is also disputed, as is the role of laryngeal features. Is a voiced stop more or less sonorous than a voiceless fricative? Where do other laryngeal features fit in? In deciding, the question is not phonetically straightforward (Clements 2009b; Parker and Lahiri 2012). For example, there is no clear phonetic basis for computing "openness" based on the side channel vs the central channel, or the oral vs nasal passage. In counting "loudness," it is unclear whether periodic noise, fricative noise, and burst noise should be counted differently. And there is very little data to determine how other laryngeal features, such as breathiness, should be counted, or where consonants with non-pulmonic sources such as ejectives and implosives fit in the hierarchy.

Miller (2012) surveys the existing typological literature on syllable phonotactics, and proposes the still-more-elaborated sonority hierarchy in Figure 7.2, including different voice qualities and airstream mechanisms, with least sonorous at the top, and established rankings indicated by a line. Due to gaps in the data (e.g., no language seems to require a phonotactic restriction on a sequence of implosive plus

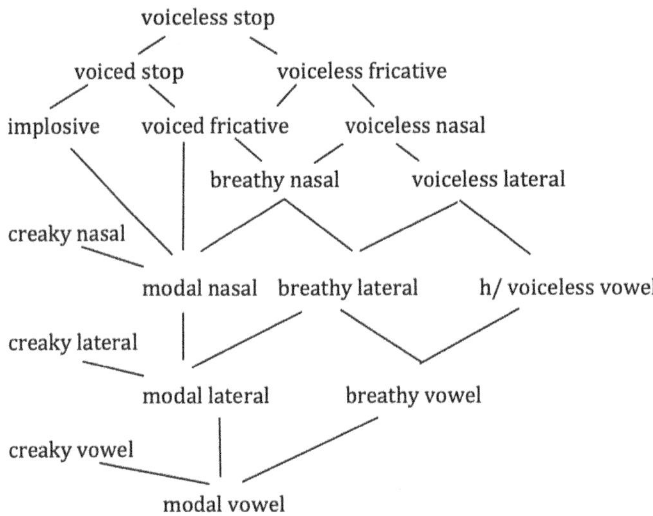

Figure 7.2 A still-more-elaborated sonority scale, including more laryngeal features, and only a partial ranking

Source: Adapted from Miller (2012).

creaky-voiced nasal) only a partial ordering is possible. Miller's ranking, however, does not include affricates, aspiration, and ejectives, as Gouskova's does.

Krämer and Zec (2020) take a different, non-phonetic, approach. They argue, based on cross-linguistic typology of onset and coda restrictions on nasals and laterals, that phonology recognizes two different kinds of nasal consonants. This is reminiscent of previous arguments, such as Kaye (1989) on two kinds of [ɛ] in Polish (Chapter 4), or two kinds of [w] (Anderson 1981; Chapter 6), that phonetic resemblance does not entail phonological identity. Nasals that occur in the coda and nucleus should be considered high sonority "vocoids" while nasals that occur in onsets should be considered low sonority consonants. Their proposed sonority scale is then Vowels > Nasal vocoids > Liquids > Nasal stops > Obstruents. In this argument, sonority is completely defined by syllable structure, rather than syllable structure being defined by a phonetically-based sonority.

The conclusion, as the papers in Parker and Lahiri (2012) explore, is that there is no final consensus on a universal sonority scale, on what phonetic properties are relevant to the phonological concept of sonority, or on exactly how the concept of sonority should be applied to defining syllable structure. Elaborated, but still incomplete and partially incompatible, sonority rankings show that sonority remains a useful concept, but one that is not yet fully understood.

7.3.3 DEFINING SYLLABLES 2: ARTICULATORY ORGANIZATION

Another line of research defines the syllable not as a particular pattern of sonority or loudness, but a particular pattern of articulatory coordination. It has been known for some decades that complex consonants have different coordinative patterns depending on syllable position (Krakow 1999). Velum lowering, for example, occurs earlier relative to oral closure for coda nasals than for onset nasals (Krakow 1989, 1993, 1999; Byrd et al. 2009). For laterals, the tongue body backing gesture occurs earlier relative to the alveolar closure when the consonant is in the coda than when it is in the onset, giving rise to the light/dark distinction mentioned above (Sproat and Fujimura 1993; Gick et al. 2006).

Recent research in Articulatory Phonology hypothesizes that articulatory differences do not just *depend on* syllable structure, but in fact *constitute* syllable structure (Browman and Goldstein 2000; Nam and Saltzman 2003; Tilsen 2016). These researchers argue that the syllable is a group of articulatory gestures that cohere in particular patterns of relative timing. Tilsen (2016) describes this coherence as "co-selection" in speech planning. Groups of co-selected gestures correspond to levels of the prosodic hierarchy, including the syllable (see also Byrd and Saltzman 1998; Byrd et al. 2000; and the discussion of the phonological phrase below). Chapter 9 delves into the details of how Articulatory Phonology defines gestures and gestural coordination, but here we can note that the idea that the syllable is basically a pattern of articulatory cohesion explains why people have to *say* a word to count its syllables, and why similar sequences, as "Joy" and "Joey" can be organized into one syllable or two.

Browman and Goldstein (1988), based on X-ray microbeam data, found that an onset consonant begins simultaneously with the vowel gesture, while a coda consonant follows the vowel in sequence. Onset consonants in clusters were also found to have greater overlap between the consonant gestures themselves. Later research (Browman and Goldstein 2000; Nam and Saltzman 2003) related these patterns to general mathematical models of stable coordination, or "coupling" (see the discussion in Chapter 9), linking syllable structure patterns to theories of how humans learn and produce patterns of rhythmic coordination in general. Onset consonants use the more stable "in-phase" or simultaneous coordination, which provides an articulatory explanation for why CV syllables are cross-linguistically preferred to VC syllables (Jakobson 1968).

A number of recent studies have used electromagnetic articulography (EMA) tracking techniques to investigate how consonant sequences in different languages are timed to each other and to the vowel, confirming the hypothesis of a stable articulatory coordination for the syllable, as well as for asymmetries between onset and coda. Languages examined have included Arabic (Gafos 2002), Berber (Goldstein et al., 2007; Ridouane 2008, 2017; Ridouane and Fougeron 2011; Ridouane et al. 2014), English (Marin and Pouplier 2010), French (Kühnert et al. 2006), Georgian (Chitoran et al. 2002; Hermes et al. 2011), Italian (Hermes et al. 2013), Romanian (Marin 2013), and Slovak (Pouplier and Beňuš 2011).

Following this hypothesis, patterns of articulatory coordination can be used as diagnostic phonetic evidence for syllable stucture (Chitoran et al. 2002; Goldstein et al. 2007; Chitoran 2016). Chitoran et al. (2002) for example, argue that complex consonant sequences in Georgian are in fact onset clusters based on their patterns of overlap, and Goldstein et al. (2007) argue based on different patterning that sequences in Berber are *not* complex onsets. In principle, patterns of articulatory coordination could address Hyman's (2011b) question: are syllables universal? We would hope that the three kinds of evidence—from phonotactic patterns, sonority profiles, and articulatory coordination—would all converge, as Chitoran (2016) argues that they do, but as Krämer and Zec (2020) argue, that may not be the case.

7.4 STRESS

Stress is defined as relative "prominence." A stressed syllable stands out as more prominent than the syllables that surround it, though exactly how to define prominence in phonetic terms has been "elusive" (as first stated fifty years ago by Lehiste 1970: 106). Much work in phonology has been devoted to describing the typology of stress systems, and the details of patterns of stress in different languages. A few of the most influential works on the phonological description of stress systems include Trager and Smith (1951), Chomsky and Halle (1968), Hyman (1977), Liberman and Prince (1977), Hayes (1985, 1995), Halle and Vergnaud (1987), Kager (1989), and McCarthy and Prince (1993), among many others. Overviews of phonological approaches to stress can be found in Kager (2007) and Gordon and van der Hulst (2018). For the most part, phonological accounts of stress have focused on explicating the often-complicated patterns of alternating stressed and unstressed syllables. Hayes (1995), for example, spends four pages on the phonetic correlates of stress, and the rest of the 450-page book on phonological analysis, based primarily on language descriptions where stress is transcribed.

At the phonology/phonetics interface, there are three important questions about stress to be asked: 1) How are stress patterns determined? 2) How are stress patterns produced (that is, what are the acoustic correlates of stress)? and 3) How are stress patterns perceived? These three questions are addressed in the following sections.

7.4.1 HOW ARE STRESS PATTERNS DETERMINED?

In some languages, there is no "pattern" to stress, meaning that the choice of which syllable in a word is stressed is an idiosyncratic property of that lexical item. To a large extent in Russian and to some extent in English, for example, stress is unpredictable. There is no phonological reason that the word "camel" has initial stress while "canal" has final stress, or that "comedy" and "committee" have different stress patterns. There are historical reasons—most English nouns with final stress were originally borrowings from French—but nothing in the synchronic phonology predicts the pattern.

More often in English, stress is "paradigmatic": it depends on morphological information. Verbs tend to have final stress, while nouns have penultimate (second-to-final) stress, resulting in many noun/verb minimal pairs: *record, insult, reject, permit,* and so on. Certain suffixes are associated with specific stress patterns: for example "-ial" attracts stress to the preceding syllable but "-y" does not (e.g., *president/presidency/presidential*). The study and description of lexical and paradigmatic stress systems lies much more at the phonology/morphology interface than the interface of phonology and phonetics.

The third type of stress system, positional stress, gets more of the phonological attention (e.g., Hayes 1995), because there is much more interesting phonological patterning to describe. As the name implies, the choice of which syllable is stressed is determined (primarily) by position in the word. Some of the systematic properties of positional stress include the facts that stress is alternating, culminative, multi-layered and demarcative (Hayes 1995; Kager 2007).

The fact that stress is alternating is to some extent required by the definition of prominence. In order to be prominent, a stressed syllable has to be surrounded by less prominent syllables, resulting in an alternating pattern. Another way to state this principle is that stress is essentially rhythmic (Liberman and Prince 1977). To capture the alternating pattern, syllables can be grouped into binary feet, consisting (in the default case) of one stressed and one unstressed syllable (Hayes 1995).

To say that stress is culminative means that every (content) word has one and only one primary stress. Multiple feet can be built across the length of a word, each with one relatively prominent syllable, but one foot is singled out as the *most* prominent, bearing primary stress. Because stress is multi-leveled, the other feet can bear an intermediate prominence, or secondary stress. Because there must be at least one stressed syllable per word, every word must consist of at least one foot, consistent with licensing in the prosodic hierarchy. The foot that is designated as the bearer of primary stress will always be positioned at the left or right edge of the word (or at least within one syllable of it), fulfilling the demarcative property of stress, indicating word edges.

In a strictly positional stress system, the phonetic makeup of individual syllables makes no difference at all in determining the stress pattern. For example, in the Australian language Pintupi (Hammond 1986; Hayes 1995), main stress falls on the initial syllable of a word, and secondary stress on alternating syllables thereafter (excluding the final), regardless of the vowels and consonants that make up the syllables. In some positional systems, however, word position and phonetic content interact.

In "quantity-sensitive" systems, heavy syllables attract stress, and stressed syllables may become heavy. A "heavy" syllable is one that has a long vowel or, in some languages, a vowel followed by a consonant. Open syllables with short vowels are inevitably "light." A canonical example is the Amazonian language Hixkaryana (Derbyshire 1979; Kager 1999, 2007). In this language, there is no contrastive vowel length. All closed syllables are stressed, the remaining open (CV) syllables are stressed in an alternating pattern, and stressed vowels in open syllables are lengthened. The result is that all stressed syllables are heavy, and all heavy syllables are stressed. Heavy syllables inherently already bear a certain degree of prominence ("carrying power" in the terms of Jones 1966: 141) due to their extra length, and it makes sense that the two kinds of prominence would coincide to promote culminativity (Trubetzkoy 1969; de Lacy 2002, 2007).

Other than the heavy/light distinction, few phonetic properties have been argued to influence stress. The absolute number of segments in a syllable does not in general count for stress systems. In English, for example, [skrɛ] is just as light as [kɛ]. Gordon (2005) however, discusses a small set of languages for which he argues that sonorant onset consonants *do* contribute to weight. De Lacy (2007) discusses a set of languages in which vowel height is relevant: stress is attracted to low vowels and is avoided on high vowels. While rare, these examples are consistent with the principle that greater sonority (that is, greater openness, loudness, and length) is best for realizing greater prominence.

7.4.2 HOW ARE STRESS PATTERNS PRODUCED?

What is the phonetic definition of stress? Van Heuven and Turk (2018) provide a useful overview of cross-linguistic patterns of phonetic realization, while Lehiste (1970), Ohala (1977), and Hayes (1995) give some history of phonetic studies of stress. Early linguists such as Sweet and Jones characterized stressed syllables as having extra intensity, described as "comparative force" (Sweet 1877: 91) or "force of breath" (Jones 1966: 57). This extra intensity could be realized through a number of different cues, including high pitch, duration, intensity, and segment quality. Sweet writes that "There is a natural connection between force, length, and high pitch, and conversely between weak force, shortness, and low pitch. The connection between force and pitch is especially intimate" (1877: 96). Ohala (1977) used the "whole body plethysmograph," a closed chamber that measures changes in lung volume and air pressure, to show that F0 changes associated with stress were due more to the larynx than the lungs, further refuting the "chest pulse" proposal of Stetson (1928), and demonstrating that pitch changes on stressed syllables were planned, not just automatic aerodynamic by-products of increased airflow.

Fry (1955, 1958) used synthetic speech to vary loudness, duration, and pitch on minimal stress pairs in English (such as ˈpermit vs perˈmit). He found that pitch, duration, and amplitude all varied with stress, but that changes in pitch and duration were more important cues to stress than loudness, at least for English speakers. Other acoustic studies on English, including Lieberman (1967) and Nakatani and Aston (1978), confirmed these results.

While loudness, pitch, and duration all play a role in signaling prominence, different languages weight the different cues differently (Cutler 2005; van Heuven and Turk 2018). Languages that use pitch contrastively (e.g., Thai, Potisuk et al. 1996) are more likely to use length to signal stress, while languages that have a phonemic length contrast (but no tonal contrast) are more likely to rely on pitch (Hayes 1995; Prince 1980). It is also important to note that it is not always high pitch that signals stress—it could be low pitch (as in English yes–no questions), or a rising or falling pitch excursion that makes the syllable stand out.

In addition, stressed and unstressed syllables can be distinguished by different allophones, with the unstressed syllables realizing a reduced inventory, or lenited versions of segments. It is common for unstressed syllables to allow only a subset of a language's vowel inventory. Crosswhite (2001) provides numerous examples: reduction to a short, centralized schwa as in English; reduction to the "corner" vowels [i, a, u], as in Luiseño, where [e] and [o] raise to [i] and [u]; or loss of a dimension of contrast, as in Italian, where the tense/lax contrast is lost in unstressed syllables. Tonal inventories are also often reduced in unstressed syllables, or tone may be contrastive on stressed syllables only (Duanmu 1990; de Lacy 1999; Yip 2002; Nitisaroj 2006).

Consonants also evidence strong and weak versions: for example, English /t/ is aspirated at the beginning of stressed syllables, and tapped at the beginning of stressless syllables. These stressed-based reductions can be motivated both articulatorily and perceptually. Lenited consonants and less extreme vowels take less energy, while the shorter duration and lower amplitude of unstressed syllables also make distinctions harder to hear, leading to loss of some dimensions of contrast.

Overall, there is no one universal correlate or diagnostic for distinguishing stressed and unstressed syllables (van Heuven and Turk 2018). As Hayes (1995: 8) writes: "Because the relation between stress and pitch/duration is both indirect and language-specific, it is impossible to 'read off' stress contours from the phonetic record." Rather, stress must be defined at the more abstract level of rhythmic prominence. Regardless of exactly which acoustic correlate a language uses, by alternating syllables with more and less of it, a stress pattern is produced.

Recent work continues to focus on how stress is realized in different languages, and how it interacts with other systems, including vowel quality, tone, and intonation. Again, van Heuven and Turk (2018) provide an up-to-date overview of studies on the phonetic realization of stress in different languages. The volume edited by Gussenhoven and Chen (2018) collects articles that survey stress (and other prosodic) systems according to eighteen areal groupings, as well as providing useful overviews.

7.4.3 HOW ARE STRESS PATTERNS PERCEIVED?

The perception of stress must go hand in hand with the production of stress. On a language-specific basis, listeners are attuned to the dimensions of prominence that speakers use to implement it (Fry 1955, 1958). Apart from the acoustic correlates of stress, however, several interesting questions arise with respect to perception of stress. How important is the demarcative function of stress in word processing? How many levels of stress can listeners perceive? Is "stress shift" a matter of production or perception?

Cutler (2005, 2012) discusses the "demarcative" function of stress. English listeners can more easily detect and remember words when the stressed syllable is initial, signaling the start of the word, and listeners mis-hear word boundaries when stress does not fall where expected (see also Cutler and Butterfield 2003 and Cutler and Norris 2002). This demarcative function seems to be an important cue for infants to bootstrap into word identification—infants more easily remember sequences that correspond to the rhythmic patterns typical of their language (Jusczyk 1999; Cutler 2012; Newport 2016). Again, however, there are language-specific differences in the importance of stress cues to word division. Listeners pay more attention to stress cues if their language uses stress contrastively (Cutler et al. 1997; Peperkamp and Dupoux 2002; Cutler 2005; Peperkamp et al. 2010).

As was mentioned above, phonologists take it as axiomatic that stress is multileveled, but how many levels of stress can listeners perceive? In principle, the phonology could potentially generate an indefinite number of levels of stress, given the cyclic application of stress rules argued for in Chomsky and Halle (1968) and subsequent phonological work. In English two-word compounds, for example, the first member receives more stress than the second, as in ˈgreenˌhouse, or ˈmotorˌcycle, while in noun phrases the second member receives the stress: ˌgreen ˈhouse. As formalized by Chomsky and Halle (1968) the formation of compounds and phrases creates "stress subordination": in compounds, the primary stress of the rightmost member is reduced one level, to secondary. But what happens when the phrases get more complex, with repeated (cyclic) subordination? Does "motorcycle club" have four levels of stress? What about "motorcycle club safety course flyer"?

Chomsky and Halle (1968) argued for four levels of stress in English words, and the IPA allows for transcription of primary stress, secondary stress, and unstressed syllables. In speech perception, however, the crucial distinction seems to be between syllables that bear some level of stress (either primary or secondary) and those that don't (Ladefoged 1993; Fear et al. 1995; Cutler 2012). When an initial syllable is stressed, word recognition and segmentation is facilitated, but the difference between primary and secondary word stress makes no difference. Cutler (2012: 232) writes that "for English listeners it is important to know whether a syllable is strong or weak . . . but it is not usually important to compute its stress level." Instead, Cutler argues, English listeners are much more sensitive to the difference between full and reduced vowel quality.

Ladefoged (1993) suggests that lexical stress should be separated from vowel reduction on the one hand and "tonic accent" on the other. The "tonic accent" is a phrase-level, rather than a word-level, prominence. Thus, when a word such as "psycholinguistics" is said in isolation, the fourth syllable is more prominent than the first, and both the first and fourth are more prominent than the second, third, and fifth, seemingly the result of primary vs secondary stress. But when said in a phrase, such as "The psycholinguistics course was fun," the difference between primary and secondary stress on the multisyllabic word disappears, because the tonic accent moves to the final word in the phrase. "In summary," Ladefoged writes,

> we can say that English syllables are either stressed or unstressed. If they are stressed, they may or they may not be the tonic syllables, which carry the major pitch change in the tone group. If they are unstressed, they may or may not have a reduced vowel. (1993: 116)

This combination—stressed and tonic, stressed but not tonic, unstressed but unreduced, and reduced—adds up to a four-way distinction without requiring the phonological representation of multiple levels of stress. Distinctions beyond four, though they may be generated by cyclic phonological compounding, are neither produced nor perceived.

Related to the question of levels of stress is stress retraction, or the "rhythm rule." In English, a final main stress is said to retract to a syllable earlier in a polysyllabic word if the following word has initial main stress. For example, in isolation, the word ˌsixˈteen is pronounced with greater prominence on the final syllable. But when the word is followed in a phrase by a word with initial primary stress, the stress pattern changes so that the initial syllable is heard as being more prominent: ˈsixˌteen ˈgirls. The state is ˌTennesˈsee but the song is ˈTennesˌsee ˈWaltz and the football team is the ˈTennesˌsee ˈTitans. This stress retraction is argued to occur because the rhythmic nature of stress requires that clashes (two stressed syllables in a row) be avoided (Nespor and Vogel 1989). Some analysts (e.g., Liberman and Prince 1977; Selkirk 1995; Hayes 1995) have argued that in these cases the main stress moves from the final syllable to a previous one in a process of "iambic reversal." Others (e.g., Gussenhoven 1991; Vogel et al. 1995; Grabe and Warren 1995) have argued that the primary stress is simply deleted or demoted, leaving no primary stress on the word but two equal stresses. This analysis is consistent with Ladefoged's analysis of the primary stress as phrasal accent as presented above. A third analysis is that speakers don't produce a change in stress at all, but that speakers hear it (Tomlinson et al. 2014). Tomlinson et al. argue that stress shift is an "auditory illusion": listeners interpret the stress pattern of a token differently depending on the words that surround it (2014: 1048). The same acoustic pattern could be heard as weak–strong or strong–weak depending on what follows.

If stress is a prominence relation, then it is necessarily relational, and sometimes the perception of rhythmic alternation can exist even if it is not produced. Because listeners expect rhythmic alternation, they impose it, whether it was produced or not. Questions of speech perception are explored more extensively in Chapter 10.

7.5 THE PHONOLOGICAL WORD AND PHRASE

"Phrasing" refers to the way that words are grouped together. Words are not produced at a completely even pace in a monotone, as though to a metronome. Rather, words are grouped together, as evidenced by pauses, pitch rises and falls, initial strengthening, final lengthening, medial lenition, and "sandhi" ("juncture") processes of assimilation or deletion that either apply at word boundaries or do not. Such groupings are crucial not only to understanding speech production, but also perception (Cutler et al. 1997). The two relevant questions for the phonology/phonetics interface are how these groups are defined, and how they are instantiated in production and perception.

7.5.1 THE PHONOLOGICAL WORD

Words are primarily understood as lexical entries, the unpredictable combination of sound and meaning. But words can also be phonological domains: processes can apply only within words (such as vowel harmony), or only at word boundaries (such as word-final devoicing). As a phonological domain, the "word" can be either larger or smaller than the word as defined by syntax or morphology.

A clear case of the former type of mismatch comes from compounds, which count as a single word for the syntax, but often as two words for the phonology. In Igbo, for example (Zsiga 1997, 2013a), words exhibit ATR vowel harmony: all the vowels in a word must be either [+ATR] or [-ATR], and affixes alternate between [+ATR] and [-ATR] versions depending on the root to which they attach. The verbal prefix, for example, alternates between [+ATR] [e] and [-ATR] [a]: [e-si]

telling, [a-sɪ] *cooking*. Igbo has a very productive process of verbal compounding, and the parts of the compound do not necessarily agree. Further, prefixes agree with the first part of the compound, and suffixes agree with the second part, as in the example in (6):

(6) Igbo compounding and ATR harmony (Zsiga 2013a: 344–5)

àɲɪ	a - kʊ̀-fu - ole	ja
1pl.	infl. – strike – hit - perf.	3s.obj

we have kicked it away

[−ATR] [+ATR]
[phol wd] [phol wd]
a - kʊ̀ fu – ole
[syntactic word]

The boundary between phonological words occurs in the middle of the syntactic word.

An example of the latter type of mismatch, where the phonological word is larger than the syntactic word, comes from copula reduction (also as discussed in Zsiga 2013a). In (General American) English, the verb *is* can be reduced to a fricative at the end of the noun, as in "Pat's tall" or "Doug's a good friend." Because we know that voicing assimilation takes place only within words in English, "Pat's" and "Doug's" must be single words with respect to the phonology, as shown in (7).

(7) Reduced copulas in English: one word in the phonology, two in the syntax

a. non-reduced form: Pat is tall

[phol wd]	[phol wd]	[phol wd]
p^{h}æt	ɪz	t^{h}ɑl
[noun]	[verb]	[adjective]

b. reduced form: Pat's tall

[phol wd]		[phol wd]
p^{h}æt	s	t^{h}ɑl
[noun]	[verb]	[adjective]

These phrases show that the major syntactic division of the sentence (between subject and verb) takes place in the middle of a single phonological word. These mismatches, in both directions, have important implications for the mapping from morphology and syntax to phonology. For the phonology/phonetics interface, the phonological word constitutes an important level in the prosodic hierarchy.

7.5.2 DEFINING THE PHRASE

Phrasing has implications for both the relation between syntax and phonology, and between phonology and phonetics. Sweet (1877: 86–7) gave a physical definition of the phrase: the speaker produces as much speech as is possible on one breath.

> There is no word-division whatever in language itself, considered simply as an aggregate of sounds. The only division actually made in language is that into "breath-groups." We are unable to utter more than a certain number of sounds in succession, without renewing the stock of air in the lungs. These breath-groups correspond partially with the logical division into sentences: every sentence is necessarily a breath-group, but every breath-group need not be a complete sentence.

This description is useful in reminding readers that there are no spaces or necessary pauses between spoken words, but there is evidence for smaller units than what can be said in a single breath. Jones (1966: 148) elaborated that "breath groups" are delimited by actual pauses. Subsidiary "sense groups" are places where the speaker *can* pause, whether or not a pause is actually produced.

A similar understanding, that prosody is controlled by physical constraints on breathing, is seen in SPE (Chomsky and Halle 1968: 13): "Writers, unlike speakers, do not run out of breath, and are not subject to other physiological constraints on output that require an analysis into phonological phrases." Nonetheless, Chomsky and Halle recognize that phrasing can be important for phonological rules, and, as was discussed above, incorporate boundary symbols for morphemes, words, and phrases (though not syllables or feet) into their rules.

Much work in the following years and decades (see, e.g., Inkelas and Zec 1990, 1995; Kisseberth 1994; Prieto 1997; Borowsky et al. 2012; Selkirk 2011; Elfner 2018) focused on discussion of the relation between syntactic and phonological domains, and on how much information the phonology and syntax could share. For example, Kaisse (1985) argued that the formation of phonological domains must reference the c-command relation, while Selkirk (1986) argues that reference to phrasal edges are sufficient. More recent theories (e.g., Seidl 2001; Adger 2007) argue for a correspondence between phonological domains and syntactic "phases" as defined by Minimalist Theory (Chomsky 2001). "Match Theory" (Selkirk 2011; Itô and Mester 2013; Elfner 2015, 2018) argues for a set of violable "Match" constraints governing the correspondence between phonological and syntactic constituents. Selkirk (2011) and Elfner (2018) provide a thorough review of the literature on the syntax/phonology interface as it relates to the syntactic definition of prosodic domains.

7.5.3 PHONETIC SIGNATURES OF PHRASING

From the point of view of the phonology/*phonetics* interface, the important question lies in how prosodic structure influences phonological and phonetic organization and realization. What are the phonetic cues to phrasing? A pause of necessity delimits a phrase, but other effects include final lengthening, initial strengthening, medial lenition, and other sandhi processes (Frota 2012).

The application of assimilation between two words ("external sandhi") is often taken as the defining characteristic of phrase membership (Selkirk 1980; Kaisse 1985; Nespor and Vogel 1986; Inkelas and Zec 1990). Zsiga (1992), for example, argues that the failure of vowel harmony to apply in compounds (example 6, above) shows that verb roots must constitute separate phonological words, while the application of a different process of complete vowel assimilation between words (example 8, below) shows that the words must belong to the same phrase.

(8) Igbo complete vowel assimilation between words (Zsiga 1992)

ŋkɪtà	ɔʧa	ɔma	ʊnʊ̀	→	[ŋkɪt**ɔ̀ɔ**ʧ**ɔɔ**m**ʊʊ**nʊ̀]
dog	white	beautiful	your		your beautiful white dog

To take another example, a process of voicing assimilation between words in Korean (Silva 1992) is also argued to be diagnostic of phrase membership. A lax stop becomes voiced between sonorants, but only when no phrase boundary intervenes. Thus assimilation applies between object and verb (Korean word order is SOV), but not between subject and object, as shown in (9).

(9) Voicing assimilation in Korean (Silva 1992)

a. kajka	papul	meknunta	→	[kaj**g**a] [pa**b**ulme**ŋ**nun**d**a]
dog	rice	eats		the dog eats rice

As was discussed in Chapter 5, it has often been asked whether rules that apply across word boundaries are properly part of the phonology or the phonetics, and such rules have thus been central to theorizing about the phonology/phonetics interface. Both Igbo vowel assimilation (Zsiga 1993, 1997) and Korean intersonorant voicing (Jun 1995) have been shown by phonetic investigation to be variable and gradient, and both Zsiga and Jun argue that these processes should therefore be accounted for in terms of the organization of articulatory gestures rather than phonological feature change. Not all word-boundary rules are variable and gradient however: Ladd and Scobbie (2003), for example, specifically argue that Sardinian consonant assimilation at word boundaries is featural and *not* gestural. The challenges posed by rules of external sandhi are further discussed in Chapters 5, 9, and 11. It is not controversial, however, that the application of external sandhi is indicative of coherence within a prosodic phrase.

Phrasing is also indicated by other phonetic cues (Fletcher 2010 and Turk and Shattuck-Hufnagel 2014, 2020 provide overviews and extensive references). Phrase ends are often (though not universally) signaled by lengthening, of either a final segment, syllable, or word (Delattre 1966; Klatt 1976; Beckman and Edwards 1992; Wightman et al. 1992; Turk and Shattuck-Hufnagel 2000, 2007; Grabe and Low 2002; Coetzee and Wissing 2007; among others). The amount of lengthening can signal the size of the phrase: slight lengthening for smaller breaks, greater lengthening for finality.

A widely-adopted framework for transcribing the strength of phrase boundaries in speech is the system of "Tone and Break Indices" or ToBI (Beckman et al. 2005). The ToBI system allows for five levels of "breaks" to be transcribed (10):

(10) Break indices in ToBI (Beckman et al. 2005)

0: very close boundary where a word break might be expected but is not, as in "who's" or "Pat's";

1: normal word boundary;
2: perceived juncture with no intonational marking;
3: intermediate phrase;
4: full intonational phrase.

Boundary strength is instantiated both in temporal patterns and in intonational patterns (see Chapter 8 for discussion of intonation). A "level 2" boundary might be cued by final lengthening but no change in pitch and no pause. While it was originally developed for English (Beckman and Elam 1994; Beckman and Hirschberg 1994), subsequent work has expanded the ToBI system to describe numerous other languages (see e.g., Jun 2005, 2014). An online course by Veilleux et al. (2006) provides an excellent introduction with numerous examples.

Byrd and colleagues (Byrd and Saltzman 1998, 2003; Byrd et al. 2000; Saltzman et al. 2008) consider boundary lengthening from the point of view of articulatory control: a slowing of the "clock" that governs timing of articulatory gestures (see Chapter 9). Byrd and colleagues argue against a top-down prosodic account, however, where a hierarchy of discrete phonological boundary specifications determines articulatory control. Rather, they argue that slowing gives rise to gradient lengthening, and lengthening gives rise to the perception of boundary strength. Final lengthening is thus seen as an instantiation of one of the basic issues at the phonology/phonetics interface: is there a need for an abstract phonological structure to explain acoustic and articulatory patterns? Byrd and colleagues would argue no, that there is no "external clock," and that an abstract prosodic hierarchy is a reification of a more variable range of articulatory patterns that arise directly from gestural organization. Turk (2012) counters, however, that non-local effects of prosodic organization do require speech planning based on independently-defined domains.

The same question applies to other instantiations of boundary strength. In a series of articles, Keating and colleagues (Fougeron and Keating 1997; Keating et al. 2003; Keating 2006; Cho 2006; Cho and Keating 2009) document the ways that consonant articulations are strengthened in initial positions. Most consistently, initial consonants have more extensive contact and longer closure, but may also evidence greater displacement and longer voice onset time (Keating 2006 and references therein). Keating et al. (2003) demonstrate similar strengthening patterns in English, French, Korean and Taiwanese, languages with otherwise dissimilar prosodies. While most work has been done on consonant strengthening, Pan (2007) finds a similar effect on pitch excursion in lexical tone in Taiwanese Min: pitch range of initial tones is positively correlated with increasing boundary strength. All of these results show that strengthening is cumulative: the larger the domain at whose edge the consonant or tone occurs, the stronger the articulation. These effects, however, do not exactly correspond to the predictions of the prosodic hierarchy.

> Strengthening does not result in discrete phonetic categories corresponding to the domains of the prosodic hierarchy.... It cannot be claimed that there are four (or however many) categories of phonetic strength, each of which gets some

additional increment of constriction; the effect appears instead to be continuous. (Keating 2006: 177–8)

As discussed by Turk (2012), while the findings on cumulative strengthening are consistent with a phonological prosodic hierarchy, the complexity of the factors that influence segmental duration (including segment identity, syllable structure, stress, speech rate, and focus, in addition to phrasing) mean that the specific mechanisms to derive precise segmental durations from more abstract symbolic representations have yet to be developed.

The flip side of domain-initial strengthening is domain-medial weakening. Lavoie (2001) gives a survey of cross-linguistic weakening and strengthening processes. As was noted above, Jun (1995) argues that intersonorant voicing in Korean (example 9, above) comes about due to reduction of glottal opening in phrase-medial positions. Similarly, Pierrehumbert and Talkin (1992) document weakened articulations of /h/ and glottal stop in unaccented, phrase-medial position. Aspiration of English voiceless stops in foot-initial position, with tapping in foot-medial position, is a third often-cited example. While the difference between an aspirated stop and a tap seem to be clearly the substitution of a different phonological allophone, phrase-medial reduction in glottal gestures in both Korean and English is argued to be gradient and variable.

The tendency for sounds in initial or other prominent positions to be more forcefully and fully articulated, while sounds in other positions lenite, is given a phonological account in Smith (2002), who argues for phonological constraints on "positional faithfulness." Under this account, prominent positions are protected from change that other segments undergo. Keating (2006) and Turk and Shattuck-Hufnagel (2014) argue for a grounding in perception for these asymmetries: the "strong attack" of initial position both signals a prosodic break, and fosters clearer articulation at a point in speech, the onset of a new phrase, where information is least predictable.

Current work on phonological phrasing continues to investigate the determinants of timing variation in speech, and the relation of that variation to the grouping and information structure predicted by the syntax via the prosodic hierarchy (Turk and Shattuck-Hufnagel 2014, 2020; Elfner 2018). Work also continues on documenting different prosodic patterns in different languages (Hirst and Di Cristo 1998; Jun 2005, 2014; Frota and Prieto 2007; Gussenhoven and Riad 2007a, 2007b).

The larger question remains: what is the cause and effect relationship between the domains of the prosodic hierarchy and the phonetic implementation of duration, pitch, amplitude, gestural coordination, and allophonic variation? Do the domains pre-exist and determine the phonetic implementation? Or do the phonetic patterns pre-exist, and give rise to the phonological abstraction? The relationship between abstraction and implementation holds of suprasegmentals as much as segmentals, as has been discussed throughout the previous chapters, and is returned to in Chapter 11.

Chapter 8 now turns to work on intonation, the highest level of the prosodic hierarchy, and to tone.

RECOMMENDED READING

Overviews

Goldsmith, J. (2011), "The syllable," in J. Goldsmith, J. Riggle, and A. Yu (eds), *The Handbook of Phonological Theory*, 2nd edn. Malden, MA and Oxford: Wiley-Blackwell, pp. 164–96.

- How does Goldsmith define the syllable?

van Heuven, V. J. and A. Turk (2018), "Phonetic correlates of word and sentence stress," in C. Gussenhoven and A. Chen (eds), *The Oxford Handbook of Language Prosody*. Oxford: Oxford University Press.

- Why is it that stress can not be simply read off the phonetic record?

Turk, A. and S. Shattuck-Hufnagel (2014), "Timing in talking: what is it used for, and how is it controlled?" *Philosophical Transactions of the Royal Society B: Biological Sciences, 369*(1658), 20130395. doi: 10.1098/rstb.2013.0395.

- How do Turk and Shattuck-Hufnagel answer the question in their own title?

Exemplary research

Something old

Fry, D. B. (1958), "Experiments in the perception of stress," *Language and Speech, 1*: 120–52.

- What do Fry's experiments teach us about the phonology of stress?

Something new

Keating, P., T. Cho, C. Fougeron, and C.-S. Hsu (2003), "Domain-initial strengthening in four languages," in J. Local, R. Ogden, and R. Temple (eds), *Phonetic Interpretation: Papers in Laboratory Phonology VI*. Cambridge, UK: Cambridge University Press, pp. 143–61.

- What do the results of this paper tell us about the prosodic hierarchy?

Tomlinson Jr., J. M., Q. Liu, and J. E. F. Tree (2014), "The perceptual nature of stress shifts," *Language Cognition and Neuroscience, 29*(9): 1046–58. doi: 10.1080/01690965.2013.813561.

- Is stress shift "real"? If so, what is it?

Opposing views

Krämer, M. and D. Zec (2020), "Nasal consonants, sonority and syllable phonotactics: the dual nasal hypothesis," *Phonology, 37*: 1–37.

vs

Chitoran, I. (2016), "Relating the sonority hierarchy to articulatory timing patterns: a cross-linguistic perspective," in M. J. Ball and N. Müller (eds), *Challenging Sonority: Cross-linguistic Evidence*. Sheffield, UK and Bristol, CT: Equinox Publishing, pp. 45–62.

- Compare and contrast the three approaches to defining sonority: based on acoustics, based on articulatory timing, and based on phonological distribution. Are they incompatible?

Byrd, D. and E. Saltzman (2003), "The elastic phrase: modeling the dynamics of boundary-adjacent lengthening," *Journal of Phonetics, 31*(2): 149–80.

vs

Turk, A. (2012), "The temporal implementation of prosodic structure," in A. C. Cohn, C. Fougeron, and M. K. Huffman (eds), *The Oxford Handbook of Laboratory Phonology*. Oxford: Oxford University Press, pp. 242–53.

- What are the differences in the ways these two papers approach the definition of the "phrase"?

QUESTIONS FOR FURTHER DISCUSSION

1. What are "suprasegmentals?" What makes them "above" the segmental aspects of speech?
2. Bolinger and others have suggested that suprasegmentals have often been treated as being at the fringes of phonology proper. How much influence does the alphabet have in making us think of suprasegmentals as less important?
3. Why are syllables difficult to define? How does the sonority hierarchy help us explain both the clear cases and the difficult cases of what constitutes a syllable? What issues remain in defining a universal sonority hierarchy?
4. How does a gestural approach define the syllable? Are the sonority-based and gestural approaches incompatible? How might cross-linguistic experimental investigation of patterns in articulatory coordination help us determine whether syllables are universal?
5. What does it mean to say that stress is "essentially rhythmic"? Is stress easier or harder to define than the syllable?
6. Why have rules of "external sandhi" been central to theorizing about the phonology/phonetics interface? What's special about them?

CHAPTER 8

INTONATION AND TONE

8.1 THE LINGUISTIC USES OF PITCH

Intonation and tone are grouped together in this chapter as two linguistic systems that are considered to be fundamentally based on pitch contrasts (though neither is limited entirely to pitch, as discussed below). They differ in that *tone* is a system of lexical contrast, and *intonation* a system of sentence-level contrast. Because of their similarities, however, the linguistic descriptions of both tone and intonation face similar issues of representation and realization, and confront the phonology/phonetics interface with similar problems. Analysis of the continuously changing "pitch track" and its relation to the segmental string presents a particular case of the problem of relating the continuous and the discrete, the physical and the abstract.

It is worth noting, at the outset, that the terms "pitch" and "F0" are not exactly equivalent. A "pitch track" should more accurately be called an "F0 track." *F0* is the abbreviation for fundamental frequency, the rate at which a periodic vibration repeats. In the case of the human voice, the vibration is that of the vocal folds, which can range in speech from a low of 50Hz for a large male up to around 400Hz for a small child. Operatic sopranos can hit notes above 1,000Hz (C_6, two octaves above middle C, is 1,046Hz), and young healthy ears can hear notes as high as 20,000Hz, but this is far outside the range of normal speech. *Pitch* is the way that F0 is perceived. Due to the structure of the human cochlea, human ears are much more sensitive to F0 changes in the lower ranges than in the higher, so it takes a much bigger change in F0 in the higher ranges to create the perception of an equivalent change in pitch. Over the whole range of human hearing, the relation of F0 to pitch is logarithmic, but in the range of vocal fold vibration in normal speech (not opera) the relationship is close to linear. Because F0 can be easily measured from an acoustic recording, and because in the relevant range F0 and pitch are close to equivalent, phonetic studies of tone and intonation often report F0 directly, without making the mathematical transformation to the pitch scale, and the terms "pitch" and "F0" are loosely used interchangeably. That practice will be followed here, but see Johnson (2013a) for further discussion.

Ladd (2008: 4) defines intonation as "the use of suprasegmental features to convey . . . sentence-level pragmatic meanings in a linguistically structured way." All three parts of the definition are important: suprasegmental, sentence-level, and linguistically structured. For Ladd, the term "suprasegmental" covers F0, intensity and duration, although in practice much of what is discussed in the intonation literature is based on pitch patterns alone. For example, Arvaniti (2012: 265) defines intonation as "the linguistically structured and pragmatically meaningful modulation of pitch." Crucially, intonation "conveys meanings that apply to phrases or utterances as a whole, such as sentence type or speech act, or focus and information structure," excluding the lexical contrasts that are the province of stress or tone (Ladd 2008: 6). Just as crucially, linguistic intonation, such as the difference between statement and question, must be distinguished from the "paralinguistic" expression of the speaker's emotional state, such as anger, fear, or surprise.

In contrast to intonation, tone is defined as a *lexically*-contrastive pitch pattern (Pike 1948; Yip 2002; Gussenhoven 2004). Cross-linguistically, the majority of languages (up to 70 percent according to Yip 2002) use pitch to distinguish lexical items. The major Indo-European languages are thus in the minority in being non-tonal, and the Eurocentric attitude that lexically-contrastive pitch is an "esoteric, inscrutable, and utterly unfortunate accretion" on an "otherwise normal" language (as Welmers 1973: 73, only slightly tongue-in-cheek, describes attitudes he had encountered) is completely backwards.

Tone contrasts may consist of different pitch levels (from two to as many as five), or of pitch movements of varying direction, slope, and shape (Maddieson 1978; Yip 2002; Gussenhoven 2004). In some languages, the pitch differences are accompanied by voice quality changes (see the discussion in Section 8.3.4 below). As with intonation, tonal phonology seeks to discover the categorical units that underlie the various patterns of pitch and laryngeal state, and to characterize the relationship between abstract units of contrast and their phonetic realization. Tonal phonology needs to account not just for the set of tone contrasts, but for tone alternations as well.

Another term that is often used in describing linguistic pitch contrasts is that of "pitch accent languages," which include Swedish, Japanese, Serbian, and Croatian. These languages are usually considered, following McCawley (1978), to be a subset of tone languages in that pitch is lexically contrastive. They differ from more typical tone languages, however, in that pitch contrasts are sparsely specified, with generally only one syllable in the word bearing contrastive pitch. In these analyses, the contrast lies either in the syllable selected to bear the tone, or in the shape of the pitch pattern associated to the selected syllable (e.g., late vs early peak, or falling vs rising). Phonetically-informed phonological analyses of pitch accent languages along these lines include Pierrehumbert and Beckman (1988) for Japanese, Bruce (1977, 1990) and Kristoffersen (2000) for Swedish/Norwegian, and Smiljanić (2004), Zsiga and Zec (2013) and Zec and Zsiga (2019) for Serbian and Croatian. Taking a different approach, Morén-Duolljá (2013) and Köhnlein (2016) argue that the North Germanic pitch accent languages (specifically Swedish and Franconian) are not underlyingly specified for contrastive tone, but for contrastive metrical foot structure, to which *intonational* tones align, producing differing pitch contours. Generally speaking, languages described as having "pitch accent" have some characteristics of tone systems (words differ in pitch pattern) and some characteristics of intonational or stress systems (pitch specifications interacting with metrical structure, one prominent syllable per word). The interactions of metrical prominence and tonal prominence are further discussed in de Lacy (2007), Köhnlein (2016), and Zec and Zsiga (2019).

Continuing the discussion in Chapter 7 of the prosodic hierarchy, in this chapter we turn first to intonation. Section 8.2.1 begins with a discussion of the difference between linguistic and paralinguistic uses of pitch. Section 8.2.2 discusses intonational patterns as "tunes," and 8.2.3 discusses how autosegmental-metrical theory breaks down those tunes into H and L autosegments. Section 8.2.4 addresses the question of how autosegments can be aligned to the segmental string. The second half of the chapter, Section 8.3, turns to the analysis of tone. After a brief overview of how tones are "marked" both orthographically and phonologically in 8.3.1, Section 8.3.2 addresses features for tone, and 8.3.3 discusses approaches to modeling tone that don't involve distinctive features. Finally, Section 8.3.4 raises the issue of the interaction between F0 and voice quality, and ends with a discussion of tonogenesis, the diachronic process by which tonal contrasts develop.

8.2 INTONATION

8.2.1 LINGUISTIC AND PARALINGUISTIC

Intonation has provided many opportunities for discussion of the relation between the physical and the grammatical in language. On the one hand, changes in pitch, loudness, and tempo convey emotional information that is outside the system of language. A listener can hear anger or fear in a voice speaking in an unknown language (Scherer 2000), or for that matter in a dog's low-pitched growl or high-pitched whine (Morton 1977; Ohala 1996). On the other hand, speakers manipulate the same parameters of pitch and loudness to convey discourse meanings that are arbitrary, language-specific, and linked to specific grammatical categories. Examples of language-specific associations between pitch patterns and grammatical categories or discourse meanings include the facts that in English yes/no questions end with a rising contour, but in Hungarian and Chickasaw, yes/no questions end with a fall (Ladd 2008: 81; Gussenhoven 2004: 54). Within English, fronted wh-questions ("What are you doing?") have falling intonation, but questions with the wh-word in situ ("You're doing what?") have rising. Positive tag questions ("You aren't going, are you?") have rising intonation, while negative tag questions ("You're going, aren't you?") have falling. Are the linguistic code and the emotional code just two separate systems of communication that happen to use the same physical parameters, or are the two related?

While some aspects of intonation are clearly arbitrary and language-specific, as noted in the preceding paragraph, there are also typological near-universals, broad generalizations that hold across many languages (Ladd 2008: 80). Most languages associate high pitch with questions, though the exact realization of pitch raising will differ: it might be a final rise, a suspension of the gradual lowering tendencies found in statements, or the realization of the whole contour in a higher register (Inkelas and Leben 1990; Jun 2014). Languages also tend to place pitch peaks on prominent or focused words. Can it be accidental that we also describe questions as indicating topics that are "up in the air" or expand our pitch range to signal excitement and engagement?

For some linguists who study intonation, the grammatical and emotional aspects of the use of pitch cannot be separated. Bolinger (1986: 34) writes "The affective, attitudinal, emotive side of intonation is inextricably intertwined with the grammatical side. There is no way to set aside what is 'ideophonic' just to make the grammarian's task easier." (See also the discussion of the "parallel encoding model" of Xu (2005) in Section 8.3.3 below.) Ladd (2008) argues, however, that the two *must* be disentangled, if the grammarian's task is even to be possible.

Gussenhoven (2004: Chapter 5) provides a roadmap for understanding the relationships between the biological, affective, and grammatical aspects of intonation. Building on work by Lieberman (1967), Ohala (1983) and others (see Gussenhoven 2004 for references), Gussenhoven discusses three "biological codes" that provide a physical ground for the linguistic use of intonation: the frequency code, the effort code, and the production code. The frequency code equates size and pitch: big things produce low pitch and small things produce high pitch. That biological fact can then be utilized by humans to convey an attitude or affect associated with size even in the absence of the physical correlate. Low pitch signals power, authority, and finality; high pitch signals vulnerability and uncertainty. That signal becomes grammaticized when the continuous variable takes on a categorical linguistic structure and association to linguistic meaning: a low boundary tone marks a statement and a high boundary tone marks a question.

The other two codes, Gussenhoven argues, follow similar paths. In the effort code, greater effort results in larger movements. Used affectively, a wider pitch range means greater excitement. Grammaticized, high tones mark focused words. Finally, the production code is based on inhalation and exhalation in the breath group. At the beginning of the breath group, airflow is higher and so is pitch, and the converse is true at the end. Thus high tone marks new topics, and low tone marks the ends of utterances.

Gussenhoven argues that these biological form/function relations explain universal tendencies in intonation. "In the absence of any motivation to the contrary, the intonational morphemes of a language will reflect the universal form-function relations" (Gussenhoven 2004: 79–80). Once grammaticized, however, intonational patterns lose their direct connection to the physical functions that gave rise to them, and can be subject to the normal processes of language change, resulting in a certain degree of arbitrariness in intonational mappings and meanings, such as the reversal of rising and falling patterns in Chickasaw. Intonation is thus just one instantiation of the general principles governing "natural" and "unnatural" patterns in phonology, as discussed at length in Chapter 6. Most languages show an intonational pattern that is grounded in universal biological principles, but once grammaticized, sometimes language change can result in unusual or marked patterns in specific languages. Intonation remains an interesting case, however, because the original physical grounding of the linguistic pattern (low pitch representing anger, or expanded pitch range indicating excitement, for example) remains present and available for speakers to use alongside the linguistic form.

Also parallel to other phonological features, the grammaticization of intonation involves reinterpreting the continuous physical scale of pitch into units that are discrete and categorical. Evidence that intonational patterns are categorical include the association of specific patterns with specific grammatical or discourse meanings as described above. Gussenhoven (2004) calls these abstract units "intonational morphemes."

On this view, intonational patterns can be represented as discrete units that are associated with the segmental string in language-specific ways. The affective cues, Ladd argues, using the familiar parameters of the phonology/phonetics interface, are quantitative modifications in the implementation of categorical units (2008: 35). The non-linguistic part is "purposeful variation in phonetic implementation" (Gussenhoven 2004: 58), while the structural part is part of the grammar. The question for intonational phonology to solve, as for segmental phonology, is what those structural units are.

8.2.2 TUNES

Describing the intonation patterns of English has been a part of describing English pronunciation for the past several centuries. Crystal (1969: 20–5) provides an overview of descriptive work to that date, starting with John Hart's "Orthographie," published in 1569, which included detailed instructions for when to use different kinds of punctuation marks. The full title was "An orthographie, conteyning the due order and reason, how to write or paint thimage of mannes voice, most like to the life of nature." The question of how to "write the image of the voice" and transcribe intonational contours has continued through the following four and half centuries.

Some writers on intonation have used the technique of moving the printed letters to represent the ups and downs of the voice, as in Figure 8.1, with examples from Bolinger (1986), and Ladefoged (1993).

Making things simpler on the typesetter, a line over and under the text can indicate relative pitch, as in Figure 8.2, from Pike (1948: 16), indicating eleven different ways to say "What's your name?" in Romanian. Note the very specific meanings associated with each variant.

Another option is to use a sort of musical scale to indicate the "tune" to which words can be set, as in the examples in Figure 8.3, from Jones (1966) and Halliday

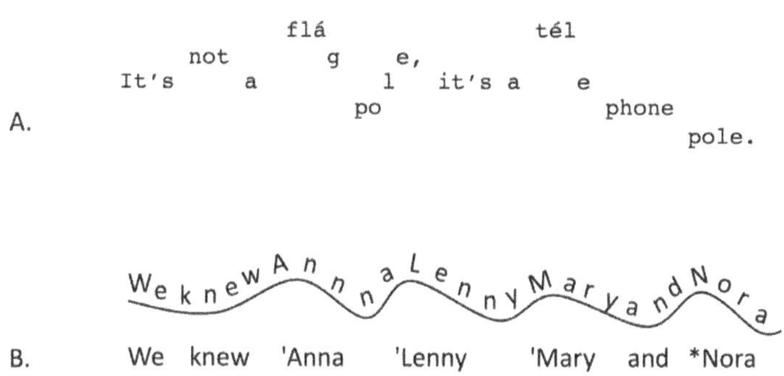

Figure 8.1 Transcribing intonation with visual placement of typography

Source: Adapted from A. Bolinger (1986: 144), B. Ladefoged (1993: 111).

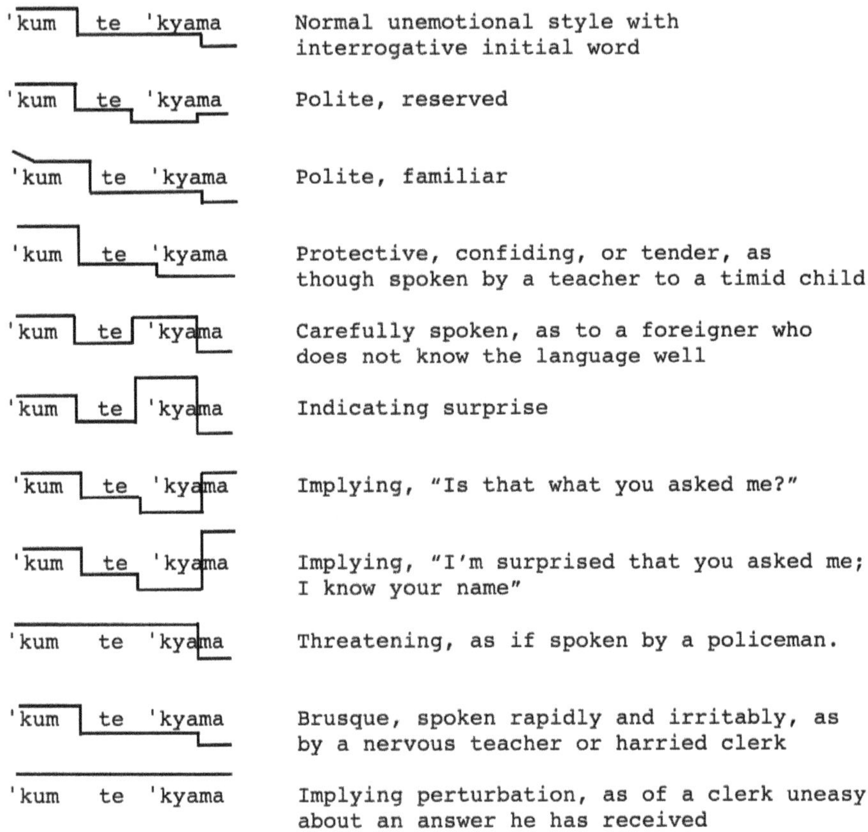

Figure 8.2 Transcribing intonation with lines: eleven ways to say "What's your name?" in Romanian

Source: Adapted from Pike (1948: 16).

(1970). In the figure, Jones's "simple fall" corresponds to Halliday's Tune A, "simple rise" to Tune B, and "emphatic fall-rise" to Tune C.

Each of these sources describes intonation in terms of "tunes" that the speaker chooses for each sentence or phrase, in the same way that one chooses a tune to set a verse to. Unlike a musical tune, however, "the melodies themselves have meanings" (Pike 1948: 15). But like a tune that is composed of notes, as pointed out by Arvaniti (2012: 265), the music cannot be interpreted as a "gestalt." The melody as a whole does not "simply shrink or stretch to fit the duration of the utterance." Rather, specific points in the tune coordinate with specific points in the segmental string, and the different points can coordinate independently. Ladd (2008: Chapter 2) makes the same point. (See also Arvaniti et al. 2006; Arvaniti and Ladd 2009). The English "calling contour," for example, is a recognizable tune involving a fall from high to mid. The fall, however, always takes place following the stressed syllable. Even though "Elizabeth" and "Marianna" both have four syllables, and both names can be called

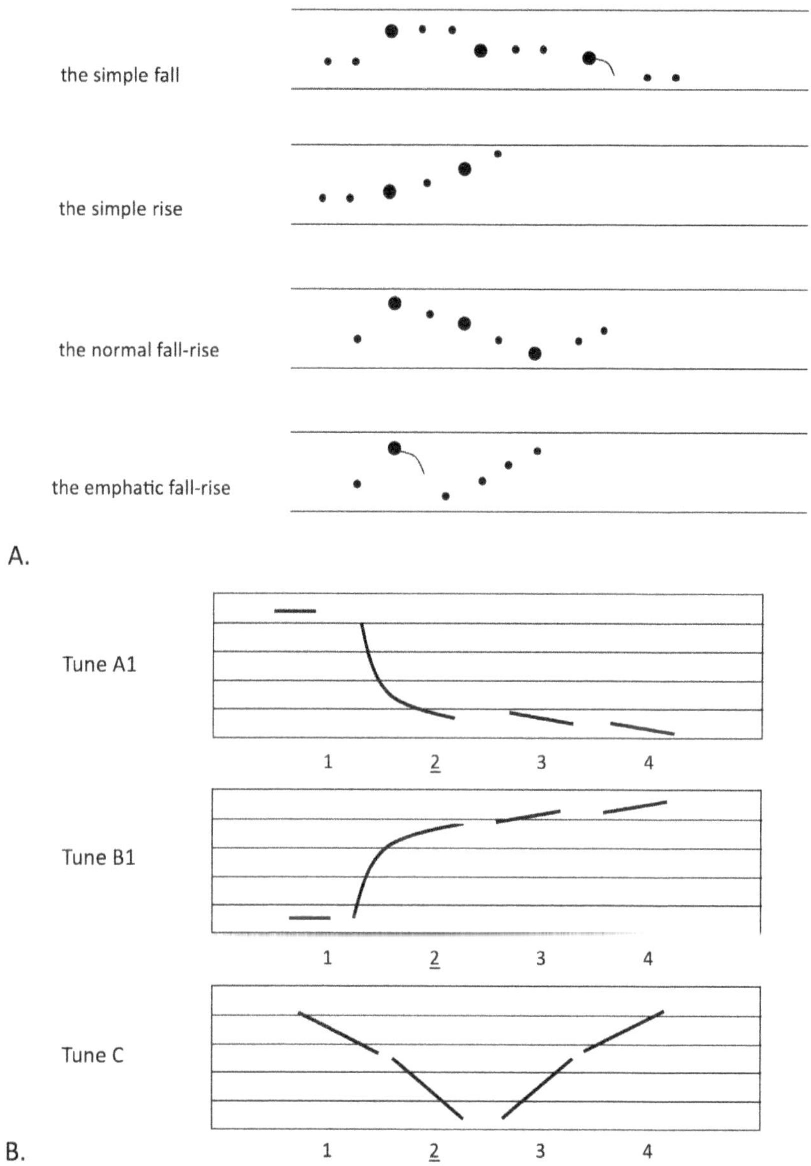

Figure 8.3 Transcribing intonation as tunes

Source: Adapted from A. Jones (1966: 152). B. Halliday (1970: 69).

with the same tune, the fall will occur after the second syllable on the former name and after the third syllable on the latter name, following the stress pattern. In order to better describe how tunes coordinate with segments, researchers on intonation break down the sentence-level patterns such as those illustrated in Figures 8.1–8.3 into smaller components. They disagree, however, on what those components are.

Halliday (1967, 1970), for example, assumes that English intonation is composed of five different "tones" (falling, high rising, low rising, falling–rising, and rising–falling) that can be combined into seven basic "melodies" (A–G), each of which can have variations in pitch height and slope. Each specific tune is described as carrying a specific meaning, from "neutral" to "special contrast" to "friendly warning," and so on (1970: 21–35). Bolinger (1951, 1958, 1986) describes English intonation in terms of contour shapes. Bolinger (1986) lists three basic "profiles": *A* is a high pitch followed by a jump down, *B* consists of a jump up, and *C* starts low and stays level or rises slightly. More complex tunes are combinations of these profiles. A is associated with assertion and some degree of finality, B indicates connection and incompletion, and C indicates backgrounding or reassurance, with many variations and possible combinations to produce specific gradations of meaning: for example, B+AC is "apt to be a disapproving comment" (p. 179) whereas CB+C is used as a "contradiction contour" (p. 183). Crystal (1969), along with others in the "British School" (which includes Palmer 1922; Armstrong and Ward 1931; Pike 1945; Gimson 1970; O'Connor and Arnold 1973; and Cruttenden 1997) describes English intonation in terms of falling and rising "nuclear tones" that associate to the accented syllable of a tone group. The various nuclear tones can be preceded by different pre-nuclear tones and followed by different "tails," resulting in melodies that can be described as "high rise," "low rise," "stylized fall-rise," and so on.

While these approaches to describing the tunes of English are similar, they differ both in the details of implementation and in the basic units that are assumed. Are there three profiles, five nuclear tones, or seven basic melodies? What parts are invariant and crucial to the linguistic meaning? All the descriptions agree that the location and direction of pitch change are important. But is the slope crucial, as in Jones's normal vs emphatic fall-rise? What about a "jump" or "step" vs continuous change, as in Halliday's system? Are high-fall and low-fall two discrete categories, or variation on a single category? Is it really possible to distinguish Halliday's 4–2–4 and 4–1–4 contours (as Liberman and Pierrehumbert 1984 ask)? In general, these approaches rely on native-speaker introspection for the tune divisions and meanings, and typographic devices as illustrated in Figures 8.1–8.3 to convey to readers (often language learners) the intended contours. Without phonetic analysis, it is difficult to provide further theoretical critique.

More phonetically-based approaches to defining intonational tunes include t'Hart and colleagues (t'Hart and Collier 1975; Collier and t'Hart 1981; t'Hart et al. 1990) on Dutch, Fujisaki (1983, 2004) and Fujisaki and Hirose (1984) on Japanese, Gårding (1983) on Swedish and other languages, and Grabe et al. (2007) on dialects of English. These approaches aim to provide a precise phonetic model of intonational contours, with the goal of synthesizing intonation by rule. The sources cited here work on the phonological assumption that tunes can be broken down into a series of accents. Each accent has a particular defined F0 shape, and the shapes are then connected via formulas for interpolation and overall declination. Grabe et al. (2007), for example, used a corpus of 700 English sentences from different dialects, and identified seven different types of pitch accent. The authors then fitted a polynomial equation to the F0 contours of each pitch accent, and proposed four coefficients for the equation that characterized the shape of the accent and differentiated the accent types. The goal of the study was to relate a particular linguistic label (regardless of the phonological model used) to a specific phonetic interpretation.

Other phonetic models (e.g., Xu and Xu 2005) work on the principle of "overlay" or super-position, rather than a linear string of accents. In an overlay model, a longer-term phrase contour is generated by rule as the base, and then shorter-term contours for accents are overlaid. (Xu's model for incorporating tone and intonation together is discussed in Section 8.3.3.) Cooper and Sorensen (1981), in another phonetically-explicit model, directly link F0 implementation to syntactic structure, without the intervention of phonological structure, though with mixed results (see the critique in Kaisse 1982).

As Grabe et al. (2007) point out, there has been a disconnect between different approaches to studying intonation. The descriptive tunes of the British School are widely used for teaching. Phonetic models are used for speech synthesis. But what is often missing from both of these approaches is an ability to address the phonology/phonetics interface, and an explicit modeling of how the linguistic structures of intonation are realized in speech. For linguists who are interested in describing both the phonology and phonetics of intonation, and the mapping between them, the most widely accepted representation is the autosegmental-metrical theory.

8.2.3 AUTOSEGMENTAL-METRICAL THEORY

Autosegmental-metrical theory (henceforth AM) expresses intonational contours as linear combinations of pitch accents that associate to prominent syllables, and boundary tones that associate to the edges of phrases. The theory builds on autosegmental work on tone (e.g., Leben 1973, 1978; see the discussion in Chapter 4), which was applied to intonation by Liberman (1985), Bruce (1977), and Pierrehumbert (1990a). The representations proposed by Pierrehumbert (1990a) have become the most widely accepted, though modified by subsequent research over the following decades. Overviews and introduction to AM theories of intonation include Ladd (1992, 2008) and Gussenhoven (2004, 2007).

The AM approach is autosegmental in that the units of contrast are H and L autosegments, the same as those proposed for the analysis of tone. (It is an advantage of the AM model that the same primitives are used for the analysis of both tone and intonation, facilitating a description of their interaction.) The theory is metrical in that the autosegments associate to metrical units, either prominent syllables or prosodic boundaries. The autosegments are sparsely specified in a linear string, with interpolation between specified points. Because of the sparse specification, the alignment of "tune to text" is an important part of the theory.

Pitch accents consist of a "starred accent," an H or L associated to a stressed syllable, with an optional associated second tone. For the simple H* and L* tones, linear interpolation to the next specified point determines F0 movement to and from the accented syllable. For the bitonal accents, a rise or fall is a necessary part of the accent type. The inventory of AM pitch accents (from Ladd 2008: 91–100) is given in Table 8.1.

Table 8.1 Inventory of AM pitch accents

H*	peak on accented syllable
L*	valley on accented syllable
L* + H	rise from the accented syllable
L + H*	rise to the accented syllable
H* + L	fall from the accented syllable
H + L*	fall to the accented syllable

Source: Ladd (2008).

Nuclear accent label	Stylization	British tradition description
H*L, %		Fall
H*L, H%		Fall-rise
H*, H%		High rise
L*H, %		Rise-plateau
L*H, H%		Rise
L*, H%		Late rise
L*H, L%		Rise-plateau-fall

Figure 8.4 AM accent, stylization, and "British School" tradition for a set of accents in a corpus of English dialects

Source: Adapted from Grabe et al. (2007: 285).

Boundary tones are marked with "%" rather than "*." H% marks a final rise, and L% the absence of a final rise (Ladd 2008). Different instantiations of the theory differ as to whether different kinds of boundary tones are needed or not. Some (e.g., Gussenhoven 2004) allow for an intermediate level phrase boundary to be marked as H^- or L^-. Over time, the inventory of accents has tended to grow less rather than more profuse, as not all of the accent combinations proposed by Pierrehumbert (1990a) have proved to be necessary. In most cases, the combination of one or more pitch accents plus a boundary tone corresponds to the "tunes" described by researchers using other approaches (Section 8.2.2).

Some examples will be useful. Figure 8.4 (from Grabe et al. 2007) shows stylized contours of the accent types (pitch accent plus boundary tone) found in their study, along with the corresponding "British School" description. Figure 8.5 gives some example utterances from Gussenhoven (2004) showing an utterance, a stylized contour, and an AM transcription.

Note in Figure 8.5 the association of starred tones to stressed syllables and boundary tones to the right edge of the phrase. Note also the linear interpolation between

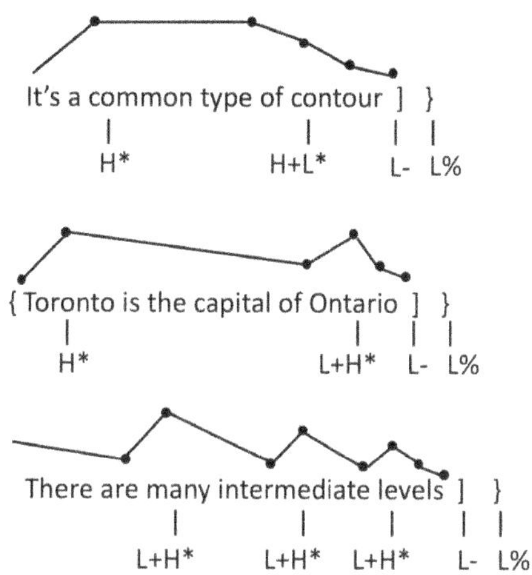

Figure 8.5 Examples of AM transcriptions from Gussenhoven
Source: Adapted from Gussenhoven (2004: 131, 135).

specified syllables that results in a plateau in Figure 8.5A and a long fall in Figure 8.5B. The figure also illustrates the use of the bitonal pitch accents, and the difference between a fall to the stressed syllable in Figure 8.5A and a rise to the stressed syllable in Figure 8.5C. Also evident in Figure 8.5C is the phenomenon of downstep: every time an L intervenes between two H tones, the second H tone is realized at a lower pitch.

Work by Pierrehumbert and colleagues in the 1980s (Liberman and Pierrehumbert 1984; Pierrehumbert and Beckman 1988; Pierrehumbert and Steele 1989) laid out the arguments in favor of the AM approach for representing intonational contrast, and modeling phonetic implementation, while separating linguistic and paralinguistic aspects. Liberman and Pierrehumbert (1984) use acoustic data from English to argue, first, for a distinction between a categorical choice of "tune" and quantitative variation in level of "prominence," and second, for an analysis of intonation in which "tunes can be decomposed into sequences of elements that are aligned with the text" (1984: 159). Their goal was to reduce the analytical degrees of freedom, including multiple tone levels and different kinds of rises and falls, needed by an approach such as Halliday's (1970; Figure 8.3B) to describe the intonational contrasts of English. One of their experiments asked participants to read a set of utterances at different levels of "overall emphasis or excitement" (p. 169). The sentences had identical strings of words, but varied in whether a particular word would count as "answer" or "background," as in (1):

(1) The questions from Liberman and Pierrehumbert (1984)
  a. What about Manny? Who came with him?
     Anna came with Manny. [Answer–Background (AB)]
  b. What about Anna? Who did she come with?
     Anna came with Manny. [Background–Answer (BA)]

Liberman and Pierrehumbert then compared the F0 height on "Anna" and "Manny." Their conclusion was that while pitch range varies continuously as a function of the level of excitement, the relation between the accented words stays constant over the varying pitch ranges, and can be predicted depending on whether the word order is Answer–Background (AB) or Background–Answer (BA). In both orders, the two words are analyzed as having L+H* pitch accents. In the AB order, the two words are phrased together and the second accent is downstepped, whereas in the BA order no downstep occurs. Crucially, a set of complex intonational contours can be mathematically modeled from a small set of primitives (L and H) and from general principles relating them, including predictable downstep and declination. These predictable relations between linguistic elements hold over quantitatively varying pitch ranges.

By separating out the phonological processes of accent choice and downstep from quantitative manipulation of pitch range, Liberman and Pierrehumbert lay out a particular division of labor between phonology and phonetics (as was discussed in Chapter 2). They argue for a model with a minimal number of phonological features, which are then subject to quantitative phonetic interpretation. Phonology is simplified (reducing a dozen or so tunes to two H and L autosegments) by moving implementation to the phonetics, allowing the analysis to "remove from the phonology the burden of representing those sound patterns for which its natural descriptive mechanisms are inappropriate" (1984: 229). Liberman and Pierrehumbert also advocate for this "style of research" (p. 232), a laboratory phonology approach that attempts to account both for phonology and phonetics, using phonetic evidence to argue for phonological analyses. The success of their study was an important impetus for further work, not only in intonation but in all areas of phonology.

Pierrehumbert and Beckman (1988) continue in this style, providing a comprehensive analysis of "Japanese tone structure," both contrastive patterns and phonetic implementation, using the AM model of pitch accents and boundary tones, along with quantitative modeling of pitch range, declination, and interpolation over unspecified syllables. Pierrehumbert and Steele (1989) further advance the model, providing experimental evidence for categorical alignment of accents to syllables. Pierrehumbert and Steele synthesized a continuum of intonational contours, with peak alignment changing in 20ms increments across a syllable boundary. When experimental participants were asked to repeat the contour they heard, they did not produce a distribution that matched what they heard, but a bimodal distribution, linking the accent to one syllable or the other. The bimodal timing pattern is further evidence that speakers plan their utterances in terms of autosegments associated to syllables, not "gesalt" tunes that can gradiently expand and contract over words or phrases.

These researchers argue for the advantage of the AM model on both phonological and phonetic grounds. The set of contrastive linguistic patterns is modeled with a small set of elements, and the linear string of pitch accents and boundary tones provide an explicit basis for computing F0 implementation. A further advantage of the AM model is that the same primitives, H and L autosegments, are used for the description and analysis of both tone and intonation.

Inkelas and Leben (1990) was one of the first papers to examine the interaction of tone and intonation. The language of investigation is Hausa. Inkelas and Leben propose a representation for tone that includes a "register" node that allows the basic tonal pattern to be raised or lowered for the purpose of marking intonational contrasts, such as that between question and statement. This type of interaction is only possible when both tone and intonation use the same primitives.

Inkelas and Leben note a number of different phonological tone rules, such as low-tone raising, question raising, tone assimilation, downstep and emphasis raising, that are bounded by the intonational phrase as the highest level of the prosodic hierarchy (see Chapter 7). These rules serve as diagnostics for division of utterances into phrases, and evidence from rule application shows that, consistent with general theories of the prosodic hierarchy, intonational phrasing is usually consistent with syntactic phrasing, but not always completely determined by it. (They cite the version of the prosodic hierarchy proposed by Selkirk 1986 and Nespor and Vogel 1986, but the data is equally consistent with Selkirk 2011). Inkelas and Leben note, for example, that intonational phrase breaks generally correspond to major syntactic breaks, but an emphasized word always begins a new intonational phrase, regardless of its position in the syntax. They also note (1990: 18) that the intonational phrase almost always corresponds to "a coherent semantic unit," consistent with early works that described intonation, such as Sweet's *Handbook of Phonetics* (published 1877) and Jones's *The Pronunciation of English* (published 1909), which argued for the importance of "breath groups" and "sense groups" (see the discussion in Chapter 7).

Research on intonation in the thirty years since 1990 has taken the basics of the AM model and applied it to the intonational patterns of numerous typologically diverse languages. While the model was developed for English (e.g., Pierrehumbert 1990a), it was never intended to be relevant to English only, and early work applied it to Japanese (Pierrehumbert and Beckman 1988) and Hausa (Inkelas and Leben 1990), as described above.

One vehicle for the description and analysis of diverse languages has been the ToBI, "Tone and Break Indices," transcription system (Beckman 2005; Veilleux et al. 2006). The aim of the system is to provide a consistent method for annotating different intonational patterns, using AM pitch accents and boundary tones, along with different levels of "break indices" (described in Section 7.5) to mark the strength of prosodic boundaries. While ToBI was first proposed in the 1990s as a systematic method for annotating the intonational patterns of English (Beckman and Elam 1994; Beckman and Hirschberg 1994), it has now been applied to a wide range of languages (Gussenhoven 2004; Jun 2005, 2014) with typologically diverse prosodies. Extensive work has been published on the major European languages, as well as on Korean and Japanese, but Jun's (2014) volume also includes AM analyses of languages as areally and typologically diverse as Arabic (Chahal and Hellmuth 2014), Basque (Elordieta and Hualde 2014), Georgian (Vicenik and Jun 2014), Tamil (Keane 2014), and the Australian language Dalabon (Fletcher 2014). While this diversity of work represents progress, and the general applicability of the AM model has made the expansion from English possible, much work still remains to be done in understanding cross-linguistic patterns of intonation.

Despite its broad general applicability, or maybe because of it, issues still remain in deciding exactly how to apply AM labels to an F0 trace, even in English. Is an accent simple or bitonal? A downstepped H or a mid tone? Jun and Fletcher (2014: 514ff.) provide practical advice on "deciding tonal categories," in ToBI transcription, but the fact that such advice is needed illustrates that the decision is not always straightforward. Ladd and Schepman (2003), for example, address at length the problem of whether a "sagging transition" between two high tones (as in forename–surname combinations such as "Nora Nelson") should be transcribed with an intervening L or not. They argue that the timing of high and low points indicates an L target, and that the distinction between H* and L+H* cannot be maintained for the transcription of English. Grabe et al. (2007) on the other hand, did find consistency in intonational labeling. As described above, Grabe et al. used mathematical modeling to distinguish the shapes of different F0 patterns in English, and they found that the groups created by the mathematical models corresponded well with the ToBI transcriptions provided by "hand-labeling." That was the case, at least, for six of the seven patterns found in the data: the model could not find a significant difference between final rises labeled L*H H% vs L* H% by human annotators.

Hualde and Prieto (2016) note the strengths and broad applicability of the AM model and ToBI transcription, but also note that it can be difficult to apply the conventions consistently across different languages. They discuss several examples of cases where very similar F0 patterns have been given different AM analyses by different authors, and other cases where identical transcriptions have different realizations. The difficulties arise, they argue, because the simplicity and generality of H and L representations require a language-specific interpretation. Thus two levels of ToBI transcription are needed: both a more abstract phonological and a more concrete phonetic level, corresponding to broad and narrow IPA for segments. Even when there is only one parameter, F0, the translation from phonological category to phonetic implementation is not automatic, and not necessarily simple.

8.2.4 TIMING OF PITCH ACCENTS

Much recent work on the phonology and phonetics of intonation, particularly within AM theory, has looked more carefully at specifying the relationship between H and L autosegments and the timing of peaks and valleys in phrases, words, syllables, and segments. It is an axiom of AM theory that tones are associated to prominent syllables and to boundaries, but given that syllables are long and have constituent parts (consonant and vowel, onset and rhyme, moraic structure), while boundaries have no extent in time independent of the segments on either side of them, the question of exactly how tones are associated to the segmental string remains an important one. Overviews of this area of research include Prieto (2011), D'Imperio (2012), and Arvaniti (2012) and references therein.

Like the choice of accent type and choice of syllable association, alignment can be contrastive. Prieto et al. (2005), for example, argue that (among other instances) in Neapolitan Italian the difference between a narrow-focus statement and a yes–no question is determined by whether the LH accent is aligned with the left or right edge of the accented syllable. In other cases alignment can be a matter of language-specific implementation. Prieto et al. (2005) argue for different general alignment patterns in Catalan vs Italian. Ladd et al. (2009) find systematic differences in peak alignment between English and Dutch, and between different dialects of English.

D'Imperio (2012), in her overview of different cross-linguistic alignment patterns, argues for a difference between phonological association to a syllable or boundary (as described by AM theory) and actual temporal alignment to a particular "anchor," which could be the left or right edge of a segment, mora or syllable, or to a different syllable entirely. Arvaniti et al. (2000), for example, argue that, in Greek, LH accents are phonologically associated to an accented syllable, but the L of the bitonal accent is in fact consistently aligned with the end of the *preceding* syllable, and H consistently aligned with the onset of the *following* syllable. Cross-linguistically, peak delay is quite common (Silverman and Pierrehumbert 1990). As is the case in tonal phonology (see the next section), H and L autosegments are not necessarily realized on the syllable with which they are associated in the phonology. For both tone and intonation, tone shift is usually to the right (Hyman 2007), unless the end of the phrase or "tonal crowding" prevents it. Peak delay may be the result of inertia, and the time needed from initiation of a movement to the time the target is reached (Xu 1999), and categorical rightward shift is the phonologization of that tendency.

Tonal crowding is the timing effect that proximate specifications have on each other (Silverman and Pierrehumbert 1990; Grice 1995; Arvaniti et al. 2000, 2006; Grice et al. 2000). There is a physical limit on the number of pitch targets that can be met within a given time span, and in cases where pitch accents and boundary tones compete for association to the same syllable or mora, or where lexical and intonational tones interact, that limit can be exceeded. In such cases, tones can be reduced, deleted, or realigned. One common realignment pattern is for tones to be realized earlier if another tone follows too closely, counter the general tendency for tone delay. Because it is the opposite of the expected phonetic effect, this early realization is evidence that tonal timing is carefully planned. Like a calligrapher spacing out letters on a page, a speaker plans tonal implementation and leaves enough room for all targets to be reached. On the principle that contrast must be preserved (Lindblom 1990), it would be expected that lexical tone would take precedence over intonational tone in cases where both cannot be realized, but Zsiga and Zec (2013) provide evidence that, in Belgrade Serbian, an intonational L can force retraction of a lexical H one syllable to the left when both compete for association with a final short vowel, resulting in neutralization of lexical contrast.

As was noted above in Section 8.2.3, it is only when tone and intonation are represented with the same primitives, H and L autosegments, that such interactions between tone and intonation can be analyzed. Ability to model both lexical tone and sentence-level intonation, and the interactions between them, is a signature strength of the AM model. It is to the representation of lexical tone that we now turn.

8.3 TONE

8.3.1 TRANSCRIBING TONE: MARKS AND MARKEDNESS

Linguistic tone is the use of pitch to create lexical contrast (Pike 1948); that is, changing the pitch changes the meaning of the word. For example, the Cantonese syllable

Table 8.2 Systems for transcribing tone

			Africa	Asia	Central America
High tone	H	acute accent	á	55/5	1
Low tone	L	grave accent	à or unmarked a	11/1	5
Mid tone	M	level accent	ā or unmarked a	33/3	3
Fall from high to low	HL	acute plus grave	â	51	15
Rise from low to high	LH	grave plus acute	ǎ	15	51

Source: Yip (2007: 231).

[jau] means "worry" if said with a high level pitch, and "have" if pitch rises from low to mid over the course of the syllable, as well as four other pitch patterns with different meanings (Yip 2007: 230). Overviews of the phonology of tone, with many varied examples, include Fromkin (1978), Yip (2002, 2007), and Gussenhoven (2004).

Languages differ in the complexity of their systems of tonal contrast. The simplest tone languages have just a two-way contrast: every syllable is either high or nonhigh, as in Navajo or Setswana (Zsiga 2013a). As reported by both Yip (2007) and Gussenhoven (2004), a three-level contrast is common, as in Yoruba, but a few languages may contrast as many as four (Mambila, described by Connell 2000) or even five levels (Benčnon, described by Wedekind 1983). When contour tones such as the rises and falls of Cantonese are added, there may be as many as ten possible contrastive patterns (Gussenhoven 2004). Languages differ in the density of their tonal specifications. In some languages, every syllable can be specified with a different tone. In others, only syllables that are metrically prominent in terms of stress or length can bear the full set of tonal contrasts (de Lacy 1999, 2007).

Several different systems have been proposed for transcribing tone (Table 8.2, from Yip 2007: 231). In simpler systems, a diacritic over the vowel suffices: an acute accent for a high tone and grave for low, with a combination for falling and rising. In two-tone systems, usually high tone will be orthographically marked and low tone left unmarked. In three-tone systems, the mid tone is the unmarked one. These orthographic conventions are consistent with phonological markedness. In two-tone systems, having a high pitch is considered the distinctive mark, and lack of high pitch is the default. Cross-linguistically, high tone tends to be the more "active" phonologically, triggering assimilations and dissimilations, for example (Clements 2001; Yip 2002). When tones reassociate to other syllables in the word, high tones tend to be the ones that move, and they tend to be attracted to syllables that are otherwise prominent: stressed, heavy, or both (Bickmore 1995; Yip 2001; de Lacy 2007). In three-tone systems, both high and low tone are marked, both orthographically and phonologically, and mid tone is often considered to be the default, literally having no tone at all. Whether "mid tone" is *always* equivalent to lack of phonological specification is a matter of debate (see, e.g., the discussion of different kinds of mid tones in Hyman 2010).

For more complex systems, with up to five levels, and different kinds of rises and falls (e.g., rising from mid to high vs low to mid, as in Cantonese), a system

Figure 8.6 Tone transcription and tone letters in the IPA chart
Source: internationalphoneticassociation.org.

of numbers is often used for tone transcription, as shown in Table 8.2, to indicate relative pitch height or the starting and ending points of contours. Unfortunately, linguists working on describing Asian languages have a tradition of using 5 for the highest pitch level and 1 for the lowest, while linguists working in the Americas have the opposite tradition. Table 8.2 also includes the autosegmental notations of H and L, and the decomposition of falling and rising contours into sequences. (See Chapter 4, and the discussion of tone features below for extensive further discussion.)

A non-ambiguous, but more cumbersome, method of transcribing tone is to use "tone letters" (Chao 1930), included as part of the chart of the International Phonetic Alphabet (Figure 8.6). The tone letter consists of a vertical line that indicates the speaker's pitch range, and then additional horizontal or slanted lines indicate the tone level and any movement. There are no constraints on possible patterns that can be drawn with a tone letter. The freedom to draw any possible design leads to a central question for tone at the phonology/phonetics interface: are there any constraints on possible tone patterns?

8.3.2 TONE FEATURES

While there are indefinitely many possible pitch patterns, a given language picks only a small set to use contrastively. As with vowel systems, where the F1–F2 space is continuous and languages choose from three to fourteen divisions for vowel contrasts, languages may choose a more or less crowded inventory of tones from the continuous F0 space. The question with tones, as with vowels, is what aspects of the realization are being used to create the contrasts. What aspects of the pattern are speakers paying attention to and manipulating (Zsiga 2012)? The question can be phrased, for tones as for vowels, in terms of what aspects of tone production and perception are phonologically significant: "What are the necessary and sufficient universal tone features?" Fromkin (1978: 1). Yip (1995: 477) expands the question to require that a universal tone feature set must "meet the familiar criteria of characterizing all and only the contrasts of natural language, the appropriate natural classes, and allowing for a natural statement of phonological rules and historical change." Abramson (1979) adds the criterion of a plausible phonetic implementation. And Hyman (2010) asks "Do tones have features" at all?

As discussed in Sections 4.3 and 8.2 above, as of this writing the tonal representation accepted by most phonologists consists of H and L autosegments that are associated to a syllable or mora (the tone-bearing unit, or TBU). Phonological arguments in favor of this representation for tone include Leben (1973, 1978), Gandour (1974), Anderson (1978), Goldsmith (1979), Yip (1989, 1995, 2002), Duanmu (1994), Zhang (2002), and Gussenhoven (2004). These sources argue that H and L associated to the TBU, usually the syllable, are sufficient for capturing the set of possible contrasts and alternations in tone languages. H vs L vs no specification account for high, low and mid level tones, while combinations of HL and LH account for falling and rising contours. More complex systems require some elaboration of the tonal geometry. An intervening register node is proposed by Yip (1995) to account for additional levels and for contours such as low rising vs high rising that traverse only a part of the pitch range. Association of tones to moras rather than syllables can account for more complex timing differences (e.g., Myers 2003; Morén and Zsiga 2006) and can capture the fact that complex tones often associate only to long or heavy syllables (Ohala and Ewan 1972; Blicher et al. 1990; Zhang 2002; Yu 2006).

Further evidence for autosegmental representations comes from tonal alternations where tonal melodies can spread or contract to fit different numbers of TBUs, such that single-syllable contours alternate with HL sequences spread out over multiple syllables (Goldsmith 1979; Yip 2002). Further, to the extent that H and L autosegments account for intonational patterns, it is an advantage that a single representation can account for both tone and intonation and their interaction (Downing 1989; Inkelas and Leben 1990; Myers 1996).

Some phonetic studies support autosegmental representations as well. Because F0 is one-dimensional, simple directions to move up or down can accurately model the pitch track (e.g., Mixdorff et al. 2002 and the intonational modeling studies cited above). Some studies that have investigated the effect of changes in speech rate or varied phonetic context have argued that H and L targets align independently to the segmental string (e.g., Myers 1996 on Chichewa; Zsiga and Zec 2013 on Serbian; and Morén and Zsiga 2006, Zsiga and Nitisaroj 2007 and Nitisaroj 2007 on Thai).

Autosegmental representations of tone cannot account for all tonal classes and alternations, however, particularly in denser systems that have more than a two-way or three-way contrast (Hyman 2010, 2014; Clements et al. 2011). Hyman (2010: 69) writes, "Because of its diversity tone is hard to reduce to a single set of features that will do all the tricks." Hyman (2010) examines various binary and unary feature systems proposed for tone, including Halle and Stevens's (1971) [stiff] and [slack] vocal folds, which were intended to apply to tone as well as to voicing, and

Pulleyblank's (1986) [upper] and [raised]. He finds that no feature set correctly defines natural classes in languages with four and five levels. Further, if the feature set provides sufficient features to contrast four levels, it can be impossible to decide which feature combinations correctly apply to systems with only three levels. Is the "mid" tone [+upper, -raised] or [-upper, +raised]?

Both Hyman (2010) and Clements et al. (2011) note that tonal systems are more cross-linguistically diverse than segment inventories and that tone rules "very often prove to be phonetically arbitrary; idiosyncratic to one language; complex . . . and/or noncomprehensive" (Clements et al. 2011: 5). Tone systems also differ from segmental systems in that tone contrasts are not built out of several different intersecting phonetic parameters (voice, place, manner) as consonant systems are, but only manipulate the single parameter, F0, in various ways. According to Hyman (2010: 69), because of the "extraordinary diversity" of tonal patterns it is a futile exercise to try to find a single small set of features (especially H and L) that will account for all of them. Both Hyman (2010) and Clements et al. (2011) argue that it would be more straightforward to describe tones in terms that are relative and scalar, referring to them by integers (H, M, L = 3, 2, 1 etc.), rather than either privative or binary features.

Continuing the discussion of universal tone features, Hyman (2014) refers to H and L autosegments as "useful analytical tools for capturing certain facts." It is a mistake, however, he argues, to expect any representation, such as particular models of tone feature geometry, to account for all classes and alternations. "Autosegmental representations come close to working, but not quite all the time," Hyman writes, a fact that "is fairly typical of phonological representations and phonological analysis in general" (2014: 390). The discussion thus recalls the points made in Chapter 4 on distinctive feature theory. Mielke (2008) found that the set of SPE features could account for 92 percent of segmental natural classes in his database (see Figure 4.7). As Cohn (2011) comments, the set is "approximately correct." And the question remains, is that good enough? Whether or not tonal systems are fundamentally different from systems of segmental contrast (Hyman 2010, 2011a), the complexities of deciding on featural representations for tone mirror the complexities of other feature systems.

8.3.3 NON-FEATURAL APPROACHES

Besides the complexities of tonal phonology, there is the issue of phonetic implementation. Mapping H and L autosegments to an F0 track is far from simple, and the phoneticians who study tone do not always agree with the phonologists on the utility of autosegmental representations. "Level" tones are seldom exactly level, and "rises" and "falls" are seldom linear diagonals. In Thai, for example, the "high" tone is a scooped contour reaching a high pitch only at the end of the syllable, while the "low" tone starts mid and falls gradually. Given this complexity, Abramson (1979) argued against the representation of the contour tones of Thai as HL and LH sequences, stating that the acoustic and perceptual data "lend no phonetic plausibility" to autosegmental representations. He writes (p. 7):

> For phonetic support of the argument one would expect to be able to devise a formula by which the dynamic tones were obviously to be derived from the shape of the static tones. Even the citation forms, let alone the F0 curves of running speech, provide no acoustic basis for such a claim. It seems psychologically far more reasonable to suppose that the speaker of Thai stores a suitable tonal shape as part of his internal representation of each monosyllabic lexical item.

In line with Abramson's reasoning, a number of linguists have argued that it is both phonetically more accurate and psychologically more plausible to describe tones in terms of movement, slope, or overall shape rather than sequences of static targets. Some of these approaches to tone description precede the advent of autosegmental proposals (e.g., Sapir 1921; Pike 1948), but other later proposals argue specifically against autosegmental representations (Gandour 1978; Abramson 1978; Clark 1990; Xu 2004; Roengpitya 2007; Barrie 2007; Xu et al. 2015). Pike (1948) argued that some languages define tone contrast in terms of "static targets" while others choose "dynamic targets." Xu (2004: 774) clarifies "for a dynamic target, the movement itself is the goal."

Evidence for dynamic targets comes from both perception and production studies. Xu (1997, 1998), for example, examines the effect of speech rate on Mandarin tonal contours. He concludes that the contours expand and contract as a whole, rather than H and L targets aligning independently to the segmental string (contra other studies, see Arvaniti 2012, that argue the opposite). He concludes that the contours are "integral dynamic" units rather than a sequence of H and L. Roengpitya (2007) conducted a similar experiment for Thai, and reaches the same conclusion that tone shapes are indivisible entities. In the perceptual domain, a number of studies such as those by Gandour and colleagues (Gandour 1978, 1981, 1983; Gandour and Harshman 1978; Gandour et al. 2000) have shown that listeners judge similarities between tones based on shape and slope rather than height of the endpoints. Gandour argues that perceptual features including direction and slope of pitch change must be included as part of the definition of tonal contrasts.

Building on this research, Xu (1997, 2004, 2005, 2011; Xu et al. 2015) argues for an explicit quantitative model in which the underlying units are movements rather than points: Parallel Encoding and Target Approximation (PENTA). A diagram of how the model works (adapted from Xu et al. 2015) is shown in Figure 8.7.

PENTA is "parallel" in that all the communicative functions of pitch—lexical, discourse, and emotional included—are encoded simultaneously in the input, and contribute equally to the determination of the pitch track. Xu argues that such parallel computation better models how pitch is actually produced and perceived in real communicative settings. In this, Xu follows Bolinger (1964), who argued that the emotional and linguistic aspects of intonation should not be separated. Each function is subject to linguistic encoding, which has both universal (e.g., questions are signaled by high pitch) and language-specific (e.g., lexical tone) aspects. The separate encoding schemes jointly determine a pitch target for each syllable. Pitch targets are not points as in AM, but include position (pitch height), slope, and strength. The

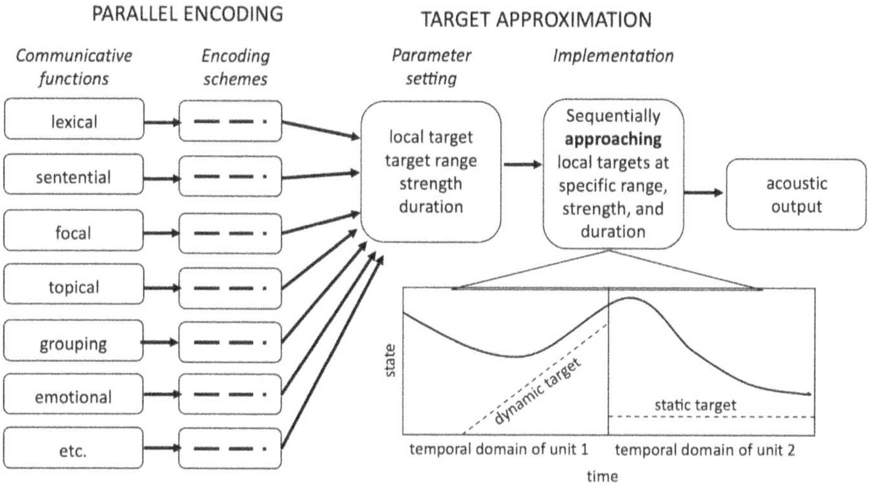

Figure 8.7 The PENTA model of tone and intonation
Source: Adapted from Xu et al. (2015: 509).

pitch targets are input to the articulatory mechanism, which produces actual acoustic output approximating the targets, weighted by "strength," as articulatory constraints allow.

The PENTA model is the strongest competitor to the AM model, and researchers continue to debate which model better captures both the phonological and phonetic aspects of tone and intonation. The models are explicitly compared in Arvaniti and Ladd (2009) arguing for the AM approach, and Xu et al. (2015) offering a counterargument in favor of PENTA. In more general terms of overall approaches to phonology and phonetics, PENTA is a functionalist approach, with emphasis on communicative function and few constraints on how linguistic encoding is accomplished. AM is a more formal approach, with an emphasis on the simplest representation of phonological contrast and alternation. Both are committed, however, to the existence of abstract categories, and to quantitative models that map those categories into F0 trajectories.

In the AM model, tonal targets are strictly acoustic. Although PENTA incorporates articulatory implementation, its targets are also acoustically defined. A different non-featural account takes the articulatory gesture as the unit of tonal contrast. This approach incorporates tone into the theory of Articulatory Phonology (Browman and Goldstein 1992a). Articulatory Phonology, as the name makes clear, assumes that articulation is the basis of all phonological contrast and alternation, and it has for the most part been applied to segmental phenomena only. Every acoustic target must be implemented by an articulatory action, however, and some recent studies have examined how the articulations that underlie tonal trajectories may be timed with other vowel and consonant gestures in order to produce the F0 patterns of contrastive tone. Tone languages that have been investigated and analyzed from an Articulatory Phonology perspective include Mandarin (Gao 2008; Yi and Tilsen

2014; Yi 2017), Thai (Karlin and Tilsen 2014; Karlin 2018), Serbian (Karlin 2018), and Igbo (Zsiga 2018b). These researchers have demonstrated the ability of this approach to model some complex patterns of tone and syllable coordination. Studies to date have been quite limited, however, and much further work is required on a wider variety of languages. Articulatory Phonology, in general, is covered in depth in Chapter 9. Before we leave the study of tone, however, one more topic must be addressed: the interaction of F0 with other segmental features.

8.3.4 INTERACTIONS OF F0 AND OTHER FEATURES

A further question related to the phonology and phonetics of tone is the interaction of pitch with other phonetic parameters and dimensions of contrast, particularly voice quality. Because contrasts in voice quality on vowels (modal vs creaky vs breathy), aspiration and voicing on consonants, and tone are all articulated by changing laryngeal configuration, it makes sense that they would interact. But the patterns of their interaction are complex, and not yet well understood.

One question is whether the definition of "tone" should be expanded to include dimensions other than pitch. In a number of languages, pitch and voice quality vary together. For example, in Burmese (Gruber 2011) and Munduruku (Picanço 2005) high-tone vowels are regularly produced with breathy voice. Low tones in Mandarin and Cantonese are regularly produced with creaky voice (Yip 2002; Yu and Lam 2014; Kuang 2017). Gussenhoven (2004) documents a tonal contrast in Yucatec Maya that is realized as glottalization in phrase-final position, but as a falling contour in phrase-medial position. Creaky voice quality is necessary to the perception of some of the tones of Northern Vietnamese (Brunelle 2009), and the contrast between breathy and modal voice is necessary for accurate tone perception in White Hmong (Garellek et al. 2013) and Green Mong (Andruski and Ratliff 2000). These and similar examples would seem to argue that voice quality is an integral aspect of tonal patterns.

Tone and voice quality are not necessarily linked, however. Glottalization can occur with both high and low pitch extrema, as the limits of regular vocal fold vibration are reached (Kingston 2005). In addition, in some languages, tone and voice quality are both contrastive, and can vary independently. In Mpi (Silverman 1997), for example, vowels contrast in "tense" and modal voice, and any of six tones can occur with either voice quality. In Jalapa Mazatek (Garellek and Keating 2011), any of three level tones can occur with either creaky, modal, or breathy voice. In order to account for the full set of contrasts in languages where tone and voice quality cross-classify, the two must be independently specified.

Further complicating an already complicated situation, the phonological relationship between tone and other laryngeal features does not necessarily remain stable over time. In the process of tonogenesis, voice quality differences result in a difference in pitch on a following vowel, which can eventually be reinterpreted as tone (Hombert et al. 1979; Kingston 2005). Numerous studies have found an effect of consonant voicing on a following vowel: pitch is slightly lower following voiced consonants and slightly higher following aspirated consonants (Halle and Stevens 1971; Ohala 1973, 1978; Sapir 1989 among others). These researchers argued that this pitch difference was an automatic effect: larynx lowering intended to promote voicing would inadvertently also lower pitch. Kingston and Diehl (1994) and Kingston (2011), however, find that explanation implausible, and argue instead that pitch changes are an intentional enhancement of the voicing difference (an example of speakers' fine-grained "phonetic knowledge," as discussed in Chapter 5). Whether planned or automatic, either way, over time the pitch difference can take over as the primary cue, and a voicing contrast can develop into a tone contrast. Recent phonetic studies that have documented tonogenesis in progress include Svantesson and House (2006), Abramson et al. (2007), Teeranon (2007), Hyslop (2009), Brunelle (2009), and Kirby (2014). In Seoul Korean, it has been well documented that the aspirated/lax consonant contrast used by older speakers has become a high and low tone contrast for younger speakers (Kim 2013; Kang and Han 2013; Bang et al. 2018). Coetzee and Beddor (2014) suggest that in Afrikaans as well, a contrast in VOT is becoming more reliably a contrast in F0. Afrikaans may be becoming a Germanic tone language.

On the one hand, linguistic pitch is simple: peaks and valleys in the F0 contour. But looked at more carefully, tone and intonation are complex, interesting, and definitely understudied. Systems of tone and intonation exemplify all the issues of the phonology/phonetics interface:

How are inventories determined?
What are the basic units/features?
What is the relationship between contrastive units and continuous implementation?
What is marked and unmarked, natural and unnatural?

Given the widespread occurrence of tone and the ubiquity of intonation in the languages of the world, the complexity of systems and contrasts that have been discussed in the previous sections, and the varied approaches to analyzing and modeling pitch systems, the biggest research imperative in the study of the phonology and phonetics of tone and intonation is to collect and analyze more data on more languages, to give us a clearer picture of what the linguistic pitch systems of the world are like.

RECOMMENDED READING

Overviews

In addition to the book-length treatments mentioned in the chapter text, these article-length descriptions provide useful overviews:

Ladd, D. R. (1992), "An introduction to intonational phonology," in G. J. Docherty and D. R. Ladd (eds), *Papers in Laboratory Phonology II: Gesture, Segment, Prosody*, Cambridge, UK: Cambridge University Press, pp. 321–34.

- What are the basic tenets of the autosegmental-metrical theory? Why the two-part name?

Prieto, P. (2011), "Tonal alignment," in M. van Oostendrop, C. Ewen, B. Hume, and K. Rice (eds), *The Blackwell Companion to Phonology*. Malden, MA: Wiley-Blackwell, pp. 1185–1203.

- Should we recognize a difference between phonological and phonetic alignment?

Yip, M. (2007). "Tone," in P. de Lacy (ed.), *The Cambridge Handbook of Phonology*. Cambridge, UK: Cambridge University Press, pp. 229–52.

- How does Yip reduce phonetic complexity to choose a small set of contrastive features?

Exemplary research

Something old

Liberman, M. and J. Pierrehumbert (1984), "Intonational invariance under changes in pitch range and length," in M. Aronoff, R. Oehrle, F. Kelley, and B. Wilker Stephens (eds), *Language Sound Structure*. Cambridge, MA: MIT Press, pp. 157–233.

- Describe how Liberman and Pierrehumbert propose to "remove from the phonology the burden of representing those sound patterns for which its natural descriptive mechanisms are inappropriate" (p. 229), and the implications of this paper beyond the representation of intonation.

Something new

Grabe, E., G. Kochanski, and J. Coleman (2007), "Connecting intonation labels to mathematical descriptions of fundamental frequency," *Language and Speech, 50*: 281–310.

- How are the findings from this paper relevant to questions of the phonological representation of intonation?

Bang, Y., M. Sonderegger, Y. Kang, M. Clayards, and T.-J. Yoon (2018), "The emergence, progress, and impact of sound change in progress in Seoul Korean: implications for mechanisms of tonogenesis," *Journal of Phonetics, 66*: 120–44.

- What are the mechanisms of tonogenesis described in this paper, and what are the implications for our understanding of the phonology/phonetics interface?

Zsiga, E. C. and R. Nitisaroj (2007), "Tone features, tone perception, and peak alignment in Thai," *Language and Speech, 50*: 343–83.

- What are the arguments for reducing the complexity of Thai tonal patterns to a small set of autosegmental features?

Opposing views

Arvaniti, A. and D. R. Ladd (2009), "Greek wh-questions and the phonology of intonation," *Phonology, 26*: 43–74.

vs

Xu, Y., A. Lee, S. Prom-on, and F. Liu (2015), "Explaining the PENTA model: a reply to Arvaniti and Ladd," *Phonology, 32*: 505–35.

- What are the objections to the PENTA model that Arvaniti and Ladd raise, and how do Xu et al. answer them? Are you convinced either way?

Yip, M. (1989), "Contour tones," *Phonology*, 6: 149–74. (See also Yip 2007 cited as an overview above.)

vs

Hyman, L. (2010), "Do tones have features?" in J. Goldsmith, E. Hume, and W. L. Wetzels (eds), *Tones and Features: Phonetic and Phonological Perspectives*. Berlin: de Gruyter Mouton, pp. 50–80.

- What arguments does Yip make in favor of tone features? What arguments does Hyman make against them? Is it OK to have a set of tone features that are "approximately correct"?

QUESTIONS FOR FURTHER DISCUSSION

1. What is the difference between linguistic, para-linguistic, and non-linguistic uses of pitch? Can you always tell which is which? How do the "biological codes" discussed by Gussenhoven relate the three levels?
2. How are intonation and tone different and how are they the same? Is it important that they should be represented in the same way?
3. Discuss the challenges in just *transcribing* tone and intonation (apart from the question of phonological features). What different systems have been proposed, and what are the benefits and drawbacks of each?
4. Should the definition of tone include other laryngeal features? Why or why not?
5. It is claimed at the end of the chapter that:

   > Systems of tone and intonation exemplify all the issues of the phonology/phonetics interface:
   >
   > How are inventories determined?
   > What are the basic units/features?
   > What is the relationship between contrastive units and continuous implementation?
   > What is marked and unmarked, natural and unnatural?

   Choose one or more of these questions, and compare and contrast the methodologies, evidence, and conclusions from segments and from suprasegmentals. Is the phonology of tone and intonation essentially the same as segmental phonology, or is it radically different? You may want to read Hyman (2011a) ("Tone: is it different?") to help organize your answers.

CHAPTER 9
ARTICULATORY PHONOLOGY

9.1 TOWARDS AN ARTICULATORY PHONOLOGY

This chapter discusses Articulatory Phonology (AP), which embodies a different approach to phonology and phonetics than the feature-based approaches discussed in earlier chapters. AP was pioneered in the 1986 article "Towards an Articulatory Phonology" by Catherine Browman and Louis Goldstein, and further developed in the decades since by Browman, Goldstein, and their students and colleagues (e.g., Browman and Goldstein 1989, 1990c, 1992a; Goldstein et al. 2006; Nam et al. 2009; Goldstein et al. 2009). Hall (2010, 2018) provides recent article-length overviews.

Unlike other phonological theories that take the segment or feature to be the basic unit of phonology, AP takes the "articulatory gesture" to be the basic unit of phonological contrast, and gestural reorganization to be the basic explanation for variation in the way sounds are realized. In AP, there in fact is no phonology/phonetics interface, no "fence" or "border," no translation from phonological units to phonetic implementation, because the units of phonology and phonetics are the same. The articulatory gesture suffices both for phonological contrast and for phonetic implementation. Whether it also suffices for phonological alternation is a current research question.

In AP, the articulatory gesture is defined as "an abstract characterization of coordinated task-directed movements of articulators within the vocal tract" (Browman and Goldstein 1989: 206). The definition is worth considering carefully. The theory, first, is based on movements of articulators: lips, tongue tip, tongue body, velum, and so on. The movements inherently take up both space and time, differing from both features and segments that are generally treated by the phonology as one-dimensional points that can be sequenced in time but have no inherent temporal extent. Importantly, articulatory gestures are "task-directed" and "coordinated." The goal of each gesture is to use a coordinated set of articulators (e.g., jaw + lower lip + upper lip) to create a particular constriction degree at a particular constriction location (e.g., bilabial stop). The characterization of these movements, however, is abstract. Goals are specified, defined as variables in a formula, but as the articulations unfold and interact with each other, the defined goals may not be met.

One way to conceptualize articulatory coordination is as a "dance" of articulators (Goldstein n.d.). Just as complicated dance routines are made up of individual steps or moves that are put together according to a particular pattern, an utterance is made up of coordinated gestures. Just as a single dance move ("plié" for instance) can involve multiple parts of the body moving in a specified way, a single gesture can recruit a set of articulators to reach the specified vocal tract state. When the dance is choreographed, it is the steps that are coordinated, not the individual muscles.

After hours of practice, a dancer will be able to execute the routine smoothly and automatically, without consciously thinking "left foot, right foot, left foot." Babies practice the routines of speech—consonants, vowels, and their coordination into syllables—through babbling in the first year of life, and these routines are then gradually enlarged into words and phrases (Browman and Goldstein 1989). For adults, the patterns of coordination are so thoroughly learned that they are executed without conscious thought, and in fact are very hard to change, as adults discover when they try to learn another language.

Goldstein uses comparison to dance to describe the "central concern" of the theory:

> The aim [of Articulatory Phonology] is to develop an explicit dynamical model of the "dance of the tongue," analyzing the observed motion of the human vocal organs during speech into "steps" (or gestures), and investigating the inventory of such steps that are used in the world's languages and the range of different dances that languages assemble from these steps. The construction of the dance in a particular language is hypothesized to be central to the cognitive activities surrounding speech, "sound structure," and reading: acquisition, perception, production, and breakdown in disease. (Goldstein n.d.)

In this view, phonology is choreography.

In the next sections, we consider the AP approach to phonology and phonetics in more detail. Section 9.2 discusses the overall gestural model and the role of gestures as units of contrast. Section 9.3 examines how AP treats alternation as gestural reorganization. Section 9.4 raises some critiques of the AP approach, specifically whether AP can or should account for categorical as well gradient assimilations and deletions, and Section 9.5 considers how constraints on gestural timing might address those critiques. Sections 9.6 concludes with challenges and directions for future research.

9.2 GESTURES AS UNITS OF CONTRAST

A flow chart for the AP gestural model, from intended utterance to speech output (from Browman and Goldstein 1992a) is shown in Figure 9.1.

First, the plan for the intended utterance is encoded in terms of gestures (not segments or features) in the "linguistic gestural model." Entries in the lexicon consist of a coordinated set of gestures associated with a meaning. The choice and coordination of gestures in an intended utterance can be represented as a "gestural score," as in Figure 9.2.

The score in Figure 9.2 diagrams the utterance [pʰæ̃n]. "Score" here is used in the sense of an orchestral score that coordinates the different instruments. Just as each instrument has its own line in an orchestral score, each articulator set has its own line in a gestural score. Articulator sets are named for the active articulator that effects the

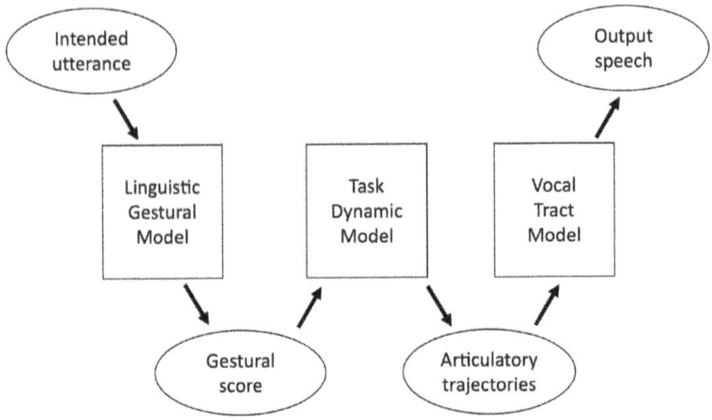

Figure 9.1 Flow chart of the Articulatory Phonology model
Source: Adapted from Browman and Goldstein (1992a).

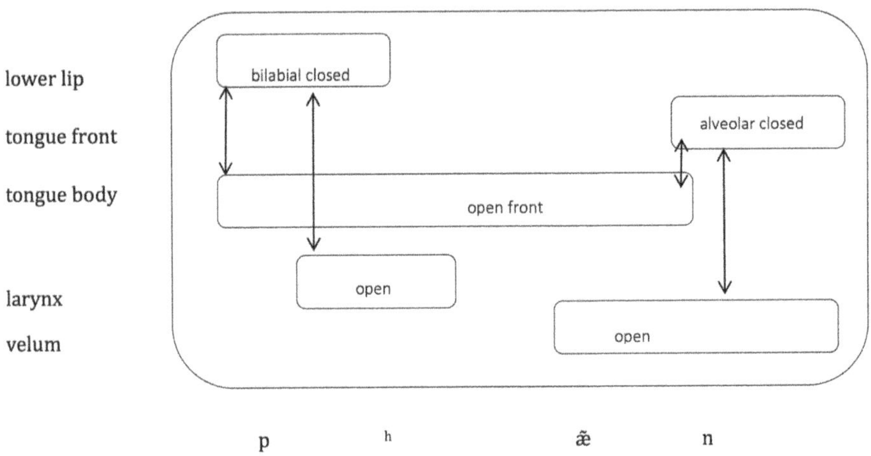

Figure 9.2 Gestural score for "pan"

constriction: lower lip, tongue front, tongue body, velum, and larynx. Each gesture, with its specified constriction location and degree, is indicated as a box in the score. Velum and larynx have only constriction degree, as they cannot move to different locations.

"Pan" consists of five gestures: bilabial closing and larynx opening corresponding to [pʰ], alveolar closing and velum opening corresponding to [n], and a single tongue body lowering gesture that constitutes [æ]. Time is on the x-axis, and the length of the box indicates each gesture's "activation interval," the time when active control of that articulator set begins and ends. The arrows in the figure show gestures that are timed directly to one another. Gestural timing, referred to as "gestural phasing," is always relative to other gestures, not to an external clock. There is no "segment tier" or other way of externally keeping track of units of elapsed time. In Figure 9.2, the score shows that gestures of the lips and tongue body begin simultaneously, that laryngeal opening begins in the middle of the bilabial gesture, that velum opening and alveolar closing reach their targets simultaneously. Thus, in some cases, such as that between alveolar closure and velum opening, the timing between units is direct and planned ("co-selected" is the term used by Tilsen 2016), while in other cases the timing is indirect, and comes about because of interaction with other gestures. For example, the velum opening gesture and the vowel are both directly timed to the alveolar closure but not directly timed to each other. Differences in gestural timing are a central concern of the theory, and are discussed in much further detail in Sections 9.3 through 9.5.

In the simplest version of AP, articulators are assumed to return to a default state when not under active control. The default is assumed to be a vocal tract position approximating [ə]: open lips, central tongue position, closed velum and modal voicing, though decisions about the default state and the necessity of release gestures have been debated (Browman 1994; Gafos 2002). Thus there is no "lip opening," "velum closing," or "voicing" gesture shown in the gestural score to return the articulators to their default states.

Note that gestures do not correspond directly to either features or segments. A bilabial closing gesture covers the information conveyed by the two features [labial] and [-continuant], but the gesture might comprise either one segment ([b]) or correspond to part of a segment ([m] or [pʰ]). Laryngeal opening might correspond to the features [spread glottis], [-voice], or both, depending on the size and timing of the opening, or it might correspond to the segment [h]. Neither features nor segments per se are referenced as units in AP. Certain "constellations" of gestures (Browman and Goldstein 1989; Tilsen 2016) might inhere more closely than others and behave as de facto units that mirror segment-type properties, but constellations are not necessarily segment-sized. A constellation might also constitute an onset, for example, or a syllable. AP does not use hierarchical nodes such as place nodes, X-slots, moras, or syllable nodes. Rather than using symbolic hierarchical representations, different constellations of gestures directly represent sets that may function together in the phonology.

The utility of gestural representation for lexical contrast can be shown by considering how the gestural scores for different words would differ. First, gestures can serve as units of contrast in terms of simple presence or absence. Take away the laryngeal opening gesture, and "pan" becomes "ban." Remove the velum opening gesture, and "ban" becomes "bad." Gestures can also contrast in the articulator sets used: "pan" recruits the lower lip and associated articulators, while "tan" recruits the tongue front. Specification of constriction location and degree can also contrast: "pan" has a bilabial closing gesture and "fan" a labiodental "critical" gesture. Fricatives are specified as having a "critical" constriction: just closed enough to constrict the airflow and cause turbulence, but not enough to cut airflow off. Vowels and glides might differ in "stiffness" (Browman and Goldstein 1989: 229), which plays a role in specifying the duration and trajectory of motion (see the description of the task dynamic model below). Another way that gestures may capture contrast is through coordination. The gestural score for "ban" has the same gestures as "mad," but they contrast in whether the velum opening gesture is associated to the labial or alveolar closure.

Crucially, in AP the same units, the articulatory gestures, suffice for both the description of phonological contrast and the exact modeling of articulator movement. The contrastive and language-specific aspects of the gestural plan are specified in the "linguistic gestural model" in Figure 9.1, and the result of that plan can be diagrammed in the gestural score. Based on the score, the gestures are converted into trajectories by the "task dynamic model."

The task dynamic model is the mathematical implementation of the gestural score. (Hawkins 1992 provides a more detailed introduction for the non-specialist linguist; see also Hall 2018.) Specific trajectories are the output of, not the input to, the model. The math of task dynamics was originally developed to model other goal-directed movements such as reaching with the hand, another type of movement that specifies a starting point and a goal, and for which a set of "articulators" would be recruited (e.g., shoulder, elbow, and fingers would all contribute), but it applies equally well to the movement of vocal tract articulators (Kelso 1982; Kelso et al. 1986; Saltzman 1986; Saltzman and Munhall 1989; Turvey 1990). The underlying math is a formula that describes a "critically-damped mass-spring oscillator": a mass attached to a spring that is distended from its equilibrium, or rest, position and then allowed to spring back. It's the springing-back movement that models the movement of the articulator. For any given articulation, variables in the mathematical formula are supplied from the linguistic gestural representation, of which the gestural score is the graphical representation. The formula (Saltzman 1986; Browman and Goldstein 1989; Hawkins 1992) is given in (1).

(1) Formula for computing an articulatory trajectory
$mx'' + bx' + k(x - x_0) = 0$
where
m = mass
b = damping of the system
k = stiffness of the spring
x_0 = equilibrium position (target)
x = instantaneous displacement
x'' = instantaneous acceleration
x' = instantaneous velocity

Certain variables specify the properties of the overall system, such as the mass and stiffness (or "springiness") of the spring (e.g., the tongue body is more massive and thus moves differently than the tongue tip; higher stiffness models the quicker movement of consonants and glides as opposed to vowels). The equilibrium position corresponds to the target—the position to which the mass will be moving after it is displaced and then released—and the displacement is the starting point. Damping specifies how quickly the movement dies down as opposed to continuing to oscillate above and below the rest position, as a real spring would do. For modeling articulation, damping is set to be "critical": the springing back movement does not overshoot the rest position, but approaches it as an asymptote as energy dies down. The resulting trajectory is a sinusoidal arc from starting point to target, with the exact shape and velocity controlled by the specified variables input to the equation.

Trajectories are computed independently for each articulator set. However, actual movement must be determined over the vocal tract as a whole. This is modeled computationally in the "vocal tract model" of Figure 9.1. The vocal tract model accounts for the ways that gestures can influence one another, hiding, revealing, or blending. It is this interaction that produces allophonic variation and connected speech alternations.

9.3 ALTERNATION AS GESTURAL REORGANIZATION

There are no phonological "rules" in AP. Rather, details of allophonic variation and positional realization are derived from interactions among gestures, particularly overlap in time. For example, it can be seen in Figure 9.2 that velum opening precedes alveolar closing. The result of this temporal coordination is nasalization that is heard on the vowel. Similarly, larynx opening extends beyond the duration of labial closure: the result is aspiration. There are no separate gestures for nasalization or aspiration: these positional variants come about as a result of the specified temporal coordination for [n] and for [p]. Other languages that have different patterns of nasalization or aspiration would have different timing specifications.

A particular type of organization is hypothesized for onset consonants: the "c-center" or consonant center, which accounts for a number of facts of positional realization, including aspiration of stops and devoicing of liquids in the English onset (Browman and Goldstein 1988; Byrd 1995). Note that in Figure 9.2, the larynx opening gesture begins at the midpoint of the labial closing gesture, and extends beyond it, resulting in aspiration. The midpoint of a single consonant is the "c-center." In a cluster of consonants, the c-center is defined as the mean of the midpoints, and Browman and Goldstein (1988) show that in English onset clusters the c-center provides the most stable point of reference for the timing of onset cluster to vowel and other gestures. Subsequent articulatory research has confirmed consistent c-center effects for onsets in English (Byrd 1995; Browman and Goldstein 2000; Marin and Pouplier 2010), as well as in Georgian (Chitoran et al. 2002; Hermes et al. 2011) and in French (Kühnert et al. 2006), and for some but not all onset clusters in Italian (Hermes et al. 2008) and Romanian (Marin 2013). C-center timing may be diagnostic of (and in fact constitute) the "onset" constituent, in that languages that lack true onset clusters, such as Moroccan Arabic and Berber, have been found to not show this kind of timing readjustment (Gafos 2002; Goldstein et al. 2007; Hermes et al. 2011; Shaw et al 2011).

C-center timing is illustrated in Figure 9.3. In an [sp] cluster (Figure 9.3a), the laryngeal opening gesture is aligned with the c-center of the cluster, so that the alveolar fricative is shifted somewhat to the left and the labial closure is shifted somewhat to the right, compared with singleton onset "pan." The result of the rightward shift of [p] is that the laryngeal gesture no longer extends past the labial closure release, and there is thus no aspiration. If instead of [sp] the cluster were [pl] (as in "plan," Figure 9.3b), given the same c-center timing, the [l] would overlap with the laryngeal

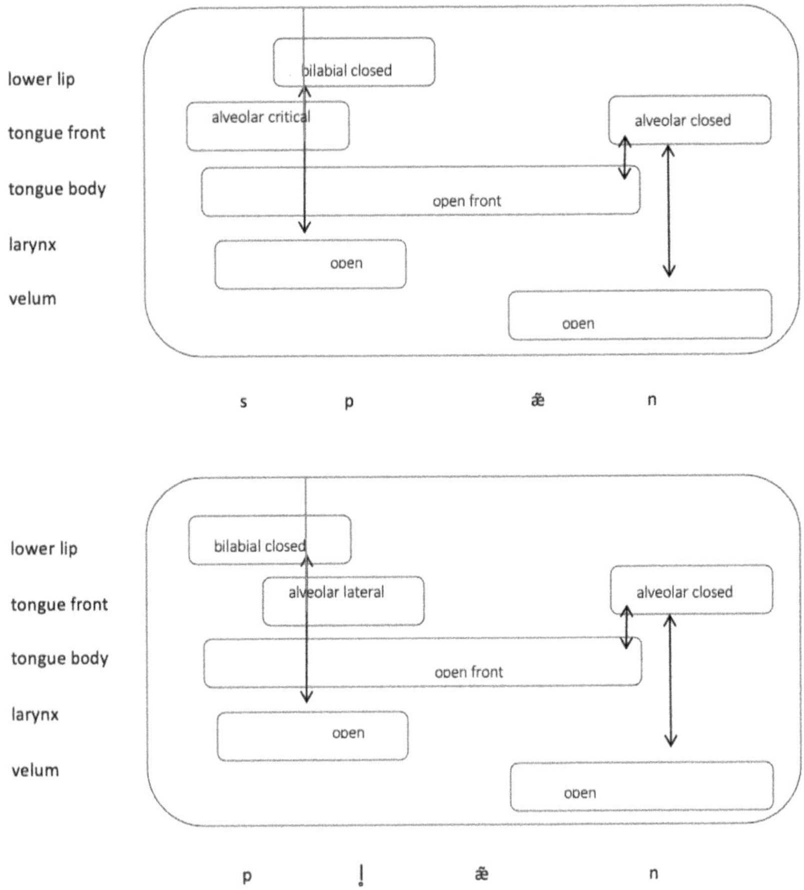

Figure 9.3 *C-center timing for "span" and "plan"*

opening gesture, resulting in the [l] being devoiced. Thus three "rules" of allophonic variation in English:

> "voiceless stops are aspirated in syllable-initial position,"
> "voiceless stops are unaspirated when they follow [s]," and
> "liquids are devoiced when they follow a syllable-initial voiceless stop"

are replaced by a single statement of gestural timing:

> "laryngeal opening is coordinated with the c-center of the prevocalic consonant sequence."

The above are all generalizations about allophonic realization. Phonological alternations, particularly alternations of "connected" or "casual" speech, are also accounted for by gestural reorganization. Browman and Goldstein (1989: 220) write:

> All changes are hypothesized to result from two simple mechanisms, which are intrinsically related to the talker's goals of speed and fluency—reduce the size of individual gestures, and increase their overlap. The detailed changes that emerge from these processes are epiphenomenal consequences of the "blind" application of these principles.

Crucially, the principles apply "blindly." There are no SPE-style rules that target a particular input configuration and specify a changed output. Rather, assimilations and deletions are seen as epiphenomena resulting from general properties of gestural reorganization.

The canonical example, from Browman and Goldstein (1990c), is the "rule" of t-deletion in clusters, exemplified by the pronunciation of the phrase "perfect memory" as [pʰɝfɛk mɛmɹi]. Browman and Goldstein collected X-ray microbeam data to reveal articulator movement in this and other phrases. In X-ray microbeam, small metal pellets are affixed to points on a talker's tongue, lips and jaw, and the positions of the pellets are tracked over time by a very focused, low intensity, X-ray beam (hence, microbeam), producing a track that shows the pellets' (and thus the articulators') movement in two dimensions over time. (This technique has now been replaced by EMA, electromagnetic mid-sagittal articulography, which produces a similar output trace, but using magnetic fields rather than even very low dose X-rays.) Figure 9.4 (Figure 19.13 from Browman and Goldstein 1990c), shows X-ray microbeam traces from two utterances of the words "perfect" and "memory," the first (Figure 9.4A) as two words in a list, the second (Figure 9.4B) as a connected noun phrase. Time is on the x-axis, vertical displacement on the y-axis. The Greek letters show the movements that correspond to the gestures: kappa stands for the tongue body gesture for [k], tau for the tongue front gesture for [t], and beta for the labial closure gesture for [m].

In the list pronunciation, note the significant overlap between the [k] and [t] gestures. Subsequent studies (Byrd 1996; Zsiga 1994, 2000) show this pattern of overlap to be typical of American English pronunciation: movement of the tongue tip toward [t] begins well before closure for [k] is released. There is a delay, however, between the [t] and the [m], consistent with a slight pause between words in a list. The acoustic consequence of this pattern of coordination is that all three consonants can be perceived. Formant transitions into the velar closure will cue the presence of the [k], even though its release will be inaudible, "hidden" behind the already-formed closure for [t]. Transitions into the [t] are similarly hidden by the [k], but since the [t] closure extends in time after the end of the closure for [k], the [t] release is heard. The [m] is not overlapped by any other consonant gesture, and is also clearly heard. Thus, the list pronunciation could be clearly transcribed [pʰɝfɛkt mɛmɹi].

In the connected phrase pronunciation, however, there is no separation between the end of the first word and the beginning of the second. Again, subsequent research

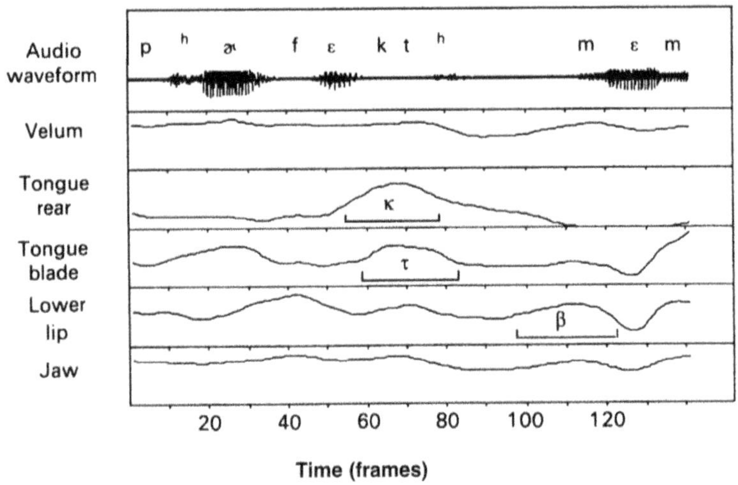

(a)

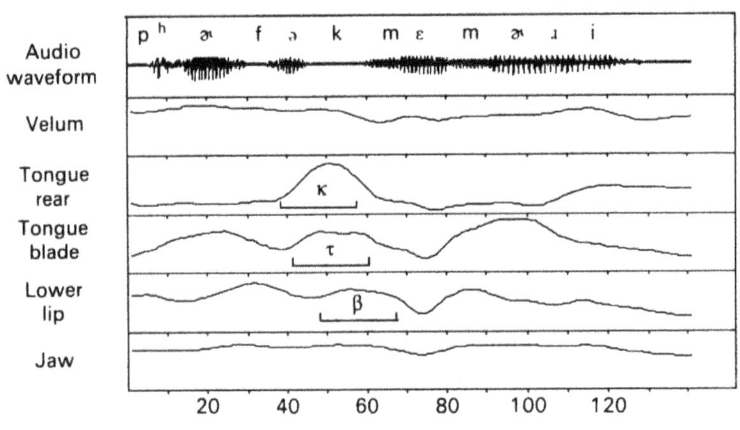

(b)

Figure 9.4 *"Perfect memory"*

Source: Browman and Goldstein (1990c: Figure 13).

finds this pattern of overlap to be typical of American English pronunciation in general. The result, however, is that the [t] is now completely inaudible: its closure transition is hidden by the [k], and its release is hidden by the now-overlapping [m]. The phrase would be correctly transcribed as [pʰɝfɛk mɛmɹi], but the X-ray micro-beam data shows that the [t] has not been deleted. The closure gesture is still there, just hidden and inaudible. Crucially, this hiding is not a fact about this particular cluster, or any three-consonant cluster, but of a very general pattern of gestural coordination in American English: a speaker of American English will overlap words in a phrase, beginning movement toward the second word before the first is complete.

The apparent deletion is an epiphenomenon resulting from this general pattern of overlap.

The degree of overlap is language-specific, but not word-specific. For example, the general pattern of between-word overlap is greater in Korean and less in Russian (Zsiga 2000, 2003, 2011; Kochetov et al. 2007). Thus, both Korean and English have noticeable assimilations and deletions at word boundaries, while Russian does not. Kochetov and Colantoni (2011) show differences in word-boundary overlap between Argentinian and Cuban Spanish, resulting in somewhat different patterns of nasal assimilation, and both these dialects show differences from the patterns observed by Honorof (1999) for Peninsular Spanish. Because it is language-specific, these patterns of overlap must be learned. But once they are learned, AP argues, there is no need to learn additional rules of positional variation—the positional variation follows from the gestural organization. Similarly, different patterns of nasalization (e.g., between Parisian and Canadian French, Cohn 1990), or different degrees of aspiration (Lisker and Abramson 1964; Davidson 2018), do not have to be learned as rules, but follow from patterns of gestural organization.

While gestural "hiding" can bring about apparent deletion, gestural "revealing" can bring about apparent insertion. For example, Browman and Goldstein (1992a) analyze epenthetic stops in words like "something" pronounced as [sʌmpθɪŋ] to be the result of gestural reorganization. If the velum closes and the larynx opens a little too soon, anticipating the upcoming voiceless fricative in the [mθ] sequence, the result will be a period of voiceless closure, heard as [p]. Similarly, if the result of increasing overlap between consonant closures can be heard as deletion, the result of *decreasing* overlap can be heard as schwa insertion. Browman and Goldstein (1992b) show that AP can model a continuum between the words "bray" and "beret" (and similar pairs), simply by increasing the temporal distance between the [b] and the [ɹ]. If the two gestures get far enough apart, listeners hear a medial vowel. Davidson (2006) analyzes schwa insertion in non-native clusters in a similar way. It is known that English speakers struggle with the initial clusters of Slavic languages, as speakers of other languages struggle with some of the initial clusters of English. Davidson uses acoustic analysis to argue that English speakers aren't actually inserting vowels into words like [gdansk]—they're just leaving more space between the consonants than would be typical of a native speaker of Polish, and the result sounds like schwa insertion. (The gestural timing differences underlying cross-linguistic consonant cluster phonotactics are further investigated in Wilson et al. 2014; Davidson et al. 2015; and Davidson and Wilson 2016.)

Gestural hiding and revealing generally take place when the overlapping gestures call on two different articulators. Each independent articulator can reach its goal, but depending on the relative timing, the acoustic result is not necessarily what was intended. Sometimes, however, two gestures might call on the same articulator at the same time. In that case, if the two gestural specifications conflict, the actual trajectory is a blend between the two competing targets. A simple example is the more palatal articulation of [k] in "key" compared to "coo." The AP analysis of this assimilation (Browman and Goldstein 1990c) is overlap between the gestures for [i] and for [k], both of which recruit the tongue body. Because the tongue body cannot be in two places at once, the velar target for the [k] is not reached, but instead is pulled forward to a more palatal position. Another example from Browman and Goldstein (1990c) is the phrase "ten things." Here, the [n] will have an alveolar target and the [θ] a dental target. When the words are spoken in connected speech and the two gestures overlap, a blend is produced, heard as assimilation of alveolar [n] to dental [n̪].

Zsiga (1995, 2000) analyzes English palatalization at word boundaries as gestural overlap. The phrase "miss you" is usually pronounced in American English to sound like [mɪʃju]. Acoustic evidence shows, however, that the fricative in such sequences is not identical to an underlying [ʃ]. There is variability between speakers and instances, but the general pattern is a gradual change in pitch from [s]-like at the beginning to [ʃ]-like at the end. This gradual change is shown in Figure 9.5 (from Zsiga 2000: 85), which compares the acoustic quality of the fricative in the phrase "press your point" to the fricatives in "press" and "pressure point." The top line of the graph (triangles)

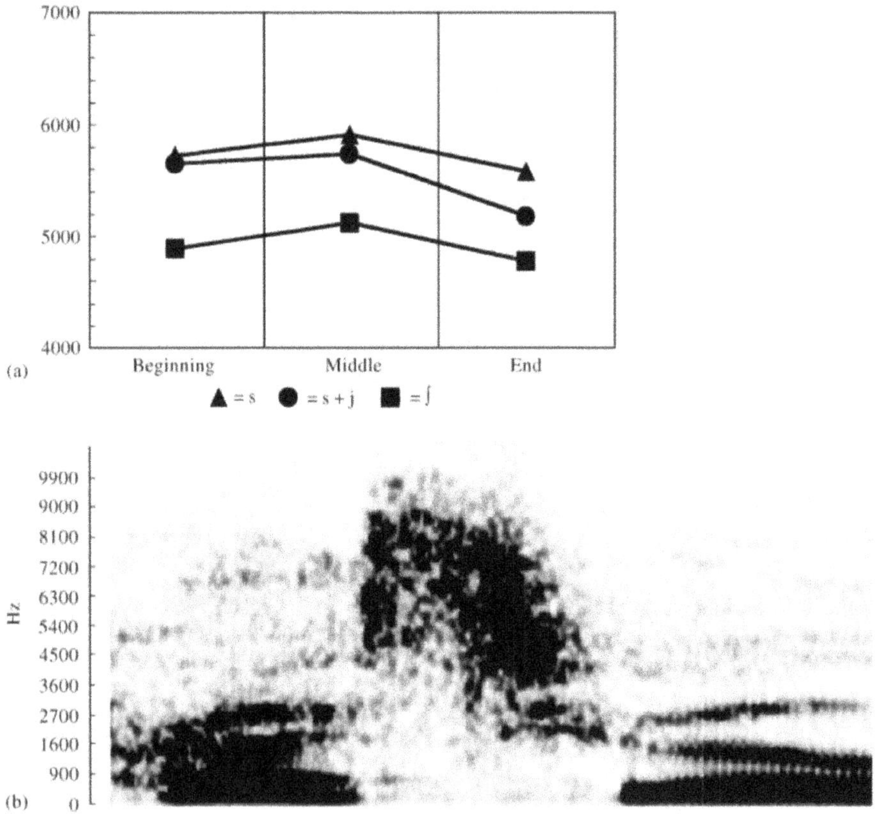

Figure 9.5 Gradient palatalization in the American English phrase "press your." (a) Comparison of centroid values in "press" (triangles), "pressure" (squares), and "press your" (circles) for one speaker. (b) Example spectrogram

Source: Zsiga (2000: 85).

shows one speaker's centroid values for [s] in "press" at three points across the duration of the fricative, all about 6,000Hz. (The centroid is a measure of where energy is concentrated in a fricative.) The bottom line (squares) shows the values for [ʃ] in "pressure," about 1,000Hz lower. The center line (circles) graphs the centroid values for the fricative in the [s#j] sequence: the values start out more like [s], but drop to an intermediate value by the end of the fricative. An example spectrogram is shown at the bottom of the figure: the change in fricative energy is clear, as well as overlap between the fricative and the high F2 and F3 formants that indicate the onset of the palatal glide. Evidence from electropalatography, which measures contact of the tongue against the palate, confirms that raising of the tongue body for [j] begins while the tongue front is still making the fricative constriction. Overlap between the two constrictions produces a configuration that results in lowered fricative noise.

Crucially, the articulatory plan—overlap at word boundaries—that produces an apparent [ʃ] in "miss you" is the same plan that produces assimilation of [n] in "ten things" and deletion of [t] in "perfect memory." AP argues that there is no need for rules of assimilation and deletion in these cases, or for a phonological constraint ranking to derive the correct output. All that needs to be learned is the gestural coordination typical of American English, and positional and allophonic variation will follow.

Numerous other studies have found evidence for a gestural account of assimilations and deletions in English (e.g., Barry 1992; Browman and Goldstein 1992a and references therein; Ellis and Hardcastle 2002; Chen 2003) and other languages including Korean (Jun 1995; Jun 1996; Zsiga 2011), Spanish and Catalan (Kochetov and Pouplier 2008), and Igbo (Zsiga 1997, as discussed in Chapter 5). These authors argue that AP gives a better account of "connected speech" or "casual speech" assimilations than a traditional feature- or segment-based phonological analysis, by showing that the output of the change is gradient and variable, as in Figure 9.5. If gestures are interacting with one another in actual space and time, overlapping more or less, such gradual changes are expected. A feature-changing approach, on the other hand, whether rule-based or constraint-based, will always change a segment from one category to another. Thus, gradient application is taken to be evidence of a gestural, rather than featural account.

The dividing line between categorical and gradient has been taken by some researchers to be the dividing line between phonology and phonetics, as discussed in Chapter 5. What the AP analysis of alternations such as [t]-deletion and palatalization adds to the debate is that an alternation may be perceived as categorical, but shown by acoustic or articulatory analysis to be gradient. The authors cited above use experimental evidence to show that at least some phonological alternations that were previously analyzed as featural can be reanalyzed as gestural. But can all?

9.4 IS ALL PHONOLOGY ARTICULATORY PHONOLOGY?

The strong version of Articulatory Phonology argues that all phonology is gestural reorganization. Gestures are never added, deleted, or changed by phonological rule. As Browman and Goldstein (1992a: 173) state, "All the lexical phonological units are present, though they may be decreased in magnitude and overlapped by other gestures. Gestures are never changed into other gestures, nor are gestures added." The first challenge to this approach is the phonetically unmotivated or "unnatural" rules that were discussed in Chapter 6, which cannot plausibly be seen as the result of gestural reorganization. AP has no account of tri-syllabic laxing in English (Kenstowicz 1994) or the morphological restrictions on [t] assibilation in Finnish that are discussed by Kiparsky (1985). Here, the analysis must be that "unnatural" rules are not phonology at all, but a matter of morphology or lexical choice. As discussed in Chapter 6, this approach is not unique to AP, but is the solution for any phonological theory that requires only phonetically natural alternations, and it remains open to the same critiques and arguments. Should unnatural rules be treated as the norm, or the exception? AP argues that they are the exception.

Another critique is that AP is strictly based in articulation, and thus has little to say about perceptual goals. Acoustic and perceptual consequences are analyzed as epiphenomena, the result of not reaching articulatory goals. Because perceptual targets are not part of the theory, any alternation that seems to have a perceptual basis, to take place to enhance perception or contrast, is not accounted for. Thus theories that take perception as basic (e.g., Liljencrants and Lindblom 1972; Steriade 1997; Flemming 2004; Boersma 2007 among many others; see Chapter 10) stand in contrast to AP, though it may well be that these theories are largely complementary rather than contradictory. Some phenomena, such as connected speech assimilation, may best be explained in terms of articulation, and others, such as vowel inventories or phonotactic asymmetries, in terms of enhancing perception. (See the discussion in 9.5 below, and in Chapters 10 and 11). Regardless of the motivation, however—whether random/historical or perceptual—if the alternation cannot be modeled as gestural reorganization, AP must assume that the different realizations are the result of allomorph selection.

A third critique of AP looks more closely at the question of gradient vs categorical application. How should the theory treat phonological alternations, such as assimilations, that seem to be clearly motivated by pressures of articulation, but are not gradient? As noted above, it is often the case that alternations that were transcribed as categorical, perhaps due to the effects of human categorical perception, or due to the limited ability of IPA transcription to record phonetic details, have proven upon acoustic or articulatory examination to not be categorical at all, and much of what has been described in the phonological literature has not been examined. But there are other cases where the acoustic and articulatory evidence *has* shown a categorical change (e.g., Holst and Nolan 1995; Honorof 1999; Ellis and Hardcastle 2002; Ladd and Scobbie 2003; Scobbie and Wrench 2003; Bradley 2007; Kochetov and Pouplier 2008; Zsiga 2011).

The problem isn't that AP can't model a categorical change at all. In the AP model, instances of what appear to be categorical deletion or assimilation are analyzed as outliers in the range of gestural variation: deletion is just the most extreme case of reduction, and complete assimilation is the most extreme case of overlap (Browman 1995). For example, Son and colleagues (2007: 515) describe the change of word-medial /pk/ → [kk] in Korean, for which they show the outcome to be indistinguishable from an underlying /kk/, as an extreme case of "lip aperture reduction." The problem is that, if all alternations are due to necessarily gradual changes in organization, the theory does not predict the existence of alternations that are consistently categorical.

In some cases, the categorical alternation is part of the lexical phonology. For example, Zsiga (1995, 2000) shows that while the palatalization of [s] to [ʃ] in "press your point" is variable and gradient (Figure 9.5 above), the [ʃ] in "pressure," arguably derived from "press," is not different from the underlying [ʃ] in a word such as "mesh." In another study, on Igbo, Zsiga (1997) shows that local vowel assimilation of [e] to [a] at word boundaries is gradient and variable, but long-distance assimilation of the same vowels from root to affix (vowel harmony) is categorical and obligatory. These publications argue that the two types of alternation, categorical and gradient, require two different representations, consistent with a traditional modular understanding of the phonology/phonetics interface. Because gestural representations have the power to capture gradience, they are not well-suited to capture alternations that are necessarily category changing.

To the extent that categorical alternations are exclusively lexical, however, as they are in a word such as "pressure," the same argument can be used as for "unnatural" rules. It can be argued that "pressure" is not derived at all, but is listed fully formed in the lexicon. Such arguments shift the question away from the phonology/phonetics interface to the phonology/morphology interface, and raise the question "What exactly is phonology accountable for?" If all categorical alternation is lexical listing and allomorph selection, and all gradient alternation is gestural reorganization, is there any space for a productive featural phonology at all? The strongest version of AP says no.

However, not all categorical alternations are lexical. Cases of categorical assimilation across word boundaries, confirmed by phonetic analysis, have been found in Sardinian (Ladd and Scobbie 2003), Spanish (Honorof 1999), and Korean (Kochetov and Pouplier 2008 for place assimilation, and Zsiga 2011 for nasal assimilation). If the alternation comes about only when words are put into phrases, it is hard to argue all the alternatives that could arise in any word combination are listed in the lexicon. (Such arguments are not impossible, however: for example, Smolensky and Goldrick 2016 argue for multiple lexical representations for the case of French liaison.) If the alternation is not lexical choice, then the alternations must have a phonological representation. However, because in the strongest versions of AP gestures are never inserted, deleted, or changed into other gestures, AP is not well suited for capturing these categorical phonological alternations (Clements 1992; Zsiga 1997; Ladd and Scobbie 2003). Ladd and Scobbie (2003: 179) argue

> that gestural overlap is on the whole *not* a suitable model of most of the assimilatory external sandhi phenomena in Sardinian, and more generally that accounts of gestural overlap in some cases of English external sandhi cannot be carried over into all aspects of post-lexical phonology. (Emphasis original)

Ladd and Scobbie argue instead for an analysis of their Sardinian data in terms of autosegmental feature spreading.

The phonetic evidence, then, is that some assimilations, deletions, insertions, and lenitions are gradient and others are categorical. Given this phonetic evidence, it must be the case that speakers' linguistic competence allows for both types of process (Scobbie 2007; Kochetov and Pouplier 2008; Pierrehumbert 2016; and references therein). Is it necessary, then, as Ladd and Scobbie (2003) argue, and as much of the literature on the phonology/phonetics interface has maintained, that gradient and categorical processes require two different kinds of representation, or can the gestural representations of AP account for both? Zsiga (2011: 292) writes "The crucial question is whether there is a theory of gestural timing and organization that is both powerful enough to account for gradient changes and constrained enough to account for changes that result in category neutralization." A more constrained theory of timing relations, formalized in terms of coupled oscillators, may provide an answer.

9.5 CONSTRAINING TIMING RELATIONS

9.5.1 TIMING RELATIONS IN FEATURE-BASED REPRESENTATIONS

Feature-based phonology allows only two temporal relations to be expressed: simultaneity and linear precedence (Sagey 1988; Zsiga 1997). In SPE notation, feature bundles are sequentially ordered, corresponding to the linear string of segments. Within each bundle, features are a-temporal. The features can be considered as labels that define the segments, determining the phonological classes to which they belong, the alternations they undergo, and their phonetic implementation (see Chapter 4). Rules can refer to preceding context or following context (linear order), or to co-occurrence within a feature bundle (phonological simultaneity), but the representation allows reference to no other temporal relation.

Autosegmental notation and feature-geometric representations (Section 4.3.5) seem to complicate the picture, by allowing many-to-one and one-to-many associations between segments and features, but linear order and simultaneity are still strictly defined. This is illustrated in Figure 9.6, which shows an autosegmental representation for the segmental string [æ̃mbi]. Four root nodes define four segments and their linear order. In addition, each feature defines its own tier, and along a tier features are linearly ordered: [+nas] precedes [-nas]. Simultaneity is defined by linking: all the features linked to a given root node (such as [labial] place and [+nasal] for [m]) are phonologically simultaneous with respect to each other, much as they are in a feature bundle, as long as they are on different featural tiers. A given feature may be linked to, and thus phonologically simultaneous with, more than one root node, but linear precedence among both roots and features must be respected. Thus [labial] can be linked to both [m] and [b], and [nasal] to both [æ̃] and [m], but the [æ̃mb] order is still respected.

In contrast, the order [æ̃bm], with double linking of both [nasal] and [labial], as in Figure 9.7, would be ill-formed because linear precedence is not respected. The single [+nas] feature cannot both precede and follow [-nas]. In general, if Feature_a precedes Feature_b on some tier, and Root_x precedes Root_y in the sequence of root nodes, then

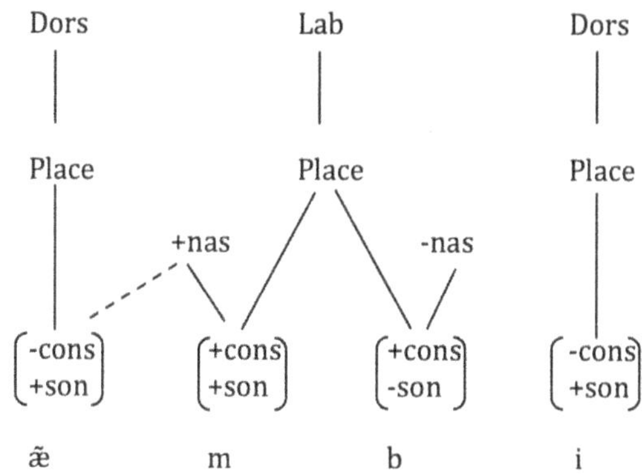

Figure 9.6 Autosegmental representation of [æ̃mbi], illustrating simultaneity and linear precedence

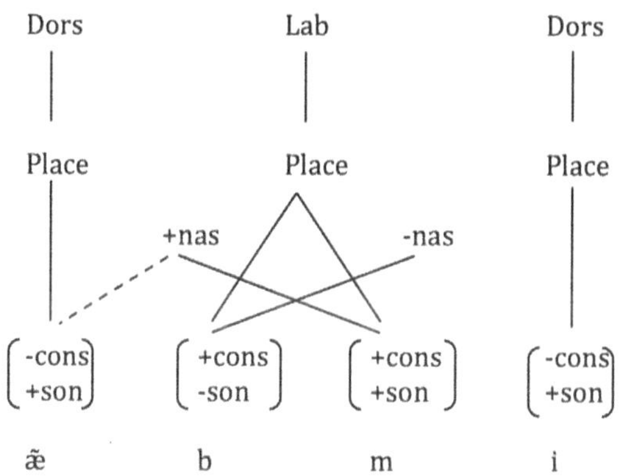

Figure 9.7 Autosegmental representation of [æ̃bmi], with crossing association lines, is ill-formed

it cannot be the case that Feature_a is linked to Root_y and Feature_b is linked to Root_x. (This is the "no-crossing constraint" of Sagey 1988.)

It is important to keep in mind that "phonologically simultaneous" does not necessarily mean phonetic realization in perfect temporal synchrony. It does mean that details of timing cannot be accessed by the phonology and are irrelevant to phonological patterning. A feature may be present or absent, linked to a root node or not, but exact degrees of overlap are not "accessible to or manipulable by phonological processes" (Sagey 1988: 112). In Figure 9.6, for example, the vowel preceding the [m] can be either nasal (association to [+nasal] present) or oral (association to [+nasal] absent), but partial nasalization cannot be represented, still less differing degrees of partial nasalization.

Thus the number of possible contrasts, and types of possible alternations, are constrained. All rules of featural assimilation are necessarily category changing: an SPE feature-changing rule, or autosegmental feature-spreading rule cannot make a preceding segment one-half or one-third nasal. This is exactly why, when an alternation is found to be gradient, that a featural account is argued to be unsuitable.

A more recent instantiation of feature-bundle notation, "Q-theory" (Inkelas and Shih 2016; Shih and Inkelas 2019), combines the linear order of SPE with the ability of feature geometry to model subsegmental ordering. In Q-theory, each segment (capital "Q") is broken down into up to three ordered subsegments (lowercase "q"). The different subsegments can have different featural specifications, allowing for the subsegmental ordering necessary for affricates ([-cont] [+cont]), pre-nasalized segments ([+nasal] [-nasal]), or contour tones (HL vs LH). While this allows for a more fine-grained ordering, features are still constrained to be either sequential or simultaneous. Q-theory does not allow one feature to spread over multiple segments, as feature geometry does. Long-distance alternations such as vowel and consonant harmony are handled by correspondence relations that require certain segments and features to match, rather than through multiple linking.

Assimilations in Q-theory *can* make a segment one-half or one-third nasal, by associating a feature to only one of the subsegmental q intervals. Thus, an important question that Q-theory has yet to address is whether three is the correct upper limit for subsegmental units. While it is always assumed that the number of subsegmental divisions is very small, to the extent that Q-theory allows a three-way or four-way contrast in nasalization, it would be subject to the same phonological over-generation critique as gestural overlap approaches are.

9.5.2 TIMING RELATIONS IN GESTURAL REPRESENTATIONS 1: CYCLES AND LANDMARKS

Because gestures have inherent extent in time, they *can* overlap more or less, and the relative timing between gestures is crucial to the account of different processes of assimilation, deletion and insertion, as described above. Gradual differences in degree of overlap result in gradual phonetic effects, such as the gradient lowering of fricative noise in Figure 9.5. It is crucial to keep in mind that, in AP, gestures are always coordinated to one another, not to an external clock. For example, in Figure 9.2 above, the gestural score for "pan" specifies that the bilabial closing gesture and the vowel gesture begin simultaneously, and that velum opening precedes alveolar closure. A gestural representation cannot encode "60ms of aspiration" or "100ms of nasalization."

But how many degrees of overlap can be defined? An infinite number? Which points in a gesture can be timed to which? As described in Section 9.2 above, the task-dynamic math underlying AP theory models each gesture as a "critically-damped oscillator." Defining the system in terms of oscillators means that the mathematical specification of a gesture is a sinusoidal function, although the critical damping assures that only one cycle of motion, movement to the target position, is ever actually realized. A cycle, of course, is 360°, and different points in the "unfolding" of a gesture can be specified as a number between 0 and 360. For example, "target" is reached at 270°, and "release" at 300°. In the earlier instantiations of AP (e.g., Saltzman 1986; Browman and Goldstein 1989, 1992a, 1992b; Saltzman and Munhall 1989), the timing of gestures to

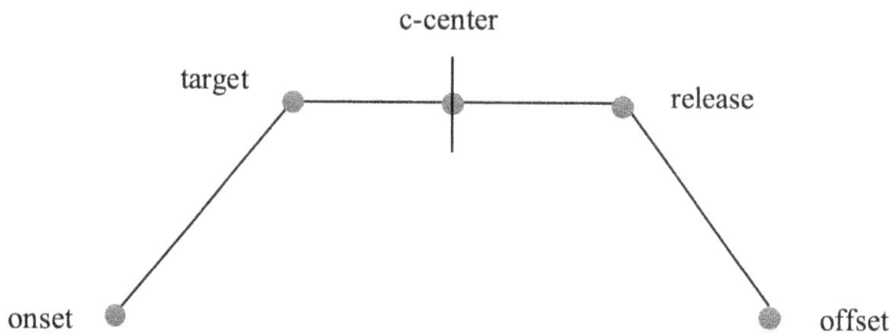

Figure 9.8 Five gestural landmarks
Source: Adapted from Gafos (2002).

one another (gestural "phasing") was defined as the specification of any points within the cycles of two gestures as being simultaneous. Gestural phasing could then be formalized as "alignment constraints." For example, Zsiga (2000) proposes an alignment constraint for consonant overlap at word boundaries in English. The proposed constraint, Align(C1,300°,C2,270°), specifies that two consonants in a sequence are timed such that the second consonant achieves closure just as the first is being released (thus ensuring that the release is inaudible, as in Figure 9.4B above). Such specifications can correctly model any degree of overlap, but this many degrees of freedom are a problem for phonological theory. If any point in one gesture can be coordinated to any point in another gesture, there are 360x360 (129,600) possible phasing relationships, not the two (precedence and simultaneity) that phonology requires.

Later work in the AP paradigm (e.g., Gafos 2002; Bradley 2007) suggest that the number of possible coordinations can be reduced by allowing alignment constraints to make reference to only five "landmarks": onset, target, c-center, release, and offset. (See Section 9.3 above for the definition of c-center. Note that onset, target, and release correspond to the three subsegmental intervals available in Q-Theory.) The five proposed landmarks are shown in Figure 9.8 (based on Gafos 2002: 271). In Figure 9.8, the gesture is schematized as movement to a target, a steady state, and movement away from the target.

Gafos (2002) models consonant timing in Moroccan Arabic in terms of these landmarks. The two different timing patterns he proposes are illustrated in Figure 9.9. In this dialect of Arabic, whether or not consonants in sequence have an audible release depends on place of articulation: homorganic sequences, such as [t-t], always have an audible release of the first consonant, while non-homorganic consonant sequences, such as [b-d], may be released or unreleased depending on speech rate. For homorganic sequences (Figure 9.9A), Gafos argues for a phasing of C2:onset to C1:offset, that is, begin C2 only after C1 is complete, which would result in audible release of C1 in all cases. For hetero-organic sequences (Figure 9.9B), he argues for a phasing of C2:onset to C1:c-center, so that movement for C2 begins in the middle of C1. In that case, release would be audible at slower speech rates but not faster.

Further, Gafos argues that these alignment constraints interact with other phonological alternations. Because of this interaction, alignment constraints must be

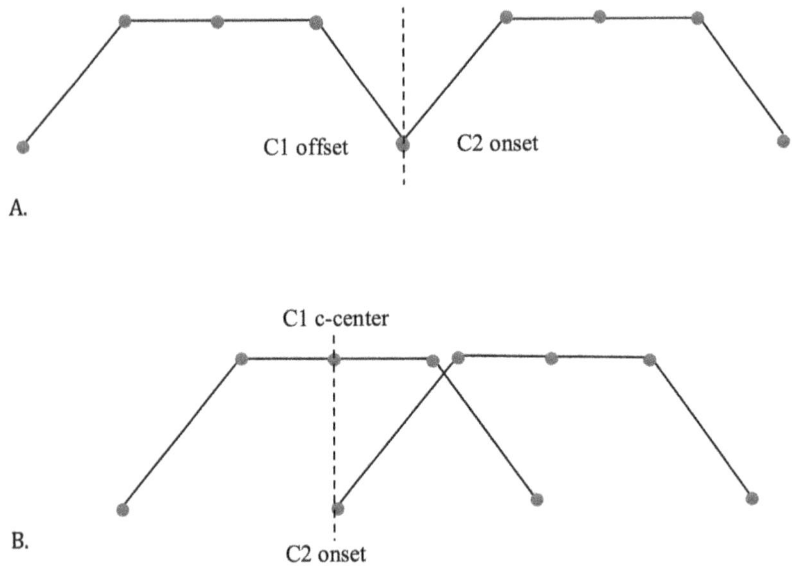

Figure 9.9 *Two patterns of landmark-based gestural coordination in Moroccan Arabic. The dashed line shows the points that are coordinated. (A) C2:onset to C1:offset results in audible release in homorganic sequences. (B) C2:onset to C1:c-center results in variable release in hetero-organic sequences*

Source: Adapted from Gafos (2002).

included in the phonological grammar, not relegated to a subsequent phonetic implementation. (See the discussion of phonetics in phonology in Chapter 6.)

If gestural coordination is limited to only five landmarks, there would be twenty-five possible patterns of coordination between two gestures, rather than 129,600. This is obviously an improvement, but still allows many more degrees of freedom than featural phonology affords. Landmark phasing captures the two different patterns of consonant coordination in Moroccan Arabic, but no language contrasts twenty-five possible degrees of nasalization or aspiration.

9.5.3 TIMING RELATIONS IN GESTURAL REPRESENTATIONS 2: COUPLED OSCILLATORS

Further work on gestural timing (Browman and Goldstein 2000; Nam and Saltzman 2003; Goldstein et al. 2006; Goldstein 2008; Saltzman et al. 2008; Nam et al. 2009) has further reduced the degrees of freedom that can be specified in gestural phasing, allowing in fact only two different essential timing modes. This later work models gestural phasing in terms of coupled oscillators, an approach first used to model other types of coordinated movements (Haken et al. 1985; Turvey 1990; Löfqvist and Gracco 1999). According to the theory of coupled oscillators, there are only two natural and easily acquired patterns of coordination: in-phase (simultaneous) and

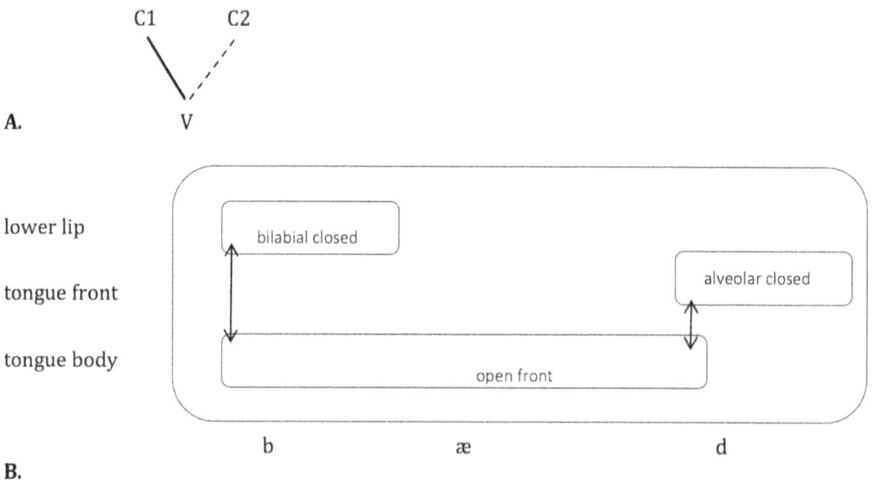

Figure 9.10 Two representations for a CVC syllable. A. Coupling graph, following Goldstein et al. (2007). B. Gestural score, following Browman and Goldstein (1986)

anti-phase (sequential), with simultaneous preferred. Other more complex rhythmic patterns, as in drumming or dancing, *can* be learned, but only with skill and practice. Under pressure of increasing speed or distraction, humans have great difficulty sustaining a complex rhythm, and revert to one of the more basic patterns (de Jong 2003).

In the coupled-oscillators version of AP, the linguistic gestural model (Figure 9.1) allows gestural timing to be specified only as an in-phase or anti-phase coupling mode. The representation can be diagrammed as a "coupling graph," as in Figure 9.10A. The figure shows three gestures in a CVC syllable, such as /bad/. The onset consonant and the vowel begin simultaneously. This is represented as an in-phase coupling, indicated by a solid line. The coda consonant follows the vowel sequentially, an anti-phase coupling, indicated by a dashed line. As was discussed in Section 7.3.3 (on articulatory approaches to syllable structure), that fact that onsets are in the preferred in-phase mode accounts for the cross-linguistic preference for CV syllables (Browman and Goldstein 2000; Goldstein et al. 2007). The corresponding gestural score is shown in Figure 9.10B.

The crucial innovation in the representation of gestural timing is that the two coupling modes correspond directly to the two types of timing that are recognized in feature-based phonological descriptions: simultaneity and precedence. While the gestural score allows multiple degrees of overlap to be specified, the coupling graph allows only two. The differences in gestural timing that give rise to gradience and variability are caused when multiple couplings interact.

Note that while a coupling graph resembles an autosegmental representation, and is similar in that lines associating units indicate temporal relations, the graphical conventions are different. In a coupling graph, a solid line indicates simultaneous timing and a dashed line indicates sequential timing. In an autosegmental representation, a

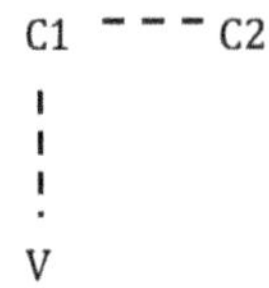

A. non-competing couplings in coda position

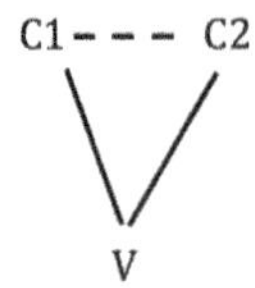

B. competing couplings in onset position

Figure 9.11 Coupling graph for consonant sequences in English codas and onsets
Source: Zsiga (2011: 295), following Browman and Goldstein (2000), Goldstein et al. (2006).

solid line indicates an existing simultaneous association and a dotted line indicates an association added in the course of a derivation. Linear sequencing of segments is indicated by sequential root nodes in an autosegmental representation; both root nodes and segments are absent from gestural representations.

Coupling graphs for different utterances are created according to language-specific principles by the linguistic gestural model. The modeling of articulatory movement then unfolds as specified in Figure 9.1 and described above (see also Goldstein et al. 2006). A gestural score is generated from the coupling graph, and articulatory trajectories are generated from the score, according to the general formulas of the task dynamic model. Gestural phasing is derived from the couplings, not independently stipulated.

Variation and gradience in the resulting movement of articulators arise not from a multitude of numerically-specific coupling modes, but from interactions between the basic modes specified across different gestures. A given gesture may participate in multiple couplings, which may not all be compatible with each other. When incompatible couplings are specified, they cannot both be achieved, and they must compete for control of the targeted articulator.

The c-center effect in onset clusters is one example of how such competition plays out. As was described above, articulatory studies (Browman and Goldstein 1988, 2000; Goldstein et al. 2007; Goldstein et al. 2009) have found different timing patterns for consonants in onsets and codas. Consonants that are behaving as an onset constituent show greater overlap with each other and with the vowel, following the c-center effect described in Section 9.3 and illustrated in Figure 9.3 above. Coda consonants are less overlapped, and are not organized around a c-center, but follow the vowel and each other in sequence. In the coupled oscillator model, this asymmetry is explained in terms of different coupling graphs for onsets and codas (Browman and Goldstein 2000; Goldstein et al. 2006, 2009).

The coda case (Figure 9.11A) is the simpler one. There are two anti-phase (sequential) couplings, indicated by the dotted lines. C1 follows the vowel, and C2 follows C1. This predicted sequential organization corresponds to what acoustic and articulatory data have shown to be true of English coda consonants.

In the onset, however (Figure 9.11B), the specified couplings are incompatible. As in the coda, the coupling mode for a consonant sequence is for C2 to follow C1 in an anti-phase coupling (dotted line). But articulatory data has shown that the specified coupling mode for an onset consonant is to be in-phase (simultaneous) with the vowel. Crucially, in Figure 9.11B, the specification of in-phase coupling with the vowel applies to *both* C1 and C2, since both are onset consonants. Thus, the coupling requirements on C2 in an onset cluster are incompatible. Since C1 is in-phase with V, C2 cannot be both anti-phase with C1 and in-phase with V.

According to the theory of coupled oscillators as implemented within AP (Goldstein et al. 2006), the solution is that the couplings *compete*. Each coupling is assigned a numerical coupling strength, according to either universal or language-specific patterns. Then, in order to produce actual trajectories, a weighted average is computed, with the result that the two consonants are neither completely sequential nor completely simultaneous, but overlapped. Different degrees of overlap may arise when different coupling patterns are present, or when different coupling strengths apply. The c-center effect, that the two gestures are offset, with the *midpoint* of the cluster timed to the vowel, is the result of the competition.

In the linguistic gestural model, which is the part of the model that expresses stored lexical contrasts, coupling graphs can indicate only in-phase and anti-phase coupling. At this level, AP and autosegmental phonology now posit the same two degrees of freedom in contrastive temporal relations. AP no longer overgenerates phonological contrasts, and typological generalizations such as onset/coda asymmetries are given a new explanation. With the addition of coupling graphs, however, AP has not lost its ability to account for the details of gradient realization, because the gestural coordinations specified in the graph still undergo implementation through the task dynamic model, with weighted averaging of competing couplings added to the effects of gestural overlap. However, the issue of categorical alternations remains.

As was noted above, AP hypothesizes that gestures are never added, deleted, or changed from one underlying specification to another. Allomorphy, such as the distribution of English [im-] and [in-] based on the initial consonant of the root, is not considered to be part of the phonology. Instead, allomorphs are indicated as part of the lexical specification of each word. All categorical alternations are handled as allomorphy, and all gradient changes are the result gestural magnitude and overlap. It follows from these hypotheses that there is no need for any statement of categorical phonological alternation in the grammar.

Zsiga (2011) argues, however, that the prohibition on adding, deleting, and changing gestures could be weakened or eliminated without doing violence to the rest of the theory. The linguistic gestural model could incorporate a module that allows for couplings to be added or deleted based on context. The specific example in Zsiga (2011) is nasal assimilation across word boundaries in Korean: /pap mekta/ "eat rice" is pronounced as [pam mekta]. Zsiga (2011) shows that nasalization is complete and categorical in Korean, even though it applies across a word boundary

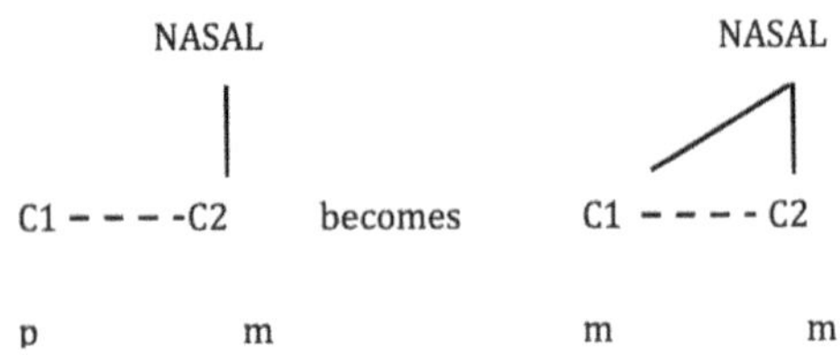

Figure 9.12 Coupling graphs for categorical nasal assimilation in Korean
Source: Adapted from Zsiga (2011).

and therefore cannot be encoded in the lexicon. (Some Korean speakers of English as a second language even apply categorical nasalization across word boundaries in English, pronouncing phrases such as "keep Matt" as "kee[m] Matt.") It is argued that the new (derived) nasal segment can be created by taking the nasal gesture originally associated to C2 and adding a new in-phase coupling with C1, as illustrated in Figure 9.12.

This approach could be used for any productive, categorical alternation in the post-lexical phonology. Under this approach, it would not be necessary to follow Ladd and Scobbie (2003) in arguing that two different representations, featural and gestural, are required for categorical and gradient assimilations. Gestural representations could handle both. The question then becomes whether gestural representations are in general the best approach to handling all kinds of alternations, whether traditionally described as phonological or phonetic.

9.6 EXTENDING THE MODEL

This chapter has argued that Articulatory Phonology offers a successful account of a large body of assimilations, insertions and deletions that occur at the interface of phonology and phonetics. AP argues that gestures can represent both phonological contrast and phonetic implementation, obviating the need for a separate featural representation or for any phonology to phonetics translation. Variable gestural overlap captures gradient assimilations. Representation in terms of coupling graphs creates a more constrained theory of phonological contrast and perhaps (following Zsiga 2011) of phonological alternation.

More research is needed in a number of areas, however. The specifics of the linguistic gestural model and task dynamic model are still being refined. These models work better for consonant gestures, for which a specific articulatory target can be defined, than it does for vowels, which many argue must be defined in acoustic terms. A truly general model of laryngeal gestures has yet to be developed. For example, can the simplifying hypothesis of voicing as a default be maintained? There has been recent promising work on incorporating gestures for tone into the model, as mentioned in Chapter 8. Gao (2008) and Karlin (2018) argue that tone gestures enter into a c-center relationship with onset consonants in Mandarin, Thai, and Serbian, but Zsiga (2018b) finds timing of tones to vowels rather than consonants in Igbo. Obviously, a sample size of greater than four would lead to improvements in the theory.

A drawback is that so much of the work in AP has been done on English. While aspects of other languages (including Arabic, Berber, Georgian, Igbo, Italian, Korean, Mandarin, Romanian, Serbian, Spanish, and Thai, see references cited above) are being modeled, more intensive work on typological differences would be welcome. For example, what is predicted about the typology of possible timing relations, and possible prosodic constituents?

The potential for correspondence between coupling graphs and autosegmental representations is obvious. There is in general a good (but not perfect) correspondence between gestures and autosegmental features: the features [labial] and [-continuant], for example, correspond to a labial closing gesture (see Zsiga 1997 for discussion). In many cases, the in-phase couplings in a coupling graph will represent the same relations as the association lines of autosegmental phonology: for example, gestures of the larynx and velum are timed to be in-phase with an oral closing gesture to create a segment-sized unit. In larger domains, gestural coupling may correspond to grouping within a prosodic domain. For example, in Figure 9.10, the presence of couplings between C and V indicate that the gestures belong to a single syllable. The hypothesis that couplings can encode higher-level prosodic constituency is further explored in Nam (2007), Nava et al. (2008), Nam et al. (2009), and Goldstein et al. (2009); see also Saltzman et al. (2008).

Different coupling strengths may indicate different prosodic levels, with smaller domains imposing a tighter coordination (Byrd and Saltzman 2003). The gestural and autosegmental approaches do not always agree on what the constituents should *be*, however. Is there a mora-sized unit? An onset constituent? What is the status of the segment? Are there discretely different kinds of phrases, as argued by the theories of the prosodic hierarchy (Chapter 7), or just gradient degrees of coupling strength? Mathematical modeling using coupled oscillators opens an important avenue for phonetic data to be used to test phonological theories of prosodic structures.

Another question is whether Zsiga (2011) is on the right track in incorporating the ability to represent categorical alternations in AP. Should the ban on adding, deleting, and categorically changing gestures be lifted? If so, how far should the ability to alter gestural representations extend? Should such changes apply to the post-lexical phonology only? Or should AP abandon the hypothesis that allomorphy is lexical choice rather than phonological derivation? This gets to the very large question of what phonology proper is responsible for, and the general relationship between phonology and the lexicon. As will be shown in Chapter 10, the role of the lexicon is a question not just for the phonology/morphology interface, but for the phonology/phonetics interface as well.

A final, overarching, question, is to what extent phonology is in fact articulatory. To what extent can a representation couched entirely in terms of articulatory units and articulatory goals account for phonological patterns? Chapter 10 now turns to the role of *perception* in phonology and the phonology/phonetics interface. The bigger problem of reconciling articulatory and perceptual approaches will be further addressed in Chapter 11.

RECOMMENDED READING

Overviews

Browman, C. P. and L. Goldstein (1992a), "Articulatory Phonology: an overview," *Phonetica, 49*: 155–80.

Hall, N. (2018), "Articulatory Phonology," in S. J. Hannahs and A. R. K. Bosch (eds), *The Routledge Handbook of Phonological Theory*. London: Routledge, pp. 530–52.

- Review the basic differences between AP and feature-based phonology.

Something old

Browman, C. P. and L. Goldstein (1986), "Towards an Articulatory Phonology," *Phonology Yearbook, 3*: 219–52.

Browman, C. P. and L. Goldstein (1990c), "Tiers in Articulatory Phonology, with some implications for casual speech," in J. Kingston and M. E. Beckman (eds), *Papers in Laboratory Phonology I: Between the Grammar and Physics of Speech*. Cambridge, MA: Cambridge University Press, pp. 341–76.

- What evidence and arguments do Browman and Goldstein provide for the inadequacy of feature-based representations to represent their data?

Something new

Davidson, L. (2006), "Phonotactics and articulatory coordination interact in phonology: evidence from non-native production," *Cognitive Science, 30*(5): 837–62.

- Describe the interaction that Davidson references in the title. How is using non-native speech important for discovering this interaction?

Shaw, J., A. I. Gafos, P. Hoole, and C. Zeroual (2011), "Dynamic invariance in the phonetic expression of syllable structure: a case study of Moroccan Arabic consonant clusters," *Phonology, 28*(3): 455–90.

- How do the authors use patterns of gestural coordination to deduce syllable structure?

Karlin, R. and S. Tilsen (2014), "The articulatory tone-bearing unit: gestural coordination of lexical tone in Thai," *The Journal of the Acoustical Society of America, 136*(4): 2176.

- Compare the proposal of Karlin and Tilsen to representations that assume the syllable or mora is the TBU.

Opposing views

Browman, C. P. and L. Goldstein (1989), "Articulatory gestures as phonological units," *Phonology, 6*(2): 201–51.

vs

Goldstein, L., D. Byrd, and E. Saltzman (2006), "The role of vocal tract gestural action units in understanding the evolution of phonology," in M. A. Arbib (ed.),

Action to Language via the Mirror Neuron System. Cambridge, UK: Cambridge University Press, pp. 215–49.

- Compare the 1989 and 2006 versions of AP. What has changed, and what has remained the same?

Ladd, D. R. and J. M. Scobbie (2003), "External sandhi as gestural overlap? Counter-evidence from Sardinian," in J. Local, R. Ogden, and R. Temple (eds), *Papers in Laboratory Phonology VI: Phonetic Interpretation.* Cambridge, UK: Cambridge University Press, pp. 164–82.

vs

Son, M., A. Kochetov, and M. Pouplier (2007), "The role of gestural overlap in perceptual place assimilation: evidence from Korean," in J. Cole and J. I. Hualde (eds), *Papers in Laboratory Phonology IX.* Berlin: De Gruyter Mouton, pp. 507–34.

vs

Zsiga, E. C. (2011), "External sandhi in a second language: the phonetics and phonology of obstruent nasalization in Korean and Korean-accented English," *Language, 87*: 289–345.

- You could also add Browman and Goldstein (1990c), cited above, and Nolan (1992) cited as recommended reading for Chapter 3. Is gestural overlap the best way to represent external sandhi?

QUESTIONS FOR FURTHER DISCUSSION

1. Why is there no phonology/phonetics interface in AP? How does AP account for lexical contrast and for what a modular approach would call lexical, post-lexical, and phonetic rules?
2. In what ways are gestures abstract?
3. Consider the words "ban" and "mad" as being composed of either three segments or four gestures each. What does thinking about these word as gestural constellations convey that the segment-based transcription does not, and vice versa? Draw each word out in an autosegmental, feature-geometrical representation. What new perspective does this representation add?
4. Draw a coupling graph and a gestural score for your name. (If it happens to be Sue or Pam you're in luck, but in that case I'd encourage you to try something a little more challenging.). Were there aspects where the gestural specification or timing were difficult to decide?
5. What would c-center timing look like for a word like "split"? How would it differ from "plan" and "span"?
6. In AP, "assimilations and deletions are seen as epiphenomena resulting from general properties of gestural reorganization." Give and discuss some examples not mentioned in the text. If you can think of examples that are not from English, all the better.
7. What are the three critiques of AP discussed in Section 9.4? How might they be answered?
8. What is the difference between in-phase and anti-phase gestural timing? How do

these gestural timing relations compare to feature-and-segment-based models? How are other, more gradient, timing patterns derived?

9. Read up on Q-theory (Shih and Inkelas 2019). How is it different from and similar to autosegmental representations? Gestural representations? What, if anything, does it do better than either?

CHAPTER 10

SPEECH PERCEPTION, EXEMPLAR THEORY, AND THE MENTAL LEXICON

10.1 HEARING AND SPEECH PERCEPTION

Chapter 9 focused on the role of articulation at the phonology/phonetics interface. In this chapter, we look at approaches that take perception as the main entry point. How is the acoustic signal translated to an image in the brain? And what can that translation tell us about what that "sound image" is like? As in the mapping from abstract symbol to articulator movement, the mapping from acoustic signal back to cognitive representation leads us to explore the relationship between the continuous and the categorical, the physical and the cognitive, the signifier and the signified, at the phonology/phonetics interface.

Section 10.1 gives a general overview of the difference between hearing and speech perception, and some of the basic transformations (normalization and cue integration) that must take place in the process of perception. Section 10.2 reviews and brings together some of the major theories that have been proposed for the ways in which perceptual factors influence phonology, particularly Dispersion Theory (Lindblom 1972; Flemming 2002), licensing by cue (Steriade 2001), and cue constraints (Boersma 1998, 2009). Section 10.3 looks at the problem of categorization, and the ways that, conversely, phonological structure influences perception. With that background, Section 10.4 addresses the question: What are the objects of perception? Of what does the "sound image in the brain" (Saussure 1959: 11) consist? The section discusses prototype models (Klatt 1989), models based on feature or cue detection (McClelland and Elman 1986; Lahiri and Marslen-Wilson 1991; Stevens 2002) and models that take the articulatory gesture to be the object of both production and perception (Liberman et al. 1967; Liberman and Whalen 2000; Galantucci et al. 2006). Section 10.5 turns to exemplar models of perception (Goldinger 1996; Johnson 1997; Frisch 2018). Exemplar models return us to the larger question of the relation between the quantitative and the categorical in linguistic representation, as discussed in the conclusion of the chapter, Section 10.6.

We begin with the important fact that hearing and speech perception are not the same thing. *Hearing* is what takes place in the ear: the chain of energy transfers that transform pressure waves in the air to vibrations of the eardrum and ossicles, to movement of the basilar membrane in the cochlea, and thence finally into electrical signals that nerves carry to the auditory cortex in the brain. (Johnson 2013a and other phonetics texts give overviews of the structures and processes involved.) *Perception* is what takes place when those signals are mapped into linguistic categories: segments, words, or sentences. As Boersma (2009: 65): defines it: "Perception is the mapping from raw sensory data to more abstract mental representations, or any step therein." Perception therefore deeply involves the questions both of how the mapping takes place, and what those representations are.

There are multiple transfers between the ear and the auditory cortex, the part of the brain where speech is processed, and the speech signal that reaches the brain is not identical to the signal that impinges on the outer ear. At every stage, aspects of the signal may be enhanced, lost, or transformed. The signal is modified even at the earliest stages of speech processing: the wrinkles of the pinna create echoes that aid in sound location, and the ear canal itself is a resonator that amplifies frequencies in the range relevant for speech (Wright et al. 1974; Ballachanda 1997).

The mapping from acoustics to perception is non-linear: equal changes in frequency and amplitude, as they would be measured by an external microphone and recorded in a waveform or spectrogram, do not result in equal responses in the perceptual system (see Stevens 2000a; Johnson 2013a). At the mechanical level in the basilar membrane, the cochlea is much more sensitive to frequencies in the lower ranges than in the higher, determining what human hearing can and cannot distinguish. Although the logarithmic nature of the acoustic to auditory mapping is close to linear at the lowest ranges of the scale (those relevant for F0), at the higher ranges relevant for speech, transformation to a logarithmic scale that better matches the response of the cochlea can make an important difference. In this range, sounds that look very similar in an acoustic spectrogram, such as click bursts or fricatives, become more distinct in auditory "cochleagrams." Figure 10.1 (from Johnson 2013a) shows that frequencies that appear to be close in the acoustic measure of kilohertz are further spread out when transformed into the Bark scale, which approximates the human auditory response.

A number of different logarithmic transformations, including the semitone scale, the Bark Scale, the Mel scale, and the equivalent regular bandwidth (ERB) scale can be used for a better representations of the auditory response than hertz. For formulas, see for example Johnson (1989), Traunmüller (1990), or Stevens (2000a).

As the auditory signal travels from the ears to the brain, it undergoes several levels of processing. Bernstein (2005) and Goldstein and Brockmole (2017) provide accessible overviews of how the different parts of the brain process speech. Abstracting away from the neuroanatomy, one model of the different stages of speech perception (Stevens 2005) is shown in Figure 10.2.

Most theories (e.g., Stevens 2002; Lahiri and Reetz 2002, 2010) assume that speech processing takes place in two stages. The first step is general auditory processing that is the same for both speech and non-speech, followed by processing that is specific to speech. These stages are discussed in the following paragraphs.

Aspects of general auditory processing include localization, attention, and extraction of specific "cue" information (Moore 1997). At this point the term "cue" is used non-technically, to mean just relevant auditory events, whatever form they might take. Localization includes integrating information from the two ears into a single stream of information. One of the first stages of perception must be "auditory

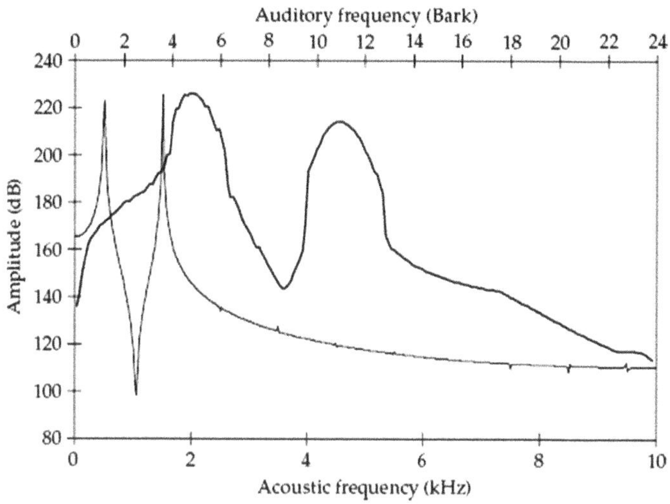

Figure 10.1 *A comparison of acoustic (light line) and auditory (heavy line) spectra of a complex wave composed of since waves at 500 and 1500Hz. Both spectra extend from 0 to 10kHz, although on different frequency scales*

Source: Johnson (2013a: 53).

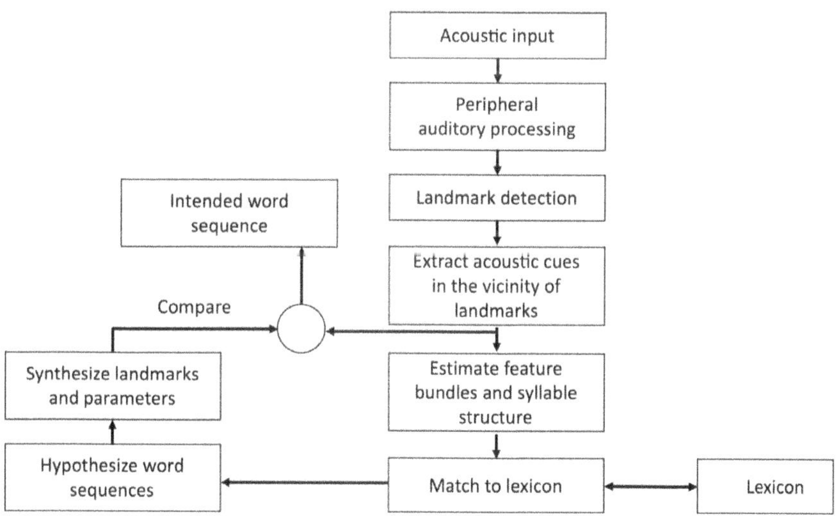

Figure 10.2 *The speech perception process according to Stevens*

Source: Adapted from Stevens (2005: 144).

scene analysis," which identifies a series of sounds that are similar, continuous and from the same localized source, so that the brain can attend to that series (Cherry 1953; Bregman 1990; McDermott 2009). Once a stream of speech has been identified and attended to, relevant cues to the linguistic content must be extracted. In most models of speech perception, the cues include formant frequencies, burst and fricative noise, F0, silent intervals, and the temporal relations between all of these. In Stevens's model in Figure 10.2, the perceptual system pays particular attention to "landmarks," which include both steady states and points of rapid transition.

The next stages become specific to speech. The cues must be normalized, integrated, and thereby mapped into linguistic representations (McQueen and Cutler 1997). Different models (see Section 10.4 below) make different assumptions about how normalization and integration take place, but a few words about the need for both will be helpful to place the more specific discussions in context.

Normalization refers to compensation or correction for differences that are irrelevant to the message per se, such as individual talker differences (size, gender, dialect), speech rate, or contextual factors. The acoustic/auditory patterns produced by different talkers diverge in many ways, and in order for these different auditory objects to be recognized as the same linguistic message, the differences have to be accounted for. Formant frequencies, for example, depend on vocal tract length, so that the second formant frequency of a large male speaker might be equivalent in hertz to the first formant frequency for a small child. F0 depends on the length and mass of the vocal folds, so 180Hz may be a high tone for one speaker and a low tone for another.

In the process of normalization, listeners interpret cues relative to each other. The information used for normalization might be simultaneous. Within a single vowel, F0, spacing of the higher formants, and voice quality differences can signal whether the talker is large or small, and thus whether a formant at 900Hz is likely to be F1 or F2. (See Johnson 2005 for some discussion of mathematical formulas that might work for this kind of normalization.) Because we use simultaneous information, we can interpret monosyllables. But listeners also take preceding context into account. A few syllables are sufficient to convey a speaker's F0 range, allowing a listener to correctly decide if 180Hz is at the top of the range and therefore a high tone or at the bottom and therefore a low tone (Gandour 1978). Ladefoged and Broadbent (1957) showed that the same synthesized vowel could be perceived as [ɪ] or [ɛ] depending on the preceding phrase. If the voice in the preceding phrase had vowels with higher formant values, the F1 of the target vowel was perceived as relatively low and thus the vowel was identified as [ɪ]. Conversely, if the preceding phrase contained vowels with lower formants, the F1 of the synthesized vowel was perceived as relatively high and was thus identified as [ɛ]. (It is worth noting that, the experiment having been conducted in 1957, Ladefoged and Broadbent used an analog speech synthesizer, the "parametric artificial talking device" at the University of Edinburgh.) Crucially, the same acoustic object was perceived to be of a different linguistic category depending on the context.

Models differ as to what happens to the talker-specific information in the process of normalization. Templatic models assume that the information is discarded, and must be ignored for the linguistic message to be identified. Of course, listeners can identify both talker and message, but for templatic models the two recognition processes are separate. Exemplar models assume that the talker-specific information is retained and used. As Nygaard (2005) poses the question: is variation noise to be discarded, or information to be used? See further discussion below.

Once cues are extracted and normalized, they must be integrated. Cue integration means that the brain creates a single object, the percept, from multiple pieces of information that it receives. Again, theories differ as to whether the integrated percept is a feature, segment, syllable, word, or something else—this is where the mapping to linguistic units takes place. Further, some theories argue that the mapping from cues to linguistic units is mediated by distinctive features; others argue for a more direct mapping. But any unit must be composed of multiple cues.

For example, formant cues must be integrated to create the perception of vowel quality. The vowels [i] and [u] are distinguished by a high vs low second formant. Changing the F2 value changes the percept. But listeners are not aware of hearing formant values: they hear vowel sounds. Formant integration can be demonstrated via "sine wave speech" (Remez et al. 1981). In sine wave speech, sinusoids of varying frequencies and amplitudes are synthesized to mirror formant trajectories. Played in isolation, the sine waves sound like strange whistles. Played simultaneously, they become recognizable (though robotic-sounding) sentences. As of this writing, demonstrations can be found at https: //www.mrc-cbu.cam.ac.uk/people/matt.davis/sine-wave-speech/ and http: //www.haskins.yale.edu/research/sws.html. Most demonstrations of speech perception phenomena are hard to describe in text: they have to be heard to be believed.

There is also integration of the auditory signal with visual and tactile input. Remez (2001) and Rosenblum (2005) give overviews of multimodal integration, including the neural links between the auditory cortex, the visual cortex, and the motor cortex. The most famous example of audio-visual integration is the "McGurk effect" (McGurk and MacDonald 1976). The effect is created by creating a mismatched video and audio. Seeing the mismatched video while hearing the audio causes the audio percept to shift. For example, seeing a video of a face articulating the lower lip against the upper teeth will cause the perception of synchronized audio [pa] to change to [fa]. Look away, and the audio is [pa] again. A BBC-sponsored YouTube demo is here: https: //www.youtube.com/watch?v=G-lN8vWm3m0. Interestingly, the McGurk effect works even if you know the trick—even the person who made the mismatched video will still have an altered perception. On the other hand, Walker et al. (1995) found an effect of familiarity. If the listeners were familiar with the faces and voices in the video (from friends or colleagues), the illusion didn't work. Knowing that the voice and face didn't match prevented the integration. The McGurk effect is a fun demo in the lab or classroom, but multimodal integration has real-world effects. Sumby (1954) found that *matching* visual information helped listeners recognize words embedded in noise. More recently, Havenhill (2018) demonstrated that American English speakers for whom the cot/caught distinction was marginal were better able to distinguish the two vowels when they were watching a video that showed the presence or absence of lip rounding. The neural links between the auditory cortex and the motor cortex are very important for the Motor Theory of speech perception (Section 10.4).

A final step in speech processing is the influence of top-down information (McQueen and Cutler 1997). "Bottom-up" processing is the incoming stream of acoustic cues. "Top-down" processing is our prior knowledge of what the communicative event is likely to be. Listeners have expectations of what they will hear, and how an ongoing utterance will be completed, and they use this information in the process of matching the signal to the linguistic message. Remez (2001: 29) quotes Weiner (1950): "In the problem of decoding, the most important information we can possess is the knowledge that the message we are reading is not gibberish." Figure 10.2 (Stevens 2005) shows the influence of top-down information as a feedback loop that checks a preliminary match against expectation. Other models (e.g., McClelland and Elman 1986) assume that frequent lexical items have a higher baseline excitation level to begin with, and thus exert an influence prior to the linguistic match being made. The exact mechanism of how and when prior expectations are integrated into the perception process is a matter of ongoing research (e.g., Sohoglu et al. 2012).

Now that this section has established a general picture of how speech perception works, the next sections turn to questions of how speech perception influences phonology, and how phonology influences speech perception.

10.2 SPEECH PERCEPTION INFLUENCES PHONOLOGY

In order for spoken communication to work, humans must be able to both articulate and hear distinctions between speech sounds. It can be argued that structuralist phonemics, which starts its analysis with a surface string of speech sounds, takes the point of view of the hearer (Wells 1947; Hockett 1961; Ladd in press). Bloch (1948: 10) writes that the linguist must be trained to correctly hear and transcribe strings of speech sounds, and that that transcribed string is the valid starting point of linguistic analysis.

> Postulate 9: An observer can be trained to make a phonetic description of the utterances of any dialect, or of a sufficient sample thereof, without the aid of laboratory devices, that will be adequate and valid for the purposes of phonological analysis. . . . [This] means that our present methods of pursuing phonological (specifically phonemic) analysis are sound: that a linguist can be trained to hear all that he needs to observe in his informant's speech; that he needs to rely only on his ears and eyes; and that his results will have scientific value even if he records only a sampling of all the possible utterances that his informant might make.

Recall also that in Postulate 11 (1948: 12, quoted in Chapter 3), Bloch argues that while phonetic analysis may show that the movements of vocal organs are continuous, speech "AS PERCEIVED" (emphasis original) is categorical: "Every phonetician has had the experience of breaking up the smooth flow of speech into perceptibly discrete successive parts." That is, for Bloch and those who follow his postulates, analysis begins with perception.

On the other hand, as detailed by Boersma and Hamann (2009), most generative models emphasize the production side of the speech chain. A generative derivation starts with an underlying cognitive representation and ends with a surface pronunciation, leaving unspecified how the ensuing conversion from acoustic signal back to cognitive representation might take place. However, there are some patterns in phonology that production cannot explain. Constraint-based models (beginning with Prince and Smolensky 2004) allow for a bi-directional mapping, and constraints that map surface and cognitive representations can work for both perception and production. Hume and Johnson (2001: 5) state that Optimality Theory "has allowed for the statement of perceptually grounded constraints that interact dynamically with constraints motivated by other principles." Different theories of phonology and of the phonology/phonetics interface have taken advantage of that ability to differing extents. The role of perception in phonology has been touched on in previous chapters, particularly the discussion in Chapter 3 on inventories and the discussion in Chapter 6 on markedness and naturalness, but we pull these discussions together here. Theories of phonology that emphasize perception begin with the premise that phonological systems are adapted for communication, and thus will evidence good design principles that make communication possible.

Dispersion Theory (Liljencrants and Lindblom 1972; Lindblom 1986; Flemming 2002, 2004, 2005) is a theory of how perception shapes inventories. (See further discussion in Chapter 3.) Dispersion Theory argues that the segments that make up a given inventory are distributed for maximum distinction in the perceptual space. "Efficient communication depends on fast, accurate perception of speech sounds, and listeners are faster and more accurate in identifying the category to which a stimulus belongs if the stimulus is more distinct from contrasting categories" (Flemming 2005: 157). Thus, for example, languages choose the "corner vowels" [i, a, u] first, as maximally distinct in the F1/F2 vowel space, and then fill in the spaces between only as inventories get larger. As Flemming notes, there is nothing difficult about articulating the high central vowel [ɨ]. The reason it does not occur more often in inventories is perceptual: the [i] vs [ɨ] contrast is less perceptually distinct than [i] vs [u], so [i] and [u] are preferred.

Flemming encodes the phonological predictions of Dispersion Theory into "minimal distance" constraints that set requirements on how close, for instance, vowels can come to each other in the F1 or F2 space. Militating against "minimal distance" is a constraint to "maximize contrast" that prioritizes larger inventories. Specific inventories come about as a function of the maximum number of contrasts that can be maintained without violating the minimum allowed distance. The constraints in Dispersion Theory operate not over individual lexical items, but over inventories.

Consistent with Dispersion Theory is the "auditory enhancement hypothesis" proposed by Diehl and colleagues (Diehl 1991; see also Diehl and Kluender 1989; Kingston and Diehl 1994, 1995; Diehl and Lindblom 2000; Kingston 2011; Kingston et al. 2014). According to this hypothesis, the features of vowels and consonants are chosen so that contrast is enhanced. For example, a contrast in backness is enhanced by rounding, the two different articulatory dimensions working together to make /i/ and /u/ more distinct. The feature of voicing is enhanced by length and pitch differences on adjacent vowels (Kingston 2011). Padgett (2001) makes a similar point: in Russian, front vowels and palatalized consonants occur together, while consonants that occur with back vowels are velarized, so that vowel and consonant distinctions are maximized together. From an articulatory point of view, the secondary enhancing features may make an articulation more difficult or complex, so the feature co-occurrence must be explained by grounding in perception instead.

Steriade's (2001) theory of "licensing by cue" expands the role of perception from inventories to phonotactics and alternations. Steriade introduces the idea of the "P-map": a cognitive representation of perceptual distinctiveness that linguistic constraints can reference. Based on the P-map, contrasts are permitted ("licensed") in positions where cues can be clearly heard, and neutralized in positions where distinctions are not clear. Licensing by cue replaces more traditional discussions of onset/coda asymmetries. According to licensing by cue, CV syllable structure is preferred not because the onset constituent has a special status, but because in general consonants are most clearly distinguished at release, through information in the burst and formant transitions. Steriade (2001) notes that retroflex consonants are an exception to the general preference for onset positions, occurring in many languages only in the coda. The reversal is explained by the unique perceptual cues to the curling back of the tongue tip, which are most clearly perceived in a *preceding* vowel. Wright (1996, 2001) provides further examples of the preservation of auditory cues acting as a determinant of phonotactic constraints on consonant sequences.

Licensing by cue extends to alternations as well. In most cases, as argued by Steriade (2001), a contrast will be neutralized in a position where it cannot be clearly heard. On this view, the ubiquitous process of nasal place assimilation (nasals assimilate in place of articulation to a following stop) is explained by the fact that nasal consonants are poor carriers of place information, because of the presence of nasal resonances that obscure formant structure. The clear release burst of the following stop predominates in the perception of the cluster, and the nasal place contrast is lost. Thus, licensing by cue provides a further grounding in perception for assimilation, in addition to a production account in terms of coarticulation and gestural overlap.

Another possibility for alternation is that weak positional contrasts can be avoided by making a change that strengthens the perceptibility of the contrast. Dissimilation, which goes against the pressure to make articulation simpler, is often argued to have a perceptual grounding. For example, in a sequence of two fricatives, one may dissimilate to a stop, making it easier to perceive that there are two different segments (Hume and Johnson 2001; Tserdanelis 2001). Hume (1998, 2001) argues that metathesis is driven by perceptual optimization. Fricatives and stops may switch position (as in [æsk] pronounced as [æks]) in order to place the stop adjacent to at least one vowel, where formant transitions can be heard. Because fricative place is strongly cued by characteristics of the noise during the fricative duration, in addition to formant transitions, there is less of a need for fricatives to be vowel-adjacent.

Attention to perceptual cues, and misperception of cues, is also implicated in sound change. As Ohala (1981) argued, sound change results if a listener either misperceives a cue, or misattributes a cue that is correctly heard. One of the clearest examples of how differences in cue attribution can cause phonological sound change is the process of tonogenesis (discussed in Chapter 8.3.4). The laryngeal configuration of a consonant will affect the pitch of a following vowel (Ohala 1973). As argued by Kingston (2005), these pitch differences should be analyzed as a perceptual enhancement of the voicing distinction. If listeners, however, take the secondary pitch cue as primary, the voicing distinction becomes a contrastive tone distinction, as is currently taking place, for example, in Seoul Korean (Kim 2013). To take a different kind of example, many studies have shown that the "voicing" cue on syllable-final consonants in English is now almost completely carried by preceding vowel length (e.g., Keating 1985; Braver 2014; Pfiffner 2019). If the listener is the source of sound change, as Ohala (1981) argued, then perception is key to shaping inventories.

One phonological model that directly incorporates perception is Boersma's (1998, 2009) model of Functional Phonology. As discussed in Chapter 6, Boersma formulates "cue constraints" that "express the speaker-listener's knowledge of the relations between continuous auditory cues and discrete phonological surface elements" (Boersma 2008: 3). Boersma (2009: 59) writes that "the phonology/phonetics interface consists of cues." In Functional Phonology, cue constraints are often formulated negatively, ruling out mappings, such as "don't perceive [F2 = 2000 Hz] as [a]." The ranking of these cue constraints will be language specific: the F1 level that cues [i] in Scottish English, for example, will cue [ɪ] in Southern British English (Escudero and Boersma 2004). Crucially, cue constraints on the phonology/phonetics mapping are interleaved with articulatory constraints, with structural constraints on phonological form, and with faithfulness constraints that implement the phonology/morphology interface, and all constraints are evaluated in parallel. Thus the effects of perception on phonology are directly modeled. If a cue constraint requiring clear formant transitions, for example, outranks a faithfulness constraint on linearity, the action of the cue constraint can directly trigger metathesis. As was discussed in Chapter 6 (see also further discussion in Chapter 11, phonologists disagree over whether such direct "functional" intervention is either necessary or desirable.

Overall, the evidence is clear that phonological inventories and alternations are shaped by both principles of perception and principles of articulation. To the extent that markedness constraints are incorporated into the grammar, they may be grounded in either perception or production. The two kinds of constraints may interact, sometimes favoring opposite outcomes, as in enhancing a contrast at the expense of additional effort, or losing a contrast to favor ease of articulation. The question remains open, however, as it is with markedness constraints in general, of whether preference for more distinct perception is encoded in the synchronic grammar, or whether it is simply the case that contrasts that are difficult to hear are lost over time. We return to this overarching question in Section 10.6 below.

10.3 PHONOLOGY INFLUENCES SPEECH PERCEPTION

The preceding section demonstrated that speech perception influences phonological structure. At the same time, the structure of the listener's phonology influences the process of speech perception. Specifically, the phonological categories of our language influence the way speech sounds are perceived. How similar or different two items sound is influenced by whether they fall into the same or different categories in the listener's phonology.

In early experiments, this influence was described as "categorical perception." Strictly speaking, the term has two parts. First, there is a sharp boundary between categories. Along any physical scale that distinguishes phonological categories, such as VOT or formant values, there will be an abrupt transition between values that characterize X and values that characterize Y. Second, there is a link between identification and discrimination. If two sounds are identified as the same category, discrimination between them is poor, but if two sounds are identified as belonging to different categories, discrimination between them is good. It has been argued (e.g., Liberman 1996; see also the discussion in Pisoni and Luce 1986) that these properties hold of speech but not of non-speech perception, such as perception of musical notes.

Categorical perception was first demonstrated with F2 transitions differentiating the syllables [ba], [da], and [ga] (Liberman et al. 1957), but it can be demonstrated with any continuum of values that differentiates linguistic categories. In one well-known experiment, Lisker and Abramson (1970) synthesized a set of syllables with a continuum of VOT values, from -150ms (a long period of voicing preceding the release burst) to +150ms (a long period of aspiration following the release burst). They then asked listeners to identify each syllable as "ba" or "pa" and, in a second task, to decide whether two syllables were the same or different. For English-speaking listeners, Lisker and Abramson found a clear category boundary at +30ms VOT. All tokens with prevoicing or with a VOT of less than +30ms were identified as "ba," while all tokens with a voicing lag of more than +30ms were identified as "pa." Fulfilling the second half of the requirement for categorical perception, two tokens from the same side of the boundary would be identified in the second task as "same," but two tokens from different sides of the boundary would be identified as "different."

In follow-up studies, Abramson and Lisker (1970, 1973) found the same categorical perception effects for speakers of Spanish and Thai, but the location of the boundary was different for each language, reflecting the language-specific phonological system. The results of their studies are summarized in Figure 10.3 (from Zsiga 2013a).

For Spanish speakers, who typically produce [b] with prevoicing and [p] with short-lag VOT, the category boundary fell at +15ms VOT. For Thai speakers, who distinguish between [b] with prevoicing, [p] with short-lag VOT, and [p^h] with longer aspiration, there was a three-way distinction. For all speakers, regardless of language, within-category distinction was poor and cross-category discrimination was good. However, the same/different judgement depended on language experience, and thus the location of the category boundary. Spanish speakers heard tokens of -30 and +20 as different, but English speakers heard them as the same, and Spanish speakers heard tokens of +20 and +50 as the same, while English speakers heard them as different. For Thai speakers, -30, +20, and +50 were all different.

Abramson, Lisker, and their colleagues concluded from their findings of categorical perception that speech is processed in an entirely different way from perception of music and other non-linguistic sounds. They argued, in fact, for the direct perception of articulatory gestures (Liberman et al. 1967, see discussion below).

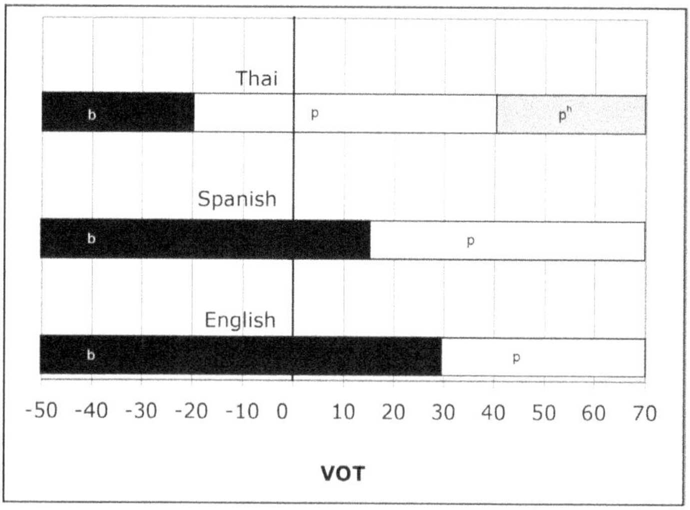

Figure 10.3 Category boundaries depend on language experience
Source: Results from Abramson and Lisker (1970, 1973).
Graph from Zsiga (2013a: Figure 9.10).

The original claims of categorical perception turned out to be too strong, however. Both the sharpness of category boundaries and the inability to distinguish within categories were called into question by further experimentation (summarized in Roussel 1996).

Further experiments by other researchers found that categorical perception is not unique to speech or even to humans. Humans *can* demonstrate categorical perception of music and other non-speech stimuli, if they are asked the right category-forming questions, such as "Was this note plucked or bowed"? (Cutting and Rosner 1974; Massaro 1987). Categorical perception was also demonstrated in animal auditory processing, including for chinchillas (Kuhl and Miller 1975), macaques (Kuhl and Padden 1983), and quail (Kluender et al. 1987).

It's not the case that humans can't hear any difference at all between tokens within a category (Massaro 1987; Boersma 2009). As Boersma notes, judging that two stimuli represent the same word is not the same as perceiving them as acoustically identical. Nonetheless, the categorical perception experiments were not completely off the mark. Research does confirm that phonemic categorization has an effect on perception, if not quite as determinative as originally argued. Speakers are still *better* at distinguishing sounds that cross category boundaries, and speakers of different languages have category boundaries at different places (see, e.g., Johnson and Babel 2010). In addition to experiments on adult native-language listeners, the effects of category formation are seen clearly in infant language acquisition and adult language learning.

Cutler (2012) provides an extensive overview of language learning from a perceptual point of view, along with a review of experimental methods for studying infant

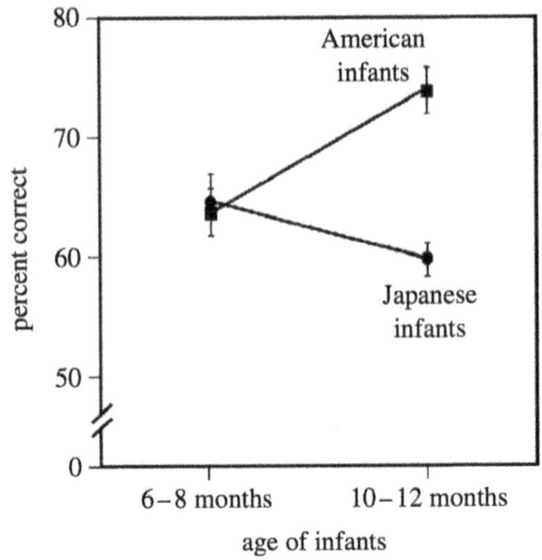

Figure 10.4 Discrimination of [ɹ] and [l] by American and Japanese infants

Source: Kuhl et al. (2006).

language learning. Infants have to be born with the ability to distinguish the sounds of whatever language environment they're born into (though see Houston 2005, who marshals evidence that their discrimination abilities are not perfect, and that some distinctions are learned only with exposure.) At some time between six and twelve months, however, babies begin to specialize: they get better at distinguishing the sounds of their own language, and begin to lose sensitivity to (or ignore) distinctions that are not important to them (Kuhl et al. 1992, 2006, 2008; Jusczyk 1997; Cutler 2012, among many others). Figure 10.4 shows one often-cited example, from Kuhl et al. (2006).

At six months of age, English-learning and Japanese-learning infants are equally good at hearing and responding to the difference between [ɹ] and [l], a distinction that exists in English but not in Japanese. (Or one might say they are equally bad—due to the nature of working with babies in experimental settings, absolute scores are above chance, but far from perfect.) By twelve months of age, the English learners have gotten better at hearing the distinction, due to continued exposure. But the Japanese learners have gotten worse. The hypothesis is that the infants are becoming attuned to the perceptual cues and contrasts that matter in their language, and are tuning out the cues and contrasts that don't. Kuhl et al. (2008) refer to this process of tuning out as "neural commitment."

It is worth noting that Houston (1999, 2005) argues that exposure to various talkers helps infants in this process of learning the categories of their language. By hearing different talkers, they are able to make generalizations about what kinds of variability are due to the talker and what kinds of variability are due to the message; that is, how to normalize. Exposure to variability helps infants form "robust representations" that hold across different talkers.

Infants at six to twelve months of age have also been found to be sensitive to transitional probabilities between segments (Jusczyk and Aslin 1995; Saffran et al. 1996). These researchers argue that infants are recognizing that frequently occurring sequences are likely to belong to words, while infrequent sequences indicate word boundaries. As they are learning to abstract over categories of sounds, they are also beginning to create the sound–meaning associations that will constitute their lexicon.

Once the contrastive categories of the lexicon are formed, they will influence perception throughout life. As has been noted by many researchers (Lado 1957; Flege 1987, 1995, 2005; Kuhl and Iverson 1995; Best 1995; Best and Tyler 2007, among many others), adult learners have great difficulty perceiving and producing contrasts between sounds in the L2 that are within the same category in the L1. Cutler (2012) cites Polivanov (1931) as one of the first linguists to have made this claim:

> The phoneme and the basic phonological representations of our native language are so tightly coupled to our perception that even when we hear words (or sentences) from a language with quite different phonology, we tend to analyze these words in terms of the phonemic representations of the native language. (Polivanov 1931: 79–80, quoted and translated by Cutler 2012: 457)

In addition to an influence of phonemic categories, L2 learners will also hear sequences of sounds in accordance with the syllable structures and phonotactic restrictions of their L1. For example, as was mentioned in Chapter 2, Dupoux et al. (1999) found that native speakers of Japanese could not perceive the difference between "ebzo" and "ebuzo." Because Japanese lacks coda consonants, listeners had no way of processing/storing the syllable /eb/, and instead heard and remembered it as /ebu/.

Different researchers have proposed different models for explaining how L1 and L2 categories interact, and exactly what it means (to use Polivanov's phrasing) to analyze a second language "in terms of the phonemic representations" of the first. Flege's (1987, 1995, 2005) Speech Learning Model (SLM) hypothesizes that the L1 and L2 systems occupy the same phonetic space, and thus use the same categories, causing a mutual influence, with the more frequently-used language dominating. While it's possible for new categories to be formed for the L2, generally the sounds of the L2 are forced into the nearest L1 class if they are at all similar, in a process of "equivalence classification." The result is that L2 distinctions are lost.

Kuhl and colleagues (Kuhl 1991; Kuhl and Iverson 1995; Kuhl et al. 2008) use the metaphor of a magnet to describe the formation and maintenance of L1 categories, in the "Native Language Magnet" (NLM) theory. According to NLM, as the infant is repeatedly exposed to instances of phonetic units, these instances are grouped into categories that share phonetic characteristics. As experience builds up, the categories are reinforced. As noted above, Kuhl et al. (2008) refer to this reinforcement as "neural commitment": physical changes in the brain that strengthen certain neural connections at the expense of others. Due to this neural commitment, the categories act as "magnets." Instances of non-native phones that are similar to the native phones are processed in terms of the pre-existing categories. Kuhl et al. (2008) argue that the infants that make the greatest neural commitments to native language categories, and in consequence show the worst discrimination on non-native (and thus to them unimportant) categories, will make the fastest progress in native language acquisition.

Best's (1995) Perceptual Assimilation Model (PAM) also addresses classification, and offers a more nuanced set of possibilities than the SLM, depending on how different the sounds of the two languages are. PAM offers six possibilities:

1. Single-category assimilation. Two contrasting sounds in the L2 are assimilated into one L1 category, and the contrast is lost.
2. Two-category assimilation. Two contrasting sounds in the L2 are assimilated into two different L2 categories. The distinction is maintained, though possibly not in exactly the same way as it would be for a native speaker of the L2.
3. Category goodness difference. In this case, the two L2 sounds are assimilated into the same L1 category, but one of the two is judged to be a typical example of the L1 category, while the other is marginal or unusual. In this case, the contrast can still be perceived, though with more difficulty than in two-category assimilation.
4. Uncategorized vs categorized. This is similar to the two-category assimilation, in that the two L2 sounds are definitely perceived to be different, but only one is close enough to an L1 sound to count as an instance of that category.
5. Both uncategorizable. In this case, neither L2 sound fits an L1 category. Depending on how different the two L2 sounds are, the learner may or may not be able to reliably perceive the difference between them.
6. Nonassimilable. This case arises if the L2 sounds are completely outside the L1 phonetic space. They are so different from L1 sounds that there is no interference from L1 phonology, and the sounds are easily discriminated.

The possibilities of a category-goodness difference and of L2 phones being uncategorizable or unassimilable are not outcomes considered by the SLM, which hypothesizes that all L2 phones are fit into L1 categories.

Best et al. (1988, 2001) tested the predictions of the PAM by asking English speakers to attempt to distinguish some of the contrastive sounds of Zulu. They found that English speakers were in fact excellent at perceiving the differences between the Zulu click sounds. Best et al. attribute this good performance to the clicks being non-assimilable, and so far outside the L1 phonetic space that there was no L1 interference. The voiced and voiceless lateral fricatives of Zulu were matched to two different English categories, /ʒ/ and /ʃ/, and were also easily discriminated. Zulu contrasts a voiceless aspirated /kʰ/ and ejective /kʼ/, a category goodness case. For English speakers, both are heard as a kind of "k" sound, the former not very different from English, the latter an unusual version. These two could be discriminated, though not as well as the laterals were. Finally, plosive /b/ and implosive /ɓ/ were assimilated to a single English category, "b," and the English speakers could not distinguish between them. Overall, the predictions of the PAM concerning the perceptibility of different kinds of classification were well-confirmed.

Research on L2 phonology is ongoing, however, and much work remains to be done. As was noted in the discussion of L2 production, it is a drawback that so much of the work has been done on learners whose native language is English. Perceptual testing of language pairs with different kinds of inventories will provide important evidence on how cross-language perceptual classification takes place.

Importantly, theories of classification differ in what they take to be the objects of perception. PAM takes a direct realist view of perception, hypothesizing that the objects of perception are articulatory gestures. The NLM relies on the importance of the aggregate of individual acoustic events, as described by Exemplar Theory. Thus the problem of how categorization takes place leads directly to the question of what it is that categories are made of.

10.4 UNITS OF PERCEPTION

We have defined perception as the steps in "the mapping from raw sensory data to more abstract mental representations" (Boersma 2009: 65). We have looked at what some of those steps are in Section 10.1, and considered in Sections 10.2 and 10.3 how the process of perception influences phonological patterns and vice versa. Here, we consider directly the question: What is the representation to which the signal is mapped?

There is no consensus on what the units of perception are. Prototype models assume the incoming stimulus is compared to a stored idealized representation, which might correspond to a segment, demi-syllable, syllable, or word. Cue-based or feature-based models assume a level at which certain acoustic events are extracted and then assembled into a linguistic representation. The Motor Theory of speech perception argues that the objects of perception are articulatory gestures. Exemplar models (which have a section of their own, 10.5) assume the stimulus is situated in a cloud of detailed episodic traces, which are associated with category labels. We consider each in turn.

10.4.1 PROTOTYPE MODELS

Some traditional models of speech perception have envisioned the mental representation as an abstract and idealized prototype, almost a Platonic idealization, that an incoming stimulus may more or less resemble. Saussure, for example, described the mental representation as a "sound image in the brain" (1959: 11). Baudouin de Courtenay considered the phoneme as an idealized abstraction, "the psychological equivalent of physical 'sound,' " (1927; Stankiewicz 1972b: 279). In Chapter 8, we noted that Abramson (1979: 7) considered that the most "psychologically reasonable" representation for Thai tone is "to suppose that the speaker of Thai stores a suitable tonal shape as part of his internal representation of each monosyllabic lexical item." The prototype might correspond to a phoneme, a word, or an identifiable part of a word, such as a syllable or tonal tune.

In the mapping from sound to abstract prototype, all variable details that represent individual speakers or particular contexts must be removed so that the mapping to the invariant prototype can be made. Thus the process of speech perception can be seen as reversing the generative derivation, stripping away the detailed and predictable information that was added step by step by phonological and phonetic rules. This removal of information would account for categorical perception: stimuli that belong to the same category sound alike because once the mapping to the prototype has been made, all the variable information is not stored in memory. Prototype models thus consider all variation to be noise (Nygaard 2005). They are not consistent with findings that show that this low-level variation is in fact stored and used.

In one detailed and successful prototype model, Klatt (1979, 1989) proposes a specific process of "template matching" for speech recognition. In Klatt's model, each morpheme is stored in memory as a template of spectrographic information to which an incoming stimulus can be directly compared. Indexical information, related to the talker, is removed by normalization, and contextual variation is removed by the undoing of phonological rules, leaving a representation with only essential information that can be compared to the template. The incoming speech information is directly compared with the spectral templates stored in memory, without any intervening analysis into segments or features. Most other models of speech perception do propose these intermediate levels.

10.4.2 FEATURE MATCHING

Most models of speech perception, including Stevens (2002, 2005), Lahiri and Reetz (2002, 2010), McClelland and Elman (1986), and Boersma (2009) assume an intermediate stage in which specific pieces of acoustic information, or cues, are extracted from the signal. It is the cues that are then matched to featural or phonemic representations that are stored in memory.

The model of Stevens (2002, 2005), which was diagrammed in Figure 10.2, is feature-based. (Stevens 1960 was an early version of this model. We should pause for a moment to consider those dates, and acknowledge the decades of contributions that Stevens has made to the study of acoustic and auditory phonetics, not the least of which is his monumental *Acoustic Phonetics* textbook and reference work, published in 2000.) As was shown in Figure 10.2, in Stevens's model acoustic cues are extracted from the signal, and matched to featural representations stored in memory. Stevens (2005: 125) writes: "We assume that words are represented in the memory of speakers and listeners in terms of sequences of segments, each of which consists of a bundle of binary distinctive features." The mapping from cue to feature is based primarily on the acoustic definitions of the features proposed by Jakobson et al. (1952). Each feature is cued by a specific acoustic correlate: nasal murmur for [+nasal], increase in amplitude for [+vowel], lowering of F2 and F3 for [+labial]. Raphael (2005) provides more extensive discussion of the mapping from acoustic cues to features.

To locate features, listeners focus on maximally informative landmarks in the speech stream: maxima and minima of amplitude, as well as points of abrupt change (Stevens 2000b). Stevens notes, however, that the process is not as difficult or complicated as it might seem.

> In the simple examples of consonants and vowels in monosyllabic words . . . [t]he acoustic landmarks are well-defined and the acoustic cues for the various

> features can be easily extracted based on knowledge of the defining articulatory and acoustic/perceptual correlates of the features. (Stevens 2005: 134)

Focusing on landmarks, whose characteristics tend to remain stable across different kinds of variation, allows listeners to abstract away from contextual and speaker-specific variation to find the crucial phonetic cues that will allow the matching features to be found.

As discussed above, and as shown in Figure 10.2, once the features are estimated a match can be made to a lexical representation, which is itself hypothesized to be a feature-based representation. As a final step, top-down processing compares the hypothesized word sequence with the incoming signal. Thus features are the objects of perception, and the phonetics to phonology mapping consists of matching acoustic cues to stored featural representations. The crux of the matter, of course, is whether, given the amount of variation in speech, the matching process is in fact as successful as Stevens argues it is.

Another feature-based model of speech perception is the FUL model of Lahiri and Reetz (2002, 2010; see also Lahiri and Marslen-Wilson 1991, 1992). FUL stands for "featurally underspecified lexicon." As the name implies, underspecification is central to the way that FUL hypothesizes listeners handle variation, both random and contextual, in the signal.

FUL assumes a single abstract underlying representation for each lexical item, represented in terms of features. Lahiri (2000) proposes a set of twelve privative features, applicable to both consonants and vowels, that are used in the model. These features are (broadly) acoustically defined: for example, [high] is defined as F1 < 450Hz. Crucially, only marked features are present in lexical representations. Unmarked values, including for example coronal place, are left unspecified.

Similar to the process proposed by Stevens (2002) and diagrammed in Figure 10.2, the perception system "scans" for "acoustic features" which are mapped into phonological features (Lahiri and Reetz 2002: 638). Features are extracted from the signal based on acoustic characteristics and "bundled" every 20ms.

The phonological features are then mapped to the lexical representation without intermediate translation to segments or syllables. For each potential mapping, the acoustic and lexical features are coded as "match," "mismatch," or "no mismatch." Each word then gets a score depending on these matches, with higher scores corresponding to higher levels of lexical activation. Possible words are ranked according to score, and the lexical item with the highest score is chosen as the perceived word.

It is the category of "no mismatch" that takes advantage of underspecification. There is an asymmetry in the matching process from signal to lexicon. If a feature is present in the lexical representation but not present in the signal, that is coded as a mismatch, and inhibits the lexical activation. On the other hand, if a feature is present in the signal but not present in the lexical representation, it is scored as "no mismatch," which does not cause inhibition. For example, a lexical item with word-final [m] is represented with a labial feature. An incoming signal with a word-final [n] will not have that labial feature. The lack of a [labial] feature in the signal will be coded as a mismatch, and the word will not be selected. On the other hand, because of underspecification, a lexical item with word-final [n] has no specified place features. An incoming signal that contains *either* word final [n] or [m] will be coded as no mismatch, and lexical activation will not be impeded. The lexical item "in" can be matched to either "i[n] place" or "i[m] place."

Thus, the FUL model crucially relies on the phonological concepts of markedness and underspecification to account for successful perception in the presence of contextual variation. Lexically-underspecified segments can be subject to variation in a way that lexically-specified segments cannot. As with Stevens's model, however, the overall success of the system relies on successful choice of both phonological and acoustic features, and successful identification of the acoustic features in the signal. The FUL model also makes very specific predictions based on phonological markedness (such as coronal place being unspecified), which makes the model both more constrained and less powerful than Stevens's, which can either be an asset or a liability as these predictions are tested with more data.

The TRACE model of McClelland and Elman (1986) similarly relies on feature matching to increase lexical activation. (The authors write TRACE in all caps but it is apparently not an acronym.) It is implemented as a neural network, with less of an explicit connection to phonology than either Stevens (2002) or Lahiri and Reetz (2002). In TRACE, a feature layer, phoneme layer, and word layer are all interconnected, and activation spreads between layers both bottom up and top down. The activation is explicitly modeled over time. As a word is processed, the acoustic features of the initial consonant activate all compatible phonemes, which activate all compatible words. As bottom-up processing continues into the vowel, certain words will be deactivated if they do not match. In the other direction, the lexical set of activated words can exert a top-down effect, such that ambiguous features or phones will be interpreted to match the words that are already strongly activated.

The sharing of activation between levels is both the greatest strength and greatest weakness of the TRACE model. It accounts well for top-down effects, and effects of "lexical competition" between similar words: words that have many similar "competitors" are recognized more slowly than words that are not similar to many others (Vitevitch and Luce 1998, 1999). On the other hand, Luce and McLennan (2005: 592) argue that division into three separate levels is a "decidedly questionable" architecture that is both "rather inelegant and probably psychologically implausible." (See also Norris et al. 2000 for further criticism.) Luce and McLennan (2005) consider other similar connectionist models of word recognition that they argue do a better job of accounting both for lexical effects and for variation. They note, however, that all of these models assume a feature-to-word mapping in which higher levels of representation are more abstract and underspecified, and they suggest that a thorough rethinking of separation into layers might be the next step in making progress in connectionist models.

10.4.3 THE MOTOR THEORY OF SPEECH PERCEPTION

The Motor Theory of speech perception hypothesizes that the units of perception—the units stored in memory to which incoming speech is matched—are not idealized spectrograms (as proposed by Klatt) nor featural representations (as proposed by Stevens) but articulatory gestures. As Liberman and Whalen (2000: 188) state: "the phonetic elements of speech, the true primitives that underlie linguistic communication, are not sounds but rather the articulatory gestures that generate those sounds." The Motor Theory of speech perception was first proposed by Liberman et al. (1967), in the context of the discovery of categorical perception and the hypothesis that speech perception is "special" (Liberman 1970, 1996), and uses different mechanisms than other types of auditory processing. The theory is elaborated and updated in Liberman and Mattingly (1985), Liberman (1996), Liberman and Whalen (2000), and Whalen (2019). Galantucci et al. (2006) argue that even if speech processing is not as unique as originally claimed, and does share general mechanisms of auditory processing, perceiving speech still can and should be analyzed as perceiving gestures.

While the idea of hearing gestures directly may at first seem counterintuitive, proponents make an analogy to visual perception (e.g., Fowler 1996). Light is the medium by which we see, but the objects of visual perception are the objects that caused the patterns of refraction. By analogy, sound is the medium by which we hear, but the objects of speech perception are the gestures that caused the patterns of pressure variation. Just as there can be visual mirages, where the brain "sees" a pattern it interprets as water even though it was caused by something else, such as heat rising from sand, there can be auditory mirages, where the brain "hears" something it interprets as a word even though it was caused by something else, such as computer-generated speech (Fowler 1990).

Recent findings that neurons in the motor cortex of the brain are activated when listening to speech (Wilson et al. 2004; Rizzolatti and Craighero 2004) are interpreted (e.g., by Goldstein et al. 2006) as evidence of a tight perception/production link consistent with the Motor Theory. Cheung et al. (2015) point out, however, that the existence of such a link does not in itself prove that the units of perception are articulatory gestures.

As Liberman and Whalen (2000) argue (consistent with proponents of Articulatory Phonology, e.g., Browman and Goldstein 1989), having a single representation for production, perception, and phonology obviates the need for any kind of mapping from one representation to the other, thus greatly simplifying the problem of linguistic representation. Those who espouse other theories would argue that it *over*simplifies linguistic representation, not taking sufficient account of the role of perceptual cues, algebraic features, segments, or any of the other linguistic representations that have been discussed in this text. A direct counterargument to the Motor Theory, in favor of other models of speech perception, is found in Massaro and Chen (2008), a reply to Galantucci et al. (2006). The debate is not over, however, and Motor Theory continues to have both its proponents and detractors.

Each of the models discussed in this section, whether based on prototypes, features, or gestures, has assumed that stored linguistic representations abstract away from variation and detail. These models all presuppose that the mapping from speech to lexicon involves discarding, disregarding, or compensating for variation. In contrast, exemplar models, discussed in Section 10.5, assume that lexical representations retain and reference high levels of detail.

10.5 EXEMPLAR THEORY

Exemplar Theory will be the final major theory introduced in this text. Compared to other theories discussed, it is relatively new, having only gained wide acceptance in the phonology/phonetics community from the late 1990s. It has only grown in importance in the last two decades, however, providing as it does new ways of thinking about and representing the relationship between the quantitative and categorical, the abstract and the specific. Johnson (2007) and Frisch (2018) provide overviews and introductions, which I draw on here.

10.5.1 DETAILED MEMORY TRACES

The fundamental hypothesis of Exemplar Theory is that the process of speech perception does not involve discarding details in order to match a stored, abstract, lexical form. Rather, the lexical form itself consists of a "cloud" of individual, detailed memory traces termed *exemplars*. Among the rich details that are stored in memory are context-specific and speaker-specific realizations. A category label may be added to the cloud of exemplars, but the label does not replace the instances.

Evidence that talker-specific information is retained in memory comes from experiments by Goldinger (1996, 1997, 2000) and others. For example, Goldinger (1996) played a list of words to participants and then sent them home. When the participants were tested a week later, they recalled words more accurately if they were played in the same voice as the original recording. The only way to make sense of that finding is if the memory of the lexical item and the voice were stored together.

Exemplars, sometimes called episodic memory traces, are linked together in memory. Exemplars that share characteristics—semantic, phonological, visual, emotional, talker-specific, and event-specific—form clusters in a multidimensional cognitive space. These clusters form the categories that we impose on the world—speech categories among them. Any new incoming stimulus is compared to those stored in memory. If it is similar to a particular set of exemplars, it will be labeled as a member of that category, and will be stored as a member of that cloud.

Exemplar Theory does not deny categorization. Categorization consists of labeling a cluster of exemplars. The crucial difference from other theories of abstract representations such as prototypes and distinctive features is that the label is in addition to, not a replacement of, the specific instances. Categories *consist* of varied instances, not abstraction away from them.

Exemplar Theory also does not presume that everything is remembered perfectly. Johnson (1997: 146) calls this the "head-filling-up" problem. While memory traces must be individual and detailed, they fade over time, and the cloud of exemplars is continuously updated with new stimuli. The more similar a new stimulus is to exemplars in the center of the cloud, the faster and more accurately it will be categorized. If new stimuli continue to be similar to old stimuli, the category label is reinforced and the boundaries between categories sharpen; if new stimuli are not very similar, but find a place at the "edge" of the cloud, the category boundary may shift.

Johnson (1997, 2005) was among the first linguists to apply Goldinger's findings to speech perception. Johnson (1997) created a corpus of vowel exemplars from thirty-nine different speakers, recording F0, formant, and duration values for each, without making any attempt to normalize among the speakers. The tokens were labeled with vowel identity and speaker sex. Johnson then pulled out individual tokens one at a time, and tested the ability of his model to correctly characterize the token, based on a formula for computing similarity to the other tokens in the dataset. The model's accuracy was 80 percent on vowel identification (comparable to human accuracy under similar conditions) and 99.6 percent for gender identification. Johnson's simple experiment showed that speech categorization by the computation of similarity, without explicit normalization, was possible.

What remains is to extend this simple model to more complex patterns of speech recognition (and to other phonological and phonetic domains, as discussed below). One immediately successful extension was to "typicality" effects. In a number of linguistic domains, it matters how typical an example of a category a given item is. This finds a natural explanation in Exemplar Theory, where similarity or typicality can be quantified, and an item can be placed either in the center or at the edge of a cloud. Strand (2000) found, for example, that words were identified faster if they were spoken in voices that were stereotypically male or female. This is explained straightforwardly if the most typical exemplars, those in the center of the cloud, have the highest level of activation, and items that are most similar to them are identified faster.

Exemplar Theory finds a natural extension to issues in sociophonetics. The links between exemplars encompass both social and acoustic information, providing a basis for modeling their interaction. For example, a speaker's dialect features become more pronounced when discussing topics that evoke strong and favorable memories of "home" (Nycz 2018). In Exemplar Theoretical terms, the social factors cause a higher activation of exemplars that are most closely linked to the place evoked, making those instances more likely to be selected for production.

Social factors can also drive language change. There will always be variation in the way lexical items are produced. But sometimes variation can shift in a certain direction, due to language contact or other social variables. If new exemplars of a category are frequently found at one edge of the current distribution, as the cloud is updated the center will shift. This category shifting over time will be recognized as language change (Wedel 2006; Pierrehumbert 2012; Hay et al. 2015).

Exemplar models also explain patterns of category acquisition and L1/L2 interaction that were discussed above. Kuhl's NLM model explicitly incorporates the idea that categories are built up out of repeated exposure to instances. The magnet effect is parallel to the typicality effect: the dense center of the exemplar cloud contains instances that are most highly activated (in Kuhl's terms, that have the highest neural commitment), and these exert the greatest effect in perception. Pierrehumbert (2003) discusses specifically how Exemplar Theory can model creation and maintenance of categories in child language acquisition.

Byun et al. (2016) apply Exemplar Theory to link child perception and production in their theory of the A-map (for "articulation"), parallel to Steriade's (2001) P-map (for "perception"). The A-map encodes memory traces of productions, linked to the resulting acoustic output. The hypothesis is that the speaker (a child in the data examined by Byun et al.) selects an exemplar from the A-map that has been successfully executed in the past, even if this doesn't exactly match the intended output. Thus the tension between articulatory "routine" and acoustic accuracy is modeled as finding the intersection between exemplars in the A-map and the P-map.

The work of Byun et al. (2016) shows the versatility of applying Exemplar Theory to a number of different problems in phonology, not just speech perception and categorization per se. If the mental lexicon consists of clouds of detailed exemplars, a number of interesting theories and results follow. For the phonology/phonetics interface, the two most important consequences of an exemplar approach to lexical representation is accounting for gradient generalizations and for frequency effects.

10.5.2 EXEMPLAR-BASED PHONOLOGY: STATISTICS IN THE LEXICON

It has long been recognized that speakers make generalizations over the patterns in their lexicons, such that they can judge if a nonce form is a possible word or not (Halle 1962). Halle assumed the choice was categorical: a nonce form would be an actual word, a possible word, or not a possible word. More recent research, however, has demonstrated that acceptability judgements are gradient: nonce forms can be more or less word-like (Coleman and Pierrehumbert 1997; Hay et al. 2003). If listeners are computing how typical or similar a word is to others in the lexicon, those computations will be reflected in gradient judgements.

Gradient judgements are directly related to frequency. Pierrehumbert (1994) and Frisch et al. (2004) showed that acceptability judgements of non-words were highly correlated with the frequency of the words' sub-parts. A word that Halle would judge as completely unacceptable, such as "mrupation," received a surprisingly high acceptability score. Even if the onset sequence was impossible, the whole word would be judged as not bad because the rest is very common.

Frequency effects are found throughout the phonological grammar (Bybee 2001, 2007). In addition to acceptability judgements, frequency of use affects production. High frequency words, for example, are more likely to undergo reduction (Bybee 2001; Pluymaekers et al. 2005; Gahl 2008, among many others). In Exemplar Theory, frequency effects on production are modeled by the constant updating of the exemplar cloud. As more reduced forms become part of the cloud, the center will move in the direction of those forms, making them more likely in subsequent productions. Such effects are not predicted if categories consist of unchanging abstract prototypes. Pierrehumbert (2002) discusses a number of cases of "word-specific phonetics" where allophonic details such as vowel reduction differ from word to word, arguing that these details must be captured as part of the lexical representation of specific words, something that would not be possible if lexical representations consist of phonological features only.

Daland et al. (2011), however, argue for a continued need for abstraction over phonological categories, even in exemplar models of phonology. Their experiments confirm that frequency accounts for differences in acceptability judgements between attested and marginal clusters. But listeners in their experiments still distinguished between clusters with zero frequency, finding that /zr/, for example, is better than /rg/. That judgement could not be made on the basis of segment or cluster frequency alone. They argue that models of frequency need to take into account not just segment sequences, but syllable structure and features that encode markedness and sonority relations. Frequency effects, and listener generalizations based on them, can be relevant at multiple levels and over different kinds of linguistic units.

Some recent work has explored the ramifications of Exemplar Theory and lexical statistics for the traditional problems of phonology. Cole (2009), for instance, applies Exemplar Theory to vowel harmony, arguing that the patterns of vowel agreement can be understood (and learned) as a pattern of statistical co-occurrence over items in the lexicon rather than as constraints on feature spreading. Hayes and Londe (2006) also consider vowel harmony as a kind of "stochastic phonological knowledge," a phrase that only makes sense if phonological representations are objects that are amenable to statistics. "Hungarian speakers," they write (2006: 99), "know not just the legal patterns of harmony, but also the frequency of these patterns, and they actively use this knowledge in guessing the harmonic behaviour of novel stems." Benus and Gafos (2007) consider the "transparent" (non-harmonizing) vowels of Hungarian: front vowels that take back suffixes rather than conditioning vowel harmony. They find that even when pronounced in isolation, words with non-harmonizing front vowels are produced with a more retracted tongue position than words with harmonizing front vowels. In isolation, this cannot be due to coarticulation at the time of utterance. Rather, Benus and Gafos argue that the lexical representation of these words has been affected by their repeated occurrence with back vowels—the exemplar cloud has been permanently shifted by repeated productions in a more back environment. Ernestus (2006, 2011) takes a similar approach to final obstruent voicing in Dutch, arguing that the lexical representation contains probabilistic information over instances where voicing does and does not occur. Just as words are not "possible" or "impossible" but more or less likely, Dutch obstruents are not [+voice] and [-voice], just more or less likely to be voiced.

Under these approaches, the lexicon is not a static repository, but a dynamic representation that includes probabilistic information (Pierrehumbert 2012). The generalizations that language users make in learning the phonology of their language, whether contrasts, phonotactics, or alternations, are made not over static representations, but over dynamic clouds of exemplars. This way of analyzing phonological patterns allows researchers to address new problems, and look at old problems in new ways. But Exemplar Theory doesn't solve everything.

10.5.3 QUESTIONS AND DIRECTIONS FOR FURTHER RESEARCH

As has been clear from the above discussion, Exemplar Theory crucially relies on computing similarity between tokens. Yet "similarity" is not necessarily clearly defined. Frisch (2018: 555) defines similarity as "proximity in a multidimensional space of features or concepts." But what are the dimensions over which similarity is computed in this highly multidimensional space? Frisch et al. (2004) demonstrate that similarity between segments can be computed over phonological features, but work that demonstrates the relevance of phonetic details and of sociophonetics shows that featural computation is not sufficient. Frisch (2018: 556) concludes that a truly comprehensive model of what similarity means is needed, and "that work has not, to date, been done."

As implied by the discussion of features and similarity, the adoption of an Exemplar Theory approach does not automatically solve the problem of what units and dimensions are being used in our mental computations (Johnson 2007; Daland et al. 2011). Johnson (1997), for example, assumes that exemplars are vowel tokens while Johnson (2007) assumes exemplars are words. Frequency effects seem to apply at all levels, from the phonetic cue to the multi-word phrase (Bybee 2007). The conclusion must follow that humans have the "headspace" to store stochastic information about multiple levels of representation simultaneously.

A final unsolved question, and an important area of current research, is the question of how categories are formed in the first place. What parameters are chosen to connect the clouds of exemplars, and how do we decide when a cloud is dense enough to merit a category label? The answer isn't obvious, but directions for research can come from what we know about learning in children. Children, and by extension adults in situations where new categories need to be learned, are statistical learners (e.g., Newport 2016). We can unconsciously compute distributions and deviations, and generalize over those distributions. The field of statistical learning of phonology is young and growing. Some recent publications with specific mathematical implementations of how categories can be induced from statistical distributions include Maye (2000), Pierrehumbert (2003), Goldsmith and Xanthos (2009), McMurray et al. (2009), Nosofsky (2014) and Pearl and Goldwater (2016).

Exemplar Theory is not in itself a theory of the phonology/phonetics interface. Rather, it is an approach that many researchers at the interface are pursuing. As Johnson (2007: 40) writes,

> There is no one exemplar-based phonology theory, but, rather, a number of nascent models seeking to use this class of memory models to help us better understand, among other things, phonological generalizations and the coexistence of gradience and categoriality in phonological knowledge.

Overall, the question of the units of speech perception is far from solved, and remains an area of research and debate. It is clear, however, that linguistic interpretation of the speech signal involves the integration of stored and incoming information at multiple levels of analysis.

10.6 PERCEPTION AND REPRESENTATION

What has our survey of perception at the phonology/phonetic interface shown us? One issue that has been at the center of the discussion has been the relationship between categorical and gradient information. In Exemplar Theory, both kinds of

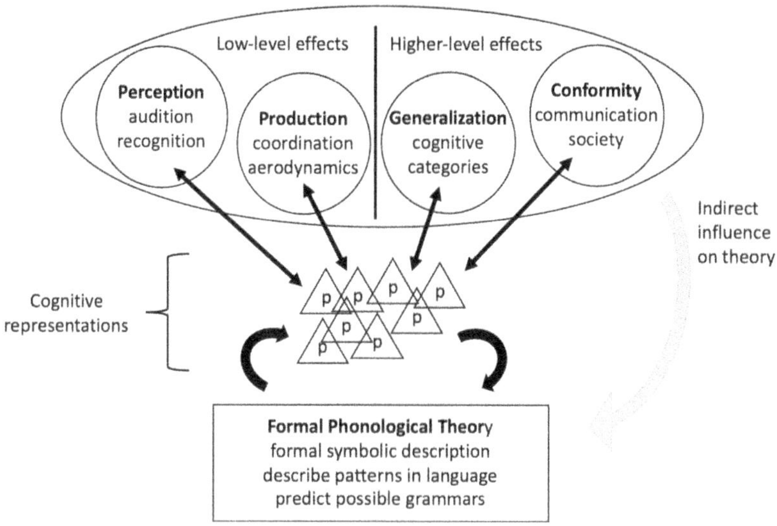

Figure 10.5 The influence of perception on phonology is indirect
Source: Adapted from Hume and Johnson (2001: 13).

information exist at the same time, and generalizations can be made at different levels of granularity (Pierrehumbert 2016).

Another perennial issue is the role of markedness in phonology. As was discussed in Chapter 6, some phonologists argue for a direct encoding of markedness constraints, including perceptual constraints, in the phonological grammar. Thus Boersma (1998, 2009) and Flemming (2002) argue for direct reference to F2 values in cue constraints or distance constraints, and these constraints directly interact with more abstract markedness and faithfulness constraints. Others, such as Blevins (2004) and Reiss (2018), argue that there is no role for synchronic markedness in the phonological grammar at all. Their arguments for a "substance-free" approach apply equally to production and perception.

Other phonologists argue for an indirect influence of perception on grammar. For Steriade (2001), the P-map, as a cognitive representation, mediates between perception and grammar. Hume and Johnson (2001) also argue for an indirect, mediated influence. Their model (2001: 13) is shown in Figure 10.5.

The central components of Hume and Johnson's model are the cognitive representations, which can be understood as exemplars. Formal phonological theory, which works through symbolic descriptions, consists of generalizations over the exemplars. (Again, the generalizations exist in addition to the exemplars but do not replace them.) Forces of perception and production, as well as the cognitive ability to generalize and the social need for effective communication, are general properties of human cognition and interaction, external to the grammar per se. (Note particularly that categorization is considered a general cognitive property and is not speech-specific.) These forces do, however, act on the cognitive representations.

Factors like ease of articulation and perceptibility of cues influence the production and perception of individual tokens, and thus the exemplar representations. Exemplars also encode, as discussed above, details of the communicative situation, which will also have an effect. Finally, the cognitive work of categorization (still not perfectly understood) influences the way the categories are created. Thus all of these factors influence grammatical patterns, but only indirectly through exemplar encoding, not directly in formal phonology.

Nonetheless, Hume and Johnson's approach is one of many. Chapter 11, concluding the text, now opens the field of survey even wider. We have looked, in the preceding ten chapters, at multiple perspectives on the phonology/phonetics interface, based on hundreds of experimental data points. Chapter 11 works towards summarizing the current state of our knowledge, and the current directions in which researchers are looking to expand the boundaries of what we know.

RECOMMENDED READING

Overviews

Johnson, K. (2013b), "Basic audition" and "speech perception," in *Acoustic and Auditory Phonetics*. Malden, MA and Oxford: Blackwell, Chapters 4 and 5.

- Describe the steps in the hearing and perception process. Which are specific to speech?

Hume, E. and K. Johnson (2001), "A model of the interplay of speech perception and phonology," in E. Hume and K. Johnson (eds), *The Role of Speech Perception in Phonology*. New York: Academic Press.

- How do Hume and Johnson define the "interplay" of the title? What is the role of speech perception in phonology?

Frisch, S. (2018), "Exemplar theories in phonology," in S. J. Hannahs and A. R. K. Bosch (eds), *The Routledge Handbook of Phonological Theory*. London: Routledge.

- List the different ways that Exemplar Theory can impact phonology.

Newport, E. L. (2016), "Statistical language learning: computational, maturational and linguistic constraints," *Language and Cognition*, 8: 447–61.

- What is statistical language learning, and what linguistic constraints is it subject to?

Exemplary research

Something old

Ohala, J. J. (1981), "The listener as a source of sound change," in C. S. Masek, R. A. Hendrick, and M. F. Miller (eds), *Papers from the Parasession on Language and Behavior*. Chicago: Chicago Linguistic Society, pp. 178–203.

- What is the evidence that the listener, not the speaker, is the source of sound change?

Ladefoged, P. and D. E. Broadbent (1957), "Information conveyed by vowels," *The Journal of the Acoustical Society of America*, 29: 98–104.

- Describe Ladefoged and Broadbent's experiment, results, and the implications for speech perception.

Something new

Kuhl, P. K., B. T. Conboy, S. Coffey-Corina, D. Padden, M. Rivera-Gaxiola, and T. Nelson (2008), "Phonetic learning as a pathway to language: new data and Native Language Magnet theory expanded (NLM-e)," *Philosophical Transactions of the Royal Society B, 363*: 979–1000.

- What do Kuhl et al. mean by "neural commitment"? How is the "magnet" effect related to exemplar theories?

Hay, J. B., J. B. Pierrehumbert, and M. Beckman (2003), "Speech perception, well-formedness, and the statistics of the lexicon," in J. Local, R. Ogden, and R. Temple, (eds), *Phonetic Interpretation: Papers in Laboratory Phonology VI*. Cambridge, UK: Cambridge University Press, pp. 58–74.

- What is the evidence for statistics in the lexicon?

Opposing views

Liberman, A. M., K. S. Harris, H. S. Hoffman, and B. Griffith (1957), "The discrimination of speech sounds with and across phoneme boundaries," *Journal of Experimental Psychology, 53*: 358–68.

vs

Johnson, K. and M. Babel (2010), "On the perceptual basis of distinctive features: evidence from the perception of fricatives by Dutch and English speakers," *Journal of Phonetics, 38*: 127–36.

- What is the influence of linguistic categories on speech perception? How has our understanding of categorical perception changed since 1957?

Galantucci, B., C. A. Fowler, and M. T. Turvey (2006), "The Motor Theory of speech perception reviewed," *Psychonomic Bulletin and Review, 13*: 361–77.

vs

Massaro, D. W. and T. H. Chen (2008), "The Motor Theory of speech perception revisited," *Psychonomic Bulletin and Review, 15*(2): 453–62.

- What is the evidence for and against the Motor Theory of speech perception? Which article do you find most convincing?

Pierrehumbert, J. B. (2002), "Word-specific phonetics," in C. Gussenhoven and N. Warner (eds), *Laboratory Phonology VII*. Berlin: De Gruyter Mouton, pp. 101–39.

vs

Daland, R., B. Hayes, J. White, M. Garallek, A. Davis, and I. Norrmann (2011), "Explaining sonority projection effects," *Phonology 28*(2): 197–234.

- What different views of the lexicon do the two articles propose? What role do exemplars play in each? Discuss the way the two articles deal with abstraction and detail.

QUESTIONS FOR FURTHER DISCUSSION

1. Think about the various "transformations" that occur in auditory speech processing, beginning with the ear canal itself, to the various neural processing steps that take place in different regions of the brain. Listen to some sine wave speech. Now

describe how each step in the speech perception process model in Figure 10.2 (from Stevens 2005) ultimately "transforms" sine waves into something you can recognize as speech.

2. Discuss some of the ways that perception influences phonology, including:

   Dispersion Theory;
   auditory enhancement;
   licensing by cue.

   Do these theories require "direct" reference of phonology to perception?
3. What is categorical perception? Discuss the evidence for and against it.
4. Why are studies of infant learners (e.g., Kuhl et al. 2008) and L2 learners (e.g., Flege 1995; Best 2001) so important to our understanding of speech perception? What do these studies provide that studies of adult native listeners can not?
5. Consider the different units of perception that are proposed in different theories discussed in this chapter—template, feature, segment, gesture. What is the evidence for each? Are they mutually exclusive?
6. How is phonological underspecification crucial to the FUL model of speech perception? How is this an example of phonology influencing speech perception?
7. What is the difference between a prototype model and an exemplar model? How does an exemplar model allow for the co-presence of categorical and gradient representation?
8. Explain Figure 10.5. In this model, how does perception influence phonology?

CHAPTER 11

CONCLUSIONS AND DIRECTIONS FOR FUTURE RESEARCH

The aim of Chapter 11 is to summarize the current state of our knowledge about the phonology/phonetics interface. In the course of ten chapters, what have we been able to conclude? To review, this first and longest section is formatted as a series of twenty questions that have been raised in the preceding chapters, and a brief summary of the possible answers that have been proposed. For details, of course, see the relevant chapters, which cannot be fully recapitulated here. These twenty chapter-based questions are followed in Section 11.2 by five overarching, more general questions that have come up repeatedly throughout the book, and apply across multiple areas of inquiry. Section 11.3 looks to the future: what avenues for future research seem most promising? Section 11.4 concludes with a bird's-eye view: what are the metaphors and graphical images that we have seen researchers use to illustrate the phonology/ phonetics interface, and how can we interpret those images from this end of our journey?

We begin with twenty questions.

11.1 QUESTIONS AND SOME POSSIBLE ANSWERS

1. *How do we define phonetics and phonology?*

Chapters 1 and 2 considered some big picture questions, beginning with the definitions of phonology and phonetics, and considering the history of the dichotomy. We started with a preliminary definition (p. 1):

> Phonology uses the methods of social science to study speech sounds as cognitive or psychological entities, which are abstract symbols that are discrete and categorical. Phonetics uses the methods of natural science to study speech sounds as physical entities, which are concrete and thus exist in actual space and time that are gradient and continuous.

The rest of the book has been about how well those definitions hold up.

2. *Are they both linguistics?*

The dichotomy between phonology and phonetics began with Ferdinand de Saussure and Jan Baudouin de Courtenay who, in defining the field of linguistics, walled off phonetic "substance" as being outside its proper scope, which concerned only the relation between form and meaning. For most of the twentieth century, linguists both structuralist and generative carried on the dichotomy, generally considering phonology to be linguistics and phonetics a mostly-uninteresting mechanistic realization of phonological categories. Yet as phonetic science advanced, and acoustic and articulatory information became both more detailed and more widely available, it became clear that phonetics could not be mechanical and automatic. Phoneticians needed phonology to know what to study, and phonologists found phonetic patterns to be of great interest as an explanation of why phonological patterns were the way they were.

The consensus towards the end of the twentieth century, as summarized, for example, by Pierrehumbert (1990b), was that phonology was about abstract mental representations, expressed in formal, algebraic rules, accessible to introspection. Phonetics, in contrast, was about measurable physical data, described by continuous mathematics, and not accessible to introspection. Both, however, are important parts of linguistics. Thirty years later, some linguists (e.g., Hale and Reiss 2008) continue to argue for a strict dichotomy along these lines, others (e.g., Port and Leary 2005) see no need for formal phonology at all, and still others (e.g., Pierrehumbert 2016) argue for a more holistic, integrated view. Most would argue that there is a role in the linguistic description of sound systems for both symbolic and quantitative levels of description, such that phonology and phonetics are distinct yet related. Chapters 3 through 10 asked what that relation looks like and how it plays out in different domains.

3. *What are the basic "building blocks" of spoken language? What are the categories and contrasts that phonology works with, and how are they phonetically defined?*

Chapter 3 looked at the question of inventories, and how they are built. The chapter began with a discussion of the "speech segment" as the default starting point for both phoneticians and phonologists, although the status of the segment as a valid unit of either phonetics or phonology has been questioned by some. Different writing systems take different units as "basic," codifying either morphemes, syllables/moras, or segments. While the syllable/mora distinction remains to be teased out, the evidence across a diverse set of languages and writing systems is that all three levels are valid. Indeed, the conclusion that "more than one level of description can be valid" is a recurring theme of the book.

Unlike writing systems developed for everyday communication, which may be more or less closely related to speech, phonetic transcription aims to accurately and explicitly represent pronunciation. Two transcription systems discussed in some detail were Bell's "Visible Speech" and the International Phonetic Alphabet (IPA). The IPA aspires to be a universal, language-neutral alphabet, able to express all possible contrastive sounds (broad transcription), and all systematic language-particular and context-dependent variation (narrow transcription). While Saussure had argued that phoneticians should avoid any reference to meaning, actual phoneticians found that in practice it was impossible to record every shade of detail, and that choices had to be made about what aspects to record. For linguists, the choice that made sense was to record only those differences between sounds that in fact created differences between words. No matter how narrow or broad, however, no transcription can ever be "language-neutral" and the IPA must be considered a tool rather than a theory.

4. *How do languages create an inventory?*

All languages select a particular inventory of sounds to combine to create words. If the inventory is too small, there aren't enough possible combinations to create a large set of lexical items. If the inventory is too large, distinctions get too hard to maintain. Dispersion Theory (Liljencrants and Lindblom 1972) argues that segments are chosen so as to be evenly spread out in the acoustic space. Quantal Theory (Stevens 1989) examines the interaction of articulation and acoustics in creating a system of stable contrasts. Hayes (1999) adds that formal symmetry and simplicity also constrain the structure of inventories, in addition to phonetic difficulty.

Phonological inventories are not chosen as random collections of segments. Rather, languages choose and reuse specific phonetic parameters, such as place of articulation, voicing, and nasality, creating symmetrical systems such as [p, t, k, b, d, g, m, n, ŋ] reusing the same places for the different consonant series. As Trubetzkoy wrote, "The phonemic inventory of a language is actually only a corollary of the system of distinctive oppositions. It should always be remembered that in phonology the major role is played, not by the phonemes, but by the distinctive oppositions" (1969: 67). The interaction of phonetic parameters and formal representations leads us to the idea of "distinctive" features based on contrast.

5. *What is universal across language modalities?*

Signed languages also have a phonology and a phonetics. Pressures of ease of articulation and the need for clear communication are equally relevant. The same question of "What are the basic building blocks of inventories?" also applies to languages in the visual mode. Signed languages are made of signs, of course, but signs are not "gestural wholes" any more than words are "vocal wholes." Several different theories of the components of signs have been proposed (e.g., Johnson and Liddell 2010; Brentari 2010), including different ways of describing handshape, location, and movement, but there is as yet no consensus on exactly what the "distinctive parameters" of signed languages are. Problems that occur in applying the techniques that work for spoken languages to signed languages include the facts that there are more degrees of freedom in the visual domain than in the articulatory domain, that there is more opportunity to express parameters simultaneously, and that iconicity plays a larger role in signed languages. Linguistic research on signed languages is much younger (and sparser) than research on spoken languages, but research on signed languages provides important opportunities for learning what is truly universal about the cognitive language faculty, and about the ways that the channel of communication constrains possible representations.

6. *What are the desiderata for a set of features?*

A large body of work at the phonology/phonetics interface has addressed the question of defining a universal set of possible "distinctive oppositions," to use Trubetzkoy's term, or distinctive features. Chapter 4 addressed the question of defining a set of distinctive features. Such work seeks to anchor possible phonological patterns of contrast and alternation in phonetic dimensions that are under sufficient articulatory control and that create sufficient acoustic difference. Clements et al. (2011: 6) suggest four criteria for feature systems:

1. phonetic motivation;
2. recurrence across languages;
3. formal simplicity;
4. comprehensiveness (accounting for the full set of data).

No feature set meets all these criteria perfectly, and they can at times be mutually exclusive, but they provide benchmarks for evaluating and comparing theories. Different theorists have emphasized different criteria in proposing their feature sets. What theories of distinctive features have been proposed?

7. What theories of distinctive features have been proposed?

A number of different theories of distinctive features have been proposed over the last 100 years. Sets of features discussed in Chapter 4 included:

- The multi-valued correlations proposed by Trubetzkoy in *Foundations of Phonology* (first published 1939), based on his analysis of the reported inventories of an impressive array of languages.
- The binary distinctive features of Jakobson et al. (1952). As required by the tenets of Information Theory, JFH proposed a very limited set of maximally efficient features, defined in terms of both articulation and acoustics.
- Halle (1959) and Chomsky and Halle (1968), who expanded on the JFH feature set to better account for natural classes in phonological rules in addition to contrastive inventories.
- Phonetically-based feature systems such as Halle and Stevens (1971) and Ladefoged (1980). Halle and Stevens mapped independently-controllable laryngeal configurations into possible phonological contrasts, while Ladefoged argued that because the needs and goals of phonology and phonetics are incommensurable, they need independent feature systems.
- Feature geometry. A graphic representation of feature groupings and dependencies, feature geometry brought phonological representation closer to phonetic representation, but in a definitely articulator-centric way. In this approach, the segment is de-emphasized, and features, independent of segments, become the basic units of phonological patterning.
- Dependency Phonology and Government Phonology, which propose a different set of unary "primes," offering an alternative to the JFH/SPE system.

Each of these systems has emphasized different criteria, some prioritizing phonetic transparency, others sacrificing phonetic specificity for greater formal simplicity or more general applicability.

8. What if there is no universal feature set?

Chomsky and Halle (1968) argued that their feature set was universal because it represented all the possible parameters that a speaker could control. Subsequent phonetic research (e.g., Keating 1985; Kingston and Diehl 1994) showed that this was not the case. Hayes (1999) suggested that features might not be genetically given, but might be universally induced from the interaction of articulatory difficulty and acoustic output that holds of every human vocal tract, combined with cognitive pressure for symmetry and efficiency. Others linguists however (e.g., Mielke 2008), argue that no feature set is in fact universally adequate, and that there is no need to require universality as a criterion: language users create whatever feature sets they need to account for the sometimes atypical data their language presents them with. That is, criterion #2 proposed by Clements et al. (2011), recurrence across languages, is eliminated, and criterion #4, comprehensiveness, is considered most important. Mielke (2008) found that any proposed universal feature set could account for no more than 92 percent of the "phonologically active classes" in his database. The open question is whether a grade of A- is passing (Cohn 2011). Debate continues on whether we should abandon the goal of finding a universal feature set and assume that individual languages/speakers create whatever features they need, or whether we should continue to refine a near-universal set to account for as much cross-linguistic patterning as possible.

9. *Does the translation from underlying representation to phonetic implementation proceed in stepwise fashion? If so, at what step is the dividing line between phonology and phonetics?*

Chapter 5 considered the concept of the derivation: the idea, proposed by Halle (1959) and Chomsky and Halle (1968), that rules apply in sequence to change abstract, underspecified, underlying representations into fully-specified phonetic representations. In addition to computational explicitness, an advantage of a derivational approach is that multiple intermediate levels, "temporary truths" as described by McCarthy (2007), can be defined. The open question is whether these intermediate levels are needed.

Derivational approaches will usually assume a strict modularity: different strata of the derivation have different properties, and different kinds of rules are assigned to different modules depending on their characteristics. Thus any given alternation has to be either a "phonological rule" or a "phonetic rule." The hypothesis is that the overall grammar will be simplified if alternations are assigned to the correct module. As Liberman and Pierrehumbert (1984: 229) write, assigning some rules to the phonetic module "removes from phonology the burden of representing those sound patterns for which its natural descriptive mechanisms are inappropriate," allowing for a simplified set of phonological features and a more constrained set of phonological rules.

The question then becomes, where exactly is the dividing line? Which are the sound patterns that are "appropriate" for phonology? Proposals for defining phonological rules included language-specificity (Chomsky and Halle 1968), structure preservation (Kingston and Diehl 1994), and categorical application (Keating 1996). Other work, however, from multiple points of view as described throughout this text, argue that neither the derivation (e.g., Prince and Smolensky 2004) nor strict modularity (e.g., Pierrehumbert 2016) should be adhered to.

10. *What is "markedness"?*

In *Foundations of Phonology* Trubetzkoy argued that for every phonological opposition, one pole would constitute the "marked" option, and the other pole would be the "unmarked" default. Theories of markedness, discussed in Chapter 6, ask what properties are universal or at least common across languages, and why. It turns out that there are very few absolute universals beyond "languages have vowels and consonants" (and even that is true only of spoken, not signed, languages, unless "movements" and "holds" are the signed correlates). Nonetheless, there are very many lopsided asymmetries (such as cross-linguistic preferences for voiceless stops and CV syllables) that phonologists and phoneticians would like to explain.

Defining "markedness" remains tricky, however (Haspelmath 2006). Relying on terms such as "easy" or "natural" risks circularity (Ohala 1990), and relying on implicational relations ("if a language has the marked pole X it will also have the unmarked pole Y") does not move toward an explanation of the pattern. Of crucial importance is the interplay of formal and functional approaches to markedness. As noted in Chapter 6, "To the extent that we can identify patterns that are common and 'natural,' do they arise because of properties of the language code itself (a 'formal' approach), properties of language use in the world (a 'functionalist' approach), or a combination?" See further discussion below.

11. What role does markedness play in phonology?
This is one specific way of asking some of the largest overarching questions of the phonology/phonetics interface. Is phonology natural? Can phonology have direct access to phonetics? Chapter 6 reviewed some of the principle approaches to defining the role of markedness in phonology.

- No role. Phonology is not "natural" (Hjelmslev 1953; Anderson 1981; Hale and Reiss 2008).
- A diachronic role only. Evolutionary Phonology (Blevins 2004) argues that pressures of markedness and naturalness play out in sound change, but should have no representation in synchronic phonology.
- A role in acquisition. Wilson (2006) suggests that phonetically natural alternations are more easily learned. Moreton and Pater (2012) argue, however, that "learning biases" have more to do with formal simplicity than with phonetic substance.
- Formal phonological constraints should encode phonetically-based markedness principles, but not refer to phonetics directly. This is the approach taken by Prince and Smolensky (2004) in Optimality Theory and by Hayes (1999) in Phonetically-driven Phonology. De Lacy (2006) and de Lacy and Kingston (2013) make explicit arguments in favor of the role of markedness in synchronic phonology.
- Phonology should have direct access to phonetic information (such as level of effort, quantitative cues, and precise timing) without the intervention of formalization. This approach, interleaving phonetic and phonological constraints, is taken by Kirchner (1998), Flemming (2002), Gafos (2002), and Boersma (1998, 2009), though from somewhat different theoretical perspectives.

Most researchers agree that markedness and phonetic grounding do play some role in phonology, but they disagree over how important and direct that role is. See further discussion in Sections 11.2 and 11.4.

12. How can we define the syllable?
Most phonologists and phoneticians agree that the syllable is an important organizational unit in spoken language, relevant for phonotactic distributions, allophonic

alternations, stress patterns, and speech timing. The syllable may be considered a foundational unit in a universal prosodic hierarchy that also includes the foot and higher levels of phrasing, as discussed in Chapter 7. The most widely accepted definition of the syllable is as a peak in sonority, defined as acoustic loudness or articulatory openness, although an exact definition of sonority, and thus of a precise sonority scale, remains a matter of debate (Parker and Lahiri 2012). Another approach (Nam et al. 2009) defines the syllable in terms of a pattern of stable articulatory coordination. The two approaches are not necessarily contradictory: Chitoran (2016) suggests that the articulatory and acoustic definitions of sonority will often make the same (correct) predictions.

13. *What is linguistic stress, and what are its phonetic correlates?*

Linguistic stress is a prominence relation between syllables. There is no one universal phonetic correlate of stress: pitch, duration, and loudness may all play a role in signaling prominence, and individual languages weight the cues differently. To the extent that there is phonological conditioning, stress patterns are largely determined by position in the word, but phonetic factors that make a syllable "heavy" (including vowel length, having a coda, and in some cases overall greater sonority) will in some languages attract the additional prominence of stress. Because prominence is necessarily relational, stress patterns rely on rhythmic alternations, and listeners may sometimes perceive a pattern of rhythmic prominence even if it is not produced.

14. *What are the phonetic cues to prosodic phrasing?*

The prosodic hierarchy proposes that syllables are grouped into stress feet, feet into words, and words into phrases. Phrasal grouping may be evidenced by assimilations across word boundaries, initial strengthening, medial lenition and final lengthening, as well as by intonational cues and pauses (Turk 2012). Some phonologists argue that the hierarchy is fixed and determined by reference to syntactic structure (e.g., Selkirk 2011). Others argue that a hierarchy of phrase types is epiphenomenal, and that gradient degrees of strengthening and lengthening give rise to the perception of phrase breaks of different strengths (e.g., Byrd and Saltzman 2003; Cho and Keating 2009). Discussion concerning the existence of an independently-defined prosodic hierarchy is thus another instance of the debate over the the usefulness of abstract, formal categories that express generalizations over phonetic facts.

15. *What are the linguistic uses of pitch? Can the linguistic and paralinguistic be separated?*

Analysis of the uses of pitch in speaking, including the formalization of pitch patterns and their relation to the segmental string, presents another case of the problem of relating the continuous and the discrete. As discussed in Chapter 8, "Tone" is defined as the use of pitch to create lexical contrast, and "intonation" is defined as the use of pitch to create discourse-level contrast. Pitch can also be used for non-linguistic communication, and some linguists (e.g., Bolinger 1986; Xu et al. 2015) have argued that the linguistic "discourse" and paralinguistic "emotional" aspects of intonation cannot be separated, although the consensus view among phonologists (Ladd 2008; Gussenhoven 2004) is that they must be.

While some aspects of intonation are arbitrary and language-specific, there are also broad generalizations that hold cross-linguistically, such as higher pitch to signal a question. Gussenhoven (2004) argues that three "biological codes," the frequency code, the effort code, and the production code, provide a physical ground from which the linguistic use of intonation abstracts. "In the absence of any motivation to the contrary," Gussenhoven writes, "the intonational morphemes of a language will reflect the universal form-function relations" (2004: 79). Thus biology forms a phonetic basis for intonational patterns, but linguistic use abstracts away from the biological conditioning.

16. How do we describe and transcribe an intonational tune?

All linguistic descriptions of intonation agree that the location and direction of pitch change are important to defining various tunes, but they differ in which aspects are considered crucial to contrastive linguistic meaning. While some phonologists who work on intonation (e.g., Cruttenden 1997) describe tunes holistically with terms such as "high fall," most researchers pursuing the relation between the phonology and phonetics of intonation adopt the autosegmental-metrical (AM) theory. The AM theory (Pierrehumbert 1990a) decomposes tunes into sequences of H and L autosegments that attach to metrically-prominent syllables or to phrase edges. Because of its phonetic explicitness, generalizability, and flexibility, AM provides consistent means of intonational transcription and modeling that has been successfully applied to many languages (Jun 2005, 2014). Issues still remain, however, in consistently applying AM labels to an F0 trace (Hualde and Prieto 2016), and in determining cross-linguistic principles of alignment between autosegments and segments (Prieto 2011).

17. What is a possible tone contrast?

As is the case with vowels and consonants, phonologists want to know and describe what is a possible inventory of tonal contrasts, what is a possible tonal alternation, and what tone patterns are marked and unmarked. Diacritics, numbers, and sketches in the form of "tone letters" can transcribe tones, but phonologists ask whether there is a set of "necessary and sufficient" tone features (Fromkin 1978; Clements et al. 2010) that will model phonological patterning. The consensus view (Yip 2002; Gussenhoven 2004) is that H and L autosegments, possibly in an elaborated feature geometry with intervening register nodes, are sufficient to capture contrasts and alternations and to allow for accurate phonetic implementation, but this view is not universal. Some advocate describing tone without a set of unary or binary distinctive features. Xu (2004) for example argues for more holistic shapes, and Hyman (2010) suggests using integers rather than features for tone levels. Some interesting new work (e.g., Karlin 2018) is being done on articulatory gestures for tone. Halle and Stevens (1971) define a set of laryngeal features that relate tone and voice quality, though much work remains to be done in studying the interaction vs independence of features for voice quality and pitch. More work on more diverse tone languages is needed to give a clearer picture of what the possible patterns of tone contrast *are*, so that theories of tone features can be effectively tested.

18. Can the articulatory gesture suffice for contrast, phonotactics, alternation, and articulator movement?

Chapter 9 gave an overview of the theory of Articulatory Phonology (AP; Browman and Goldstein 1986, 1989), which takes the articulatory gesture, "an abstract characterization of coordinated task-directed movements of articulators within the vocal tract," as the basic unit of both phonological patterning and phonetic instantiation. AP argues that lexical items contrast in the presence/absence, specification, or timing of gestures, while articulator movements derive from the inherent dynamic specification of each gesture and the coordination of gestural "constellations." Contextual variation, including allophonic generalizations and word-boundary alternations, are the result of changes in gestural overlap. There is no phonology to phonetics translation in AP, because the gesture suffices for both.

An important contribution of research in AP has been to show through acoustic and articulatory analysis that a number of alternations (such as t/d deletion, schwa epenthesis, and different kinds of assimilation) that were previously described as categorical and feature-changing are better reanalyzed as subject to gradient variation due to changes in articulatory organization. Not all alternations are gradient, however, and accounting for categorical alternations remains a challenge. Browman and Goldstein (1989) suggest that categorical differences should be accounted for through allomorph selection, though this treats what may be productive and "natural" alternations in the same way as less productive and "unnatural" ones. (Whether or not this is a problem depends on your view of what phonology is responsible for.) Some researchers (e.g., Ladd and Scobbie 2003) suggest retaining featural accounts for categorical alternations while implementing gestural accounts for gradient alternations, though this reinstitutes a phonology/phonetics split and undercuts one of the main claims of the theory. More recent developments modeling gestural coordination using a more constrained mathematics of "coupled oscillators" (Browman and Goldstein 2000) make it easier to model all-or-nothing changes, and suggest that theories of gestural organization could be expanded to capture both categorical and gradient alternations (Zsiga 2011).

Most work in AP has been on segmental alternations, especially consonant assimilations, but recent work on vowels, laryngeal contrasts, and tone (e.g., Karlin 2018; Zsiga 2018b) is expanding the empirical base. New work is also examining the possible relationship between the coupling graphs of AP and more traditional representations of autosegmental phonology, syllable structure, and larger prosodic domains (Goldstein et al. 2009; Zsiga 2011). It remains to be seen whether expanding the possibilities of changes in gestural coordination to include categorical alternations is a good idea or not.

19. What are the objects of perception?

Saussure defined the linguistic sign as linking a meaning to a "sound image in the brain" (1959). Later researchers in speech perception have asked what that "sound image" consists of. Chapter 10 described some of the multiple transfers and transformations (including logarithmic scaling, normalization, and cue integration) that the auditory signal undergoes between the ear and the auditory cortex. The chapter also described the ways in which perception is shaped, if not entirely determined, by the categories imposed by a language's system of phonological contrast (Liberman et al. 1957; Abramson and Lisker 1970). Contrastive categories are learned in early infancy (Kuhl et al. 2008), and once that "neural commitment" is made, learning new categories as an adult can be difficult (Best and Tyler 2007).

There are a number of different theories of the stored mental representation, the "sound image" to which the auditory signal is eventually mapped. Prototype approaches (e.g., Klatt 1989) assume that the incoming signal is compared against a stored idealized template. Feature-based theories (e.g., Stevens 2002; Lahiri and Reetz 2002) assume that mental representations consist of featural arrays for segments and words (as originally proposed by Jakobson et al. 1952). Cues to features are extracted from the incoming signal, and mapped into stored representations. Mapping of cues to features and features to words leads to word recognition. The Motor Theory of speech perception (Liberman et al. 1967; Liberman and Whalen 2000; Whalen 2019) hypothesizes that the objects of perception are articulatory gestures. Overall, the question of the units of speech perception remains an area of research and debate. It should be noted, however, that each of the models described here assumes that stored linguistic representations abstract away from variation and detail, and that that the mapping from speech to lexicon involves discarding, disregarding, or compensating for variation.

20. Is variation and detail noise to be discarded or information to be used?

In contrast to the models discussed above, Exemplar models, also discussed in Chapter 10, assume that lexical representations retain and reference high levels of detail. The basic hypothesis of Exemplar Theory (Johnson 2007) is that the stored lexical representation consists of a "cloud" of individual memory traces, termed *exemplars*, that retain rich speaker-specific and context-specific information. Clouds of exemplars may be mentally labeled as a category, but the label is in addition to the detailed traces in all their variability, not an abstract replacement of them.

Two of the most important successes of Exemplar Theory are accounting for frequency effects and gradient generalizations. If clouds of exemplars are updated with every new instance, the frequency with which an item is heard, and the contexts in which it is heard most frequently, will influence the representation. If listeners are computing how similar a word is to others in the cloud, those computations will be reflected in gradient judgements. On this approach, variation is information to be used, not discarded.

Challenges for Exemplar Theory include defining similarity (Frisch 2018) and determining how categories are formed and labeled in the first place. Adopting an Exemplar Theory approach does not automatically solve the problem of what stored units and dimensions are being used to compute similarity and frequency (Daland et al. 2011). As Johnson (2007) notes, Exemplar Theory is not itself a theory of the phonology/phonetics interface, but is a very promising approach to modeling the interaction of categoricality and gradience in phonological judgements.

These are some of the questions where phonology and phonetics intersect. These twenty questions have provided a review of the topics discussed in each of the ten preceding chapters of this book. They will inspire research articles and dissertations for years to come, continuing to test hypotheses and fill in the gaps in our knowledge. Yet there remain some larger-scale questions that were not confined to a single chapter, and inspire the research trajectories of entire careers. Some of these most important questions about the phonology/phonetics interface, as I see it, are addressed in Section 11.2.

11.2 SOME BIG UNANSWERED QUESTIONS

In the preceding sections, we looked at a number of important questions that arise in different domains at the phonology/phonetics interface, such as inventories, features, syllable structure, prosody, tone, and so on. From these more specific questions, there arise some big overarching issues that touch on all aspects of the phonology/phonetics interface. Here are five of them. (Or maybe just one big question asked five different ways.)

1. *Is perception or articulation more important for phonology?*

This is of course an unfair question, but one that has been lurking throughout the pages of the book. My own training has been on the articulatory side, and this may show in some of the topics I've selected to cover and in the order of presentation I've chosen, but I have tried not to let my own background tip the scales too much. Historically, as discussed in Chapter 10, structuralist models emphasized the hearer while most generative models emphasized the production side of the speech chain. Overall, the evidence is clear that phonological inventories and alternations are shaped both by forces of perception and by forces of articulation. To the extent that markedness constraints are incorporated into the grammar, they may be grounded in either perception or production. The two kinds of constraints may interact, sometimes favoring opposite outcomes, as in enhancing a contrast at the expense of additional effort, or losing a contrast to favor ease of articulation.

Theories of phonology that emphasize the role of perception in creating and maintaining inventories and alternations include Dispersion Theory (Liljencrants and Lindblom 1972), minimal distance constraints (Flemming 2005), auditory enhancement (Diehl and Kluender 1989), licensing by cue (Steriade 2001), and cue constraints (Boersma 1998, 2009). Theories that emphasize articulation include Articulatory Phonology obviously (Browman and Goldstein 1986), feature geometries that mirror vocal tract shape (McCarthy 1988), and Kirchner's (2004) constraints on articulatory effort. Some researchers (e.g., Boersma 1998, 2009; Boersma and Hamann 2009) make a definite effort to incorporate both perception and production, ending up with very complicated models, which may be what is, after all, necessary. Most phonological theories, however (e.g., Prince and Smolensky 2004), just allow for reference to both perception and production, without attempting to judge their relative importance or create a detailed model of how they interact.

2. *Is all phonology necessarily natural? If not, how abstract and "unnatural" can phonology be?*

To an extent, the answer to this question depends on what you mean by "phonology." It is certainly the case that not all contextual variation in the realization of morphemes has a clear phonetic basis (Anderson 1981). There is no synchronic phonetic motivation for tri-syllabic laxing in English, no aerodynamic reason for final consonants to voice in Lezgian, and no phonetic distinction between vowels that do and don't cause velar fronting in Icelandic. However, the less clearly "phonetic" an alternation is, the more likely that it will have other characteristics, such as lexical exceptions or lack of productivity, that suggest it might be better understood as morpheme suppletion, and thus outside the purview of phonology proper. Approaches that allow for only "natural" phonology (such as Articulatory Phonology, Browman and Goldstein 1986) assume that anything non-phonetic has to be morphology.

Those who would give a larger role to phonology in accounting for *all* predictable and productive contextual realization cannot argue that it is all "natural." For these theories, the question becomes one of emphasis, or of the preponderance of the evidence. Are phonological alternations in general phonetically based, with occasional special consideration for non-phonetic cases, or is phonology essentially independent of phonetics, with phonetic conditioning accidental, historical, and/or beside the point? Optimality Theory (Prince and Smolensky 2004) with its reliance on markedness constraints, and Phonetically-driven Phonology (Hayes 1999) with its theories of inductive grounding, lean to the former. Substance-free Phonology (Hale and Reiss 2008), Evolutionary Phonology (Blevins 2006), and Emergent Feature Theory (Mielke 2008) argue that phonology can be as abstract and far-removed from phonetics as it needs to be.

3. *Is phonology teleological? That is, is it goal-oriented? Are alternations and sound changes random, or are they optimized to meet the needs of speaker and hearer?*

This may be another way of asking the previous question, but it steps a little further away in considering that the way to ask the question is not just reference to phonetic principles in phonological rules, but more general optimization of communication. Dispersion Theory, Quantal Theory, and Phonetically-based Phonology propose specific pathways to optimizing phonological systems. Optimality Theory, obviously, holds to the tenet that phonology happens to make outputs better than inputs, on some dimension, though the theory can be flexible on what dimensions are available for reference. Evolutionary Phonology would argue that phonological systems are optimized over time (on the analogy of survival of the fittest) but that any particular synchronic innovation is random chance (on the analogy of genetic mutation). And finally, a more severe substance-free approach would hold that phonology is not teleological at all.

4. *How useful are formal representations?*

I have generally avoided using the terms "formalist" and "functionalist" in this text, as I think labeling an approach as one "ist" or the other is more likely to lead to an automatic dismissal or endorsement than to a consideration of the merits. Formalism and functionalism represent divergent ways of doing linguistics (not just phonology), turning on the issue of whether formal representations are basic to the enterprise, or a waste of time. Although the terms per se have been avoided, throughout this book the question of whether formal representations are accurate and useful ways to describe phonological patterning has come up often.

One way to think about Substance-free Phonology is that it is entirely "formalist." Phonological theory consists of formal, logic-like or algebraic rules, relations, and representations that make no reference to phonetic substance. Conversely, "functionalist" theories of phonology (e.g., Bybee 2001, 2007) eschew the constraints of formal representations and emphasize general principles of clear messaging and conservation of energy that arise because of the communicative contexts in which language is used. Like Ohala (1990), purely functionalist approaches see no added value in formal representations of communicative function. Like perception vs production, formalism vs functionalism can be viewed as a conflict that must be won by one side or the other. But at the phonology/phonetics interface, there must be room for both.

Languages do impose structure and categories on phonetic continua, and phonology references those structures and categories per se. To return to some quotes from previous chapters:

> To a phonetician whose work requires Laplace transforms or signal detection theory, treatments of phonology using the mathematically restricted resources of logic alone surely have a surprising degree of success, in terms of the coverage of the phenomena they address. This suggests that the cognitive system makes a simplified or impoverished use of the wealth and complexity of the phonetic domain—a very interesting fact about language which we have by no means gotten to the bottom of. (Pierrehumbert 2000: 9, quoted in Chapter 6)

> Not just in contemporary cognitive science, but implicitly since antiquity, the fundamental hypothesis has held that the mind is a serial computer that manipulates discrete structures built of abstract symbols . . . Going well beyond language and reasoning, modern computational work in higher visual perception, action planning, problem solving, and most other higher cognitive domains has similarly gained great explanatory leverage by analyzing cognition as arising from the manipulation of complex, abstract, symbolic structures. Such a symbolic cognitive architecture seems inescapable from the perspective of research on higher mental functions. (Smolensky and Legendre 2006: 6–7, quoted in Chapter 1)

> The only justification [for positing segments in the continuous speech stream] that can be given is a pragmatic one; such a description, based on a segmental structure imposed on the event by the analyst, has been the basis of virtually every result of note that has ever been obtained in the field of linguistic phonetics or phonology. (Anderson 1974: 6, quoted in Chapter 3)

These quotes argue that the best evidence for formal representations has been their success, coverage, and "explanatory leverage." These successes, across multiple domains, are taken as evidence that our "cognitive architecture" manipulates abstractions or symbols.

At the same time, there are equally-real phonetically-based markedness asymmetries, and clear evidence of forces of perception and production shaping phonological patterns, as has been discussed throughout this book. (Gordon 2007 gives an article-length overview of "functionalism in phonology.") Aerodynamic factors really do make voicing in obstruents difficult. Place of articulation contrasts really are better cued at release than at closure. Syllables that are already prominent through length or openness attract the additional prominence of high pitch. These phonetic forces are at work in shaping phonology, though whether the influence is diachronic or synchronic remains a matter of debate.

Whether the domain is the representation of tone and intonation, the emergence of distinctive features, the description of assimilation, or accounting for word-likeness judgements, for most researchers at the phonology/phonetics interface the question isn't whether formalism or functionalism wins, but of how formalism and functionalism interact, and how functional forces can be captured in formal terms.

5. *Are phonology and phonetics separate modules of grammar?*

If one takes a formalist approach to phonology, then phonetics *must* be a separate module. Following the tenets of modularity, aspects of sound patterning are apportioned to the module appropriate to their characteristics, according to our original definition: phonological alternations deal with cognitive entities that are discrete and categorical, represented with algebraic formalisms, while phonetic implementation deals with physical dimensions that are gradient and continuous, represented with continuous mathematics. This division of labor, which has been followed by phonologists and phoneticians for the past 100 or so years, still makes the most sense for many researchers. Theories can be more constrained and more predictive if their domains are delimited (Myers 2000), and there is no "greedy reductionist" attempt (Hayes 1999) to be a theory of everything.

For researchers such as Hayes (1999) and Myers (2000), as well as those who follow most versions of Optimality Theory, there is an influence of phonetics on phonology, but it is mediated through the process of phonologization. Phonological constraints may be based on phonetic principles (as intonational tunes may be based on emotional outcries), but they are subject to the cognitive constraints of categorization, abstraction, symmetry, and efficiency. The questions of exactly how that phonologization works, both synchronically and diachronically, at the level of the individual speaker/hearer/understander, and at the level of the social communicative system, are still in the process of being worked out.

On the other hand, if one takes a purely functionalist approach, with no use for algebraic formalism at all, there will be no need for a separate phonological module. As argued by Ohala (1990), "there is no interface between phonology and phonetics" because there is no need for a separate phonology at all. Everything interesting about linguistic sound patterns can be stated in purely phonetic terms.

There is room, however, for a more nuanced position. One can accept the need for both cognitive abstraction and physical description, without accepting a strict modularity between them. Pierrehumbert and colleagues (2000: 274), for example, as quoted in Chapter 2, assert that the cutting edge of linguistic research has "abandoned the doctrine of dualism" that requires a strict separation between mind and body. While praising, as quoted above, the usefulness of logical representations in phonology, Pierrehumbert goes on to say that "the large scale analyses made possible by recent technology, however, show that even core areas of phonology as not as categorical as more intuitive methods of data collection once suggested" (p. 9). That is (as Pierrehumbert 2016 also argues), analyses at different levels of "granularity" are needed, some more and some less abstract, but this does not necessitate a strict division between modules. As Pierrehumbert and colleagues (2000: 287) put it, again

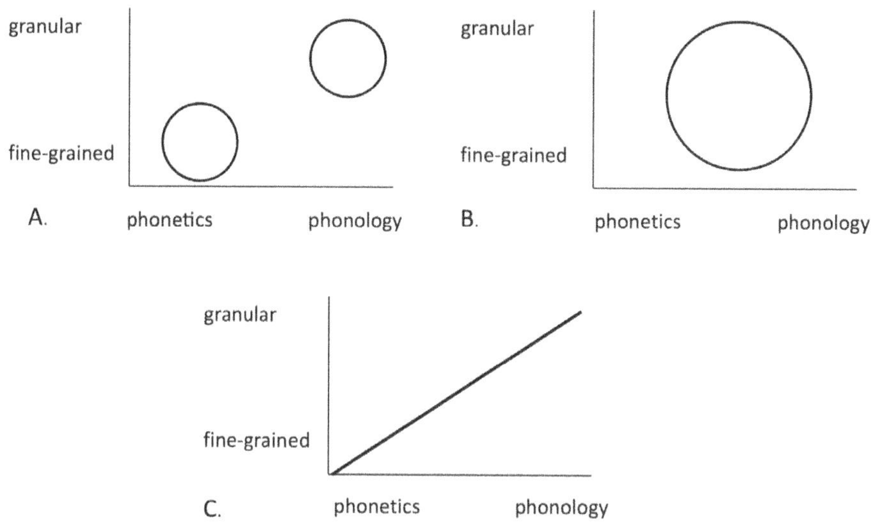

Figure 11.1 *Three different views of the phonology/phonetics interface*
Source: Adapted from Cohn (2006: 43).

as quoted in Chapter 6, "there is no particular point on the continuum from the external world to cognitive representations at which it is sensible to say that phonetics stops and phonology begins." Phonological patterns can be more or less phonetic, but there is no sharp, categorical division between modules.

These three different views of the relation between phonetics and phonology are illustrated in Figure 11.1, from Cohn (2006), and repeated from Chapter 5. In Figure 11.1A, phonetics is responsible for all "fine-grained" (that is, gradient and continuous) types of information, while phonology is responsible for all "granular" (that is, discrete and categorical) types of information, and there is no overlap between them. In Figure 11.1B, a single module is responsible for all levels of information, with no distinction between the properties of phonology and the properties of phonetics. Figure 11.1C represents the differentiated but gradient view proposed by Pierrehumbert (2000): highly granular information is handled by the phonology and highly fine-grained information is handled by the phonetics, but there are intermediate cases, and no definite modular division.

Which of these ways of viewing the interface is best and most useful continues to be one of the most important questions for both phonologists and phoneticians.

11.3 DIRECTIONS FOR FUTURE RESEARCH

Where do we go from here, to try to answer all these questions that have been raised? Some clear directions for further research have emerged from the foregoing discussion. In future research, we need to take the hypotheses that have been proposed, and test them against new data, using new technologies.

To begin with, our knowledge base needs to be expanded with more data from more diverse languages. As has been noted in several preceding chapters, English is far over-represented in our data. We need to move away from the Anglo- and Eurocentrism that characterized so much twentieth-century work, and that must still be guarded against. We must never take English to be our baseline, or assume that the structure of European languages is the default. All linguists at this end of the twenty-first century should take seriously the call to document the phonetics and phonology of endangered languages, while there are still speakers of these languages to record.

As part of increasing data diversity, studies of signed languages will make important contributions to understanding what is truly cognitively universal about human language, and how language structure is constrained by the physical channel in which it is expressed, questions at the heart of the phonology/phonetics interface.

Future studies will take advantage of new research technology. Magnetic resonance imaging (MRI) and electro-magnetic articulography (EMA) have replaced X-rays. Frame rates for all imaging techniques continue to increase, allowing for greater precision of measurement in both space and time. Phonologists and phoneticians are just beginning to take advantage of techniques to gather images and data directly from the brain, including functional (f)-MRI to measure blood flow in the brain and electroencephalography (EEG) with event-related potentials (ERP) to measure electrophysiological responses. As they did at the beginning of the twentieth century, advances in phonetic techniques are creating a wealth of new data on which to test phonological theories.

Alongside new imaging techniques are new computational tools. Corpora for different languages are becoming larger and more widely available, again providing fertile theoretical testing grounds. Increased computational power allows for more sophisticated mathematical modeling that gives us different ways of visualizing and abstracting over data (as in theories of coupled oscillators lead to new theories of gestural timing). As Pierrehumbert et al. (2000) remind us, mathematical models are also abstractions.

What would Daniel Jones, who in 1917 draped a lead chain over his tongue to X-ray its shape, think about EMA, and MRI, and portable sonography that can be carried to speakers anywhere in the world? What will researchers 100 years from now think of the images that we are able to create? What will they think of our questions, and our preliminary answers? We can only hope that our theories will be as enduring as those of Jones and his colleagues, while our understanding of and respect for speakers of languages around the world will be greater.

11.4 REVISITING THE METAPHORS

To conclude, let's circle back to the beginning.

We began in Chapter 1 with three metaphors for the phonology/phonetics interface (originally from Scobbie 2007). Some see the interface as a fence. This metaphor emphasizes modularity. Each sub-discipline must claim the acreage that properly belongs to it, and take care of its own property using its own tools. Property-line

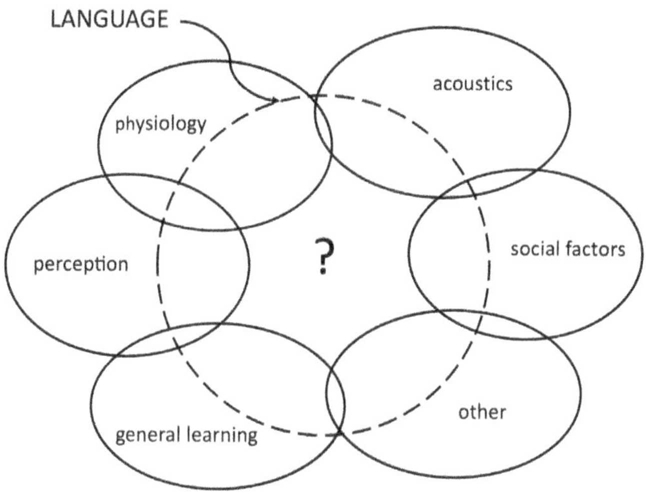

Figure 11.2 Linguistics should study what's left of language patterns when all the other explanations have been factored out

Source: Adapted from Anderson (1981: 494).

disputes inevitably ensue, either to claim greater territory, or perhaps to rake problematic foliage onto the other side of the fence. Some see the interface as a transducer. This metaphor emphasizes a derivational approach. Phonology has to be translated into phonetics. Work at the interface then emphasizes what that process of translation entails. Finally, some see the interface as a beach. There is sand and there is water, but there is also a tidal zone, where the influence of both is felt. Artificial fences or boundaries would soon be washed away, and the better approach is to study the interacting forces of water, wind, and sand in creating the overall landscape.

All good metaphors have some truth to them, and some shortcomings. They hopefully help us to think more clearly about large and important ideas. Another way to visualize big ideas is through diagrams, and here in conclusion I would like to revisit some more of the diagrams we've seen in previous chapters.

Figure 11.2, repeated from Chapter 6, diagrams Anderson's (1981) view of what linguistics (and therefore phonology) should be responsible for. This diagram, from Anderson's paper "Why phonology isn't 'natural'," represents the formalist view that phonology should account for only those aspects of contrast and alternation that are left after physiology, acoustics, perception, general learning, and so on have been factored out. The functionalist, of course, then contends that there is *nothing* left to be accounted for.

Figure 11.3, also from Chapter 6, suggests that form and function work together to determine the set of possible languages. In this diagram, from De Lacy and Kingston (2013), **C** indicates the space of possible languages allowed by cognitive constraints on grammatical formalism. **P** indicates the space of possible languages that could be arise from phonetic processes. Due to the interaction of different phonetic forces and to change over time, the languages in **P** will be more or less likely, as indicated by the

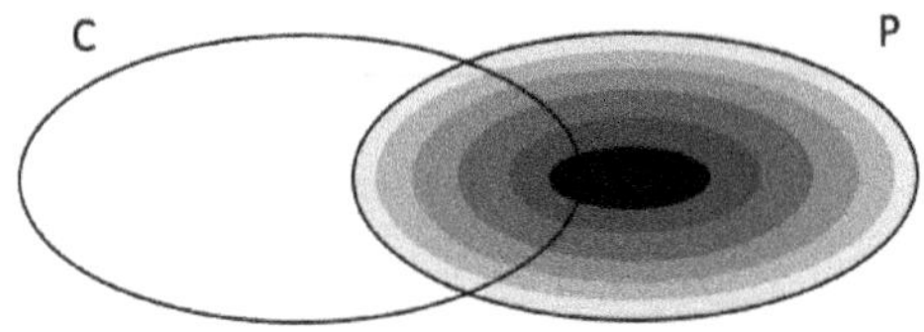

Figure 11.3 *The intersection of possible "formal" grammars ("C"), and possible "functional" grammars ("P")*

Source: Adapted from de Lacy and Kingston (2013: 343).

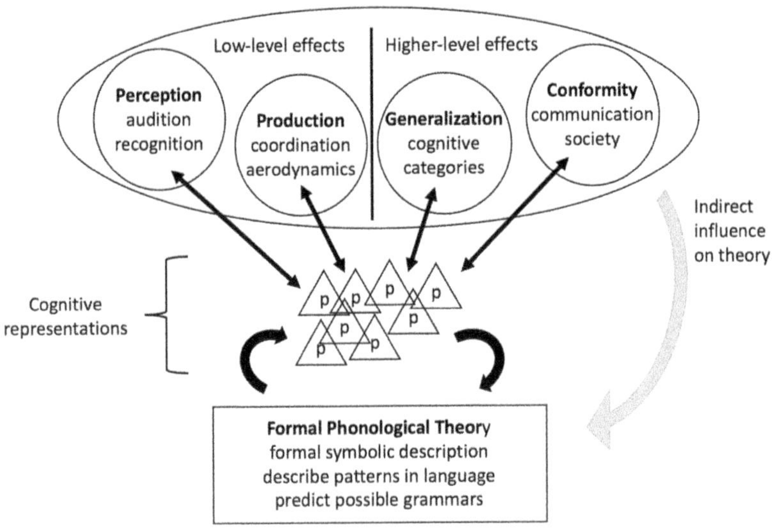

Figure 11.4 *The influence of perception on phonology is indirect*

Source: Adapted from Hume and Johnson (2001: 13).

darkness of the concentric circles. Not all languages in **C** are possible in **P**, and vice versa. Actually-occurring languages must fall in the intersection, jointly determined.

Figure 11.4, repeated from Chapter 10, gives another view of the interaction of different forces. In this figure, from Hume and Johnson (2001), the effects of functional forces (perception, production, generalization, and communicative needs) on the formal grammar are mediated by exemplar representations. Factors of language use cause differences in individual exemplar traces, which eventually cause changes in the overall shape of the cloud. Formal symbolic descriptions, which abstract over the exemplars, then respond to the changed shape. In this model, both detailed information and formal abstractions are co-present, but they are not collapsed. As in Figure 11.3, formal and functional constraints jointly determine the grammar, but the emphasis here is on the grammar of the individual speaker, rather than the set of possible grammars in the world. The challenge in pursuing this model is considering, for all the arrows, how that influence plays out in specific instances.

Figure 11.5 *Saussure's speech circuit*

Source: Saussure (1959: 11).

The final diagram, and metaphor, is the one we began with: Saussure's "speech circuit" between two interlocutors (Figure 11.5, repeated from Chapter 2).

Saussure defined five stages in every spoken communication. Repeating from Chapter 2:

- **Stage 1** is psychological. Interlocutor *A* thinks of something to say, and this mental concept "unlocks a corresponding sound image in the brain" (p. 11).
- **Stage 2** is physiological. The brain sends a signal to the speech organs to implement the sound image.
- **Stage 3** is physical. The speech organs move and send acoustic waves from the mouth of *A* to the ear of *B*.
- **Stage 4** is physiological again. Having received the acoustic information, the ear of *B* sends a signal corresponding to the sound image to his brain.
- **Stage 5** is psychological again: *B* associates the received sound image to the concept.

Nine chapters later, we can see that there are phonological and phonetic influences at every stage. Different theories emphasize different points. Algebraic phonology emphasizes Stages 1 and 5, as Saussure did: the crucial constraints on phonology are those of mental computation, and do not involve the physiological and physical at all. Articulatory Phonology, as the name states, emphasizes Stage 2. Evolutionary Phonology discounts production and perception factors in synchronic phonology, and allows only for very general cognitive principles that influence language, and therefore puts the emphasis on Stage 3, explaining phonological patterns in terms of transmission and channel bias. Finally, Exemplar Theory puts the emphasis on perception, Stage 4. Different feature theories, and different theories of markedness, seek to relate the stages to each other. All of the theories that we have discussed seek to find a path where the psychological, physiological, and physical converge. That convergence is the phonology/phonetics interface.

From this end of the book, we can look at Saussure's diagram again, and see it with deeper meaning and connection to all the theories and data we have discussed. And so we find that what T. S. Eliot wrote in "The Four Quartets" rings true:

We shall not cease from exploration
And the end of all our exploring
Will be to arrive where we started
And know the place for the first time.

It is my hope that this book will not be an end of exploring for readers, but a beginning. My hope is that that you will explore further, bringing new evidence to old debates, and asking and answering questions we haven't thought of yet.

RECOMMENDED READING

Overviews
Any of the works cited in this final chapter would be worth revisiting (or visiting, if you haven't read them yet). If you have a favorite, explain why.

Exemplary research
A fair amount of time is going to have elapsed between when I finish writing this paragraph and when you read it. Check out the latest issues of some of the journals that have been frequently cited here, for example:

Laboratory Phonology
Journal of Phonetics
Language
Language and Speech
Natural Language and Linguistic Theory
Phonology
What new research can you find that addresses the questions raised here? How does the new data move the conversation forward?

Opposing views
Find some textbooks on general linguistics, phonetics, and phonology, maybe some you've used in your own classes, and read the sections where they define phonetics and phonology. What unexamined assumptions do they make? What do they get right and what do they get wrong?

QUESTIONS FOR FURTHER DISCUSSION

1. What kinds of questions about the phonology/phonetics interface will further investigations of signed languages help us ask (and answer) that we are unable to ask/answer in examining spoken languages only? Select one of the "big unanswered questions" in Section 11.2 and think about how signed languages might help answer this question.
2. Can you think of ways some of the newer technologies ((f)-MRI, EMA, and ERP) can provide novel data on which to test phonological theories? Read up on the

technology that you find most interesting. How could you use that method to address one or more of the twenty questions in Section 11.1?

3. The book began with metaphors of the phonology/phonetics interface. Revisiting some of these metaphors after having read the intervening chapters, how do these metaphors help you to understand the different approaches and views of the phonology/phonetics interface? What truth does each of the metaphors reveal? What shortcomings? Which metaphor do you think is the best one?
4. Look at the diagram in Figure 11.1 depicting different views of understanding the relationship between phonology and phonetics. Do you think it is possible that there is some truth or accuracy in each of them, for example some domains where one depiction will be most apt and others where another depiction is more appropriate? In 100 years from now, do you think one of these will have "won out" as an approach to the interface?
5. If you could have lunch with any of the scholars cited in this book, living or not, who would it be, and why? What would you ask them if you did?
6. Consider the fact that some of the most interesting work at the phonology/phonetics interface has come about when scholars with different training and different intellectual backgrounds became collaborators, such as Jakobson, Fant, and Halle or Prince and Smolensky. Find a classmate that you've disagreed with on one or more of the topics of this book, and talk about what kind of data would be necessary to resolve your disagreement.
7. Take any question raised in this chapter, or any of the discussion questions throughout the book. Design an experiment that would move us closer to an answer. Now go carry it out, and let us all know what you learn.

REFERENCES

Abramson, A. (1978), "Static and dynamic acoustic cues in distinctive tones," *Language and Speech, 21*(4): 319–25.

Abramson, A. (1979), "The coarticulation of tones: an acoustic study of Thai," in T. I. Thongkum, P. Kullivanijaya, V. Panupong, and K. Tingsabadh (eds), *Studies in Tai and Mon-Khmer Phonetics and Phonology in Honour of Eugenie J. A. Henderson.* Bangkok: Indigenous Languages of Thailand Research Project, pp. 1–9.

Abramson, A. and L. Lisker (1970), "Discriminability across the voicing continuum: cross-language tests," in H. Bohuslav, M. Romportl, and P. Janota (eds), *Proceedings of the 6th International Conference of Phonetic Sciences.* Prague: Academia Publishing House of the Czechoslovak Academy of Sciences. pp. 569–73.

Abramson, A. and L. Lisker (1973), "Voice-timing perception in Spanish word-initial stops," *Journal of Phonetics, 1*: 1–8.

Abramson, A., P. Nye, and T. Luangthongkum (2007), "Voice register in Khmu: experiments in production and perception," *Phonetica, 64*: 80–104.

Adger, D. (2007), "Stress and phasal syntax," *Linguistic Analysis, 33*: 238–66.

Anderson, S. R. (1974), *The Organization of Phonology.* New York: Academic Press.

Anderson, S. R. (1975), "On the interaction of phonological rules of various types," *Journal of Linguistics, 11*: 39–62.

Anderson, S. R. (1978), "Tone features," in V. Fromkin (ed.), *Tone: A Linguistic Survey.* New York: Academic Press, pp. 133–73.

Anderson, S. R. (1981), "Why phonology isn't 'natural'," *Linguistic Inquiry, 12*: 493–539.

Anderson, S. R. (1985a), *Phonology in the Twentieth Century: Theories of Rules and Theories of Representations.* Chicago: University of Chicago Press.

Anderson, S. (1985b), "Generative phonology and its origins," in *Phonology in the Twentieth Century: Theories of Rules and Theories of Representations.* Chicago: University of Chicago Press, Chapter 12.

Andruski, J. E. and M. Ratliff (2000), "Phonation types in production of phonological tone: the case of Green Mong," *Journal of the International Phonetic Association, 30*: 37–61.

Archangeli, D. and D. Pulleyblank (1994), *Grounded Phonology.* Cambridge, MA: MIT Press.

Archangeli, D. and D. Pulleyblank (2015), "Phonology without universal grammar," *Frontiers in Psychology, 6*: 1229.

Armstrong, L. E. and I. C. Ward [1926] (1931), *Handbook of English Intonation.* Cambridge, UK: Heffner.

Aronoff, M. (1985), "Orthography and linguistic theory: the syntactic basis of Masoretic Hebrew punctuation," *Language, 61*: 28–72.

Arvaniti, A. (2012), "Segment-to-tone association," in A. C. Cohn, C. Fourgeron, and M. K.

Huffman (eds), *The Oxford Handbook of Laboratory Phonology*. Oxford: Oxford University Press, pp. 265–74.

Arvaniti, A. and D. R. Ladd (2009), "Greek wh-questions and the phonology of intonation," *Phonology, 26*: 43–74.

Arvaniti, A., D. R. Ladd, and I. Mennen (2000), "What is a starred tone? Evidence from Greek," in M. Broe and J. Pierrehumbert (eds), *Papers in Laboratory Phonology V*. Cambridge, UK: Cambridge University Press, pp. 119–31.

Arvaniti, A., D. R. Ladd, and I. Mennen (2006), "Phonetic effects of focus and 'tonal crowding' in intonation: evidence from Greek polar questions," *Speech Communication, 48*: 667–96.

Avery, P. and W. J. Idsardi (2001), "Laryngeal dimensions, completion and enhancement," in T. A. Hall and U. Kleinhanz (eds), *Studies in Distinctive Feature Theory*. Berlin: De Gruyter Mouton, pp. 41–70.

Bach, E. and R. T. Harms (1972), "How do languages get crazy rules?" in R. Stockwell and R. Macaulay (eds), *Linguistic Change and Generative Theory*. Bloomington: Indiana University Press, pp. 1–21.

Ballachanda, B. B. (1997), "Theoretical and applied external ear acoustics," *Journal of the American Academy of Audiology, 8*: 411–20.

Bang, Y., M. Sonderegger, Y. Kang, M. Clayards, and T.-J. Yoon (2018), "The emergence, progress, and impact of sound change in progress in Seoul Korean: implications for mechanisms of tonogenesis," *Journal of Phonetics, 66*: 120–44.

Barrie, M. (2007), "Contour tones and contrast in Chinese languages," *Journal of East Asian Linguistics, 16*: 337–62.

Barry, M. (1992), "Palatalisation, assimilation and gestural weakening in connected speech," *Speech Communication, 11*: 393–400.

Battison, R. (1978), *Lexical Borrowing in American Sign Language*. Silver Spring, MD: Linstok Press.

Battison, R. (1980), "Signs have parts: a simple idea," in C. Baker and R. Battison (eds), *Sign Language and the Deaf Community*. Silver Spring, MD: National Association of the Deaf, pp. 35–51.

Baudouin de Courtenay, J. N. (1870), *O drevne-pol'skom jazyke do XIV-go stoletija* [*On the Old Polish Language before the 14th Century*]. Leipzig, Germany: Behr and Hermann.

Baudouin de Courtenay, J. N. (1927), "Ilościowość w myśleniu językowym" ["Quantity as a dimension of thought about language"], *Symbolae gramaticae in honorem Joannis Rozwadowski*, vol. I. Kraków, Poland: Uniwersytet Jagiełłoński, pp. 3–18.

Becker, A. (2018), "The effect of iconicity on phonological and phonetic processes in American Sign Language," unpublished manuscript, Georgetown University.

Beckman, M. E. and J. Edwards (1992), "Intonational categories and the articulatory control of duration," in Y. Tohkura, E. Vatikiotis-Bateson, and Y. Sagisaka (eds), *Speech Perception, Production and Linguistic Structure*. Tokyo: OHM Publishing Co., Ltd.

Beckman, M. E. and G. A. Elam (1994), "Guidelines for ToBI labelling," <http: //www.ling.ohio-state.edu/research/phonetics/E_ToBI/> (last accessed March 3, 2020).

Beckman, M. E. and J. Hirschberg (1994), "The ToBI annotation conventions," <http://www.cs.columbia.edu/~julia/files/conv.pdf> (last accessed April 3, 2020).

Beckman, M. E., J. Hirschberg, and S. Shattuck-Hufnagel (2005), "The original ToBI system and the evolution of the ToBI framework," in S.-A. Jun (ed.), *Prosodic Typology—The Phonology of Intonation and Phrasing*. Oxford: Oxford University Press, pp. 9–54.

Bell, A. M. (1867), *Visible speech, the science of universal alphabetics: Or self-interpreting physiological letters, for the writing of all languages in one alphabet*. London, England: Simpkin, Marshall & Co.

Benua, L. (1997), *Transderivational identity: Phonological relations between words.* Doctoral dissertation, University of Massachusetts Amherst.

Benus, S. and A. I. Gafos (2007), "Articulatory characteristics of Hungarian 'transparent' vowels," *Journal of Phonetics, 35*(3): 271–300.

Berent, I. (2013), *The Phonological Mind.* Cambridge, UK: Cambridge University Press.

Berent, I., D. Steriade, T. Lennertz, and V. Vaknin (2007), "What we know about what we have never heard: evidence from perceptual illusions," *Cognition, 104*: 591–630.

Berent, I., T. Lennertz, P. Smolensky, and V. Vaknin-Nusbaum (2009), "Listeners' knowledge of phonological universals: evidence from nasal clusters," *Phonology, 26*: 75–108.

Bernstein, L. E. (2005), "Phonetic processing by the speech perceiving brain," in D. B. Pisoni and R. E. Remez (eds), *The Handbook of Speech Perception.* Oxford: Blackwell, pp. 79–98.

Best, C. T. (1995), "A direct-realist perspective on cross-language speech perception," in W. Strange (ed.), *Speech Perception and Linguistic Experience: Issues in Cross-language Research.* Baltimore: York Press, pp. 167–200.

Best C. T. and M. D. Tyler (2007), "Nonnative and second-language speech perception: commonalities and complementarities," in M. Munro and O.-S. Bohn (eds), *Second Language Speech Learning.* Amsterdam: John Benjamins, pp. 13–34.

Best, C. T., G. W. McRoberts, and N. M Sithole (1988), "Examination of perceptual reorganization for nonnative speech contrasts: Zulu click discrimination by English-speaking adults and infants," *Journal of Experimental Psychology: Human Perception and Performance, 14*: 345–60.

Best, C. T., G. W. McRoberts, and E. Goodell (2001), "Discrimination of non-native consonant contrasts varying in perceptual assimilation to the listener's native phonological system," *The Journal of the Acoustical Society of America, 109*: 775–94.

Bickmore, L. (1995), "Tone and stress in Lamba," *Phonology, 12*: 307–41.

Blevins, J. (1995), "The syllable in phonological theory," in J. Goldsmith (ed.), *Handbook of Phonological Theory*, London: Basil Blackwell, pp. 206–44.

Blevins, J. (2004), *Evolutionary Phonology: The Emergence of Sound Patterns.* Cambridge, UK: Cambridge University Press.

Blevins, J. (2006), "A theoretical synopsis of Evolutionary Phonology," *Theoretical Linguistics, 32*: 117–65.

Blicher, D., R. Diehl, and L. Cohen (1990), "Effects of syllable duration on the perception of Mandarin tone 2/tone 2 distinction: evidence of auditory enhancement," *Journal of Phonetics, 18*: 37–49.

Bloch, B. (1941), "Phonemic overlapping," *American Speech, 16*: 278–84.

Bloch, B. (1948), "A set of postulates for phonemic analysis," *Language, 24*: 3–46.

Bloomfield, L. (1926), "A set of postulates for the study of language," *Language, 2*: 153–64.

Bloomfield, L. (1933), *Language.* New York: Holt, Rinehart, and Winston.

Bloomfield, L. (1939), "Menomini morphophonemics," *Travaux du cercle linguistique de Prague, 8*: 105–15.

Blust, R. A. (1973), "A double counter-universal in Kelabit," *Working Papers in Linguistics, 5*(6): 49–56.

Boersma, P. (1998), *Functional Phonology: Formalizing the interactions between articulatory and perceptual drives.* Doctoral dissertation, University of Amsterdam.

Boersma, P. (2007), "Some listener-oriented accounts of *h*-aspiré in French," *Lingua, 117*: 1989–2054.

Boersma, P. (2008), "Emergent ranking of faithfulness explains markedness and licensing by cue," ROA-954, Rutgers Optimality Archive, <http://roa.rutgers.edu/files/954-0308/954-BOERSMA-0-0.PDF> (last accessed March 3, 2020).

Boersma, P. (2009), "Cue constraints and their interactions in phonological perception and production," in P. Boersma and S. Hamann (eds), *Phonology in Perception*. Berlin: De Gruyter Mouton, pp. 55–110.

Boersma, P. and S. Hamann (2008), "The evolution of auditory dispersion in bidirectional constraint grammars," *Phonology, 25*: 217–70.

Boersma, P. and S. Hamann (2009), "Introduction: models of phonology in perception," in P. Boersma and S. Hamann (eds), *Phonology in Perception*. Berlin: De Gruyter Mouton, pp. 1–24.

Boersma, P. and J.-W. van Leussen (2017), "Efficient evaluation and learning in multi-level parallel constraint grammars," *Linguistic Inquiry, 48*: 349–88.

Bolinger, D. L. (1951), "Intonation: levels versus configurations," *Word, 7*: 199–210.

Bolinger, D. L. (1958), "A theory of pitch accent in English," *Word, 14*: 109–49.

Bolinger, D. L. (1964), "Around the edges of language: intonation," *Harvard Educational Review, 34*: 282–93.

Bolinger, D. (1986), *Intonation and Its Parts: Melody in Spoken English*. Stanford: Stanford University Press.

Borowsky, T., S. Kawahara, M. Sugahara, and T. Shinya (eds) (2012), *Prosody Matters: Essays in Honor of Elisabeth Selkirk*. Sheffield: Equinox Publishing.

Boyce, S. E. (1990), "Coarticulatory organization for lip rounding in Turkish and English," *The Journal of the Acoustical Society of America, 88*: 2584–95.

Bradley, T. G. (2007), "Morphological derived-environment effects in gestural coordination: a case study of Norwegian clusters," *Lingua, 117*(6): 950–85.

Bradlow, A. R. (1993), *Language-specific and universal aspects of vowel production and perception: a cross-linguistic study of vowel inventories*. Doctoral dissertation, Cornell University.

Braver, A. (2014), "Imperceptible incomplete neutralization: production, non-identifiability, and discriminability in American English flapping," *Lingua, 152*: 24–44.

Bregman, A. S. (1990), *Auditory Scene Analysis*. Cambridge, MA: MIT Press.

Brentari, D. (1998), *A Prosodic Model of Sign Language Phonology*. Cambridge, MA: MIT Press.

Brentari, D. (ed.) (2010), *Sign Languages: A Cambridge Language Survey*. Cambridge, UK: Cambridge University Press.

Browman, C. P. (1994), "Lip aperture and consonant releases," in P. Keating (ed.), *Phonological Structure and Phonetic Form: Papers in Laboratory Phonology III*. Cambridge, UK: Cambridge University Press, pp. 331–53.

Browman, C. P. (1995), "Assimilation as gestural overlap: comments on Holst and Nolan," in B. Connell and A. Arvaniti (eds), *Phonology and Phonetic Evidence: Papers in Laboratory Phonology IV*. Cambridge, UK: Cambridge University Press, pp. 334–42.

Browman, C. P. and L. Goldstein (1986), "Towards an Articulatory Phonology," *Phonology Yearbook, 3*: 219–52.

Browman, C. P. and L. Goldstein (1988), "Some notes on syllable structure in Articulatory Phonology," *Phonetica, 45*: 140–55.

Browman, C. P. and L. Goldstein (1989), "Articulatory gestures as phonological units," *Phonology, 6*(2): 201–51.

Browman, C. P. and L. Goldstein (1990a), "Gestural specification using dynamically-defined articulatory structures," *Journal of Phonetics, 18*: 299–320.

Browman, C. P. and L. Goldstein (1990b), "Representation and reality: physical systems and phonological structure," *Journal of Phonetics, 18*(3): 411–24.

Browman, C. P. and L. Goldstein (1990c), "Tiers in Articulatory Phonology, with some implications for casual speech," in J. Kingston and M. E. Beckman (eds), *Papers in Laboratory*

Phonology I: Between the Grammar and Physics of Speech. Cambridge, MA: Cambridge University Press, pp. 341–76.
Browman, C. P. and L. Goldstein (1992a), "Articulatory Phonology: an overview," *Phonetica, 49*: 155–80.
Browman, C. P. and L. Goldstein (1992b), "Targetless schwa: An articulatory analysis," in G. J. Docherty and D. R. Ladd (eds), *Papers in Laboratory Phonology II: Gesture, Segment, Prosody*. Cambridge, UK: Cambridge University Press. pp. 26–56.
Browman, C. P. and L. Goldstein (2000), "Competing constraints on intergestural coordination and self-organization of phonological structures," *Bulletin de La Communication Parlée, 5*: 25–34.
Brown, J. C. (2004), "Eliminating the segmental tier: evidence from speech errors," *Journal of Psycholinguistic Research, 33*(2): 97–101.
Bruce, G. (1977), *Swedish Word Accents in Sentence Perspective*. Lund: Gleerup.
Bruce, G. (1990), "Alignment and composition of tonal accents: comments on Silverman and Pierrehumbert," in J. Kingston and M. Beckman (eds), *Papers in Laboratory Phonology I*. Cambridge, UK: Cambridge University Press, pp. 107–14.
Brunelle, M. (2009), "Tone perception in Northern and Southern Vietnamese," *Journal of Phonetics, 37*: 79–96.
Buckley, G. (2006), "Syllabic writing and syllable structure," presentation slides, April, <http://www.ling.upenn.edu/~gene/papers/syllabaries.pdf> (last accessed March 3, 2020).
Burzio, L. (2005), *Principles of English Stress*. Cambridge, UK: Cambridge University Press.
Bybee, J. (2001), *Phonology and Language Use*. Cambridge, UK: Cambridge University Press.
Bybee, J. (2007), *Frequency of Use and the Organization of Language*. Oxford: Oxford University Press.
Byrd, D. (1995), "C-centers revisited," *Phonetica, 52*: 285–306.
Byrd, D. (1996), "Influences on articulatory timing in consonant sequences," *Journal of Phonetics, 24*(2): 209–44.
Byrd, D. and E. Saltzman (1998), "Intragestural dynamics of multiple phrasal boundaries," *Journal of Phonetics, 26*: 173–99.
Byrd, D. and E. Saltzman (2003), "The elastic phrase: modeling the dynamics of boundary-adjacent lengthening," *Journal of Phonetics, 31*(2): 149–80.
Byrd, D., A. Kaun, S. Narayanan, and E. Saltzman (2000), "Phrasal signatures in articulation," in M. B. Broe and J. B. Pierrehumbert (eds), *Papers in Laboratory Phonology V*. Cambridge, UK: Cambridge University Press, pp. 70–87.
Byrd, D., S. Tobin, E. Bresch, and S. Narayanan (2009), "Timing effects of syllable structure and stress on nasals: a real-time MRI examination," *Journal of Phonetics, 37*(1): 97–110.
Byun, T. A., S. Inkelas, and Y. Rose (2016), "The A-map: articulatory reliability in child-specific phonology," *Language, 92*: 141–78.
Cairns, C. and E. Remy (2010), *Handbook of the Syllable*. Leiden: Brill.
Casali, R. (2008), "ATR harmony in African languages," *Language and Linguistics Compass, 2/3*: 496–549.
Chahal, D. and S. Hellmuth (2014), "The intonation of Lebanese and Egyptian Arabic," in S.-A. Jun (ed.), *Prosodic Typology II: The Phonology of Intonation and Phrasing*. Oxford: Oxford University Press.
Chao, Y. (1930), "A system of 'tone letters'," *Le Maître Phonétique, 45*: 24–7.
Chebanne, A. (2000), "The Sebirwa language: a synchronic and diachronic account," *Pula: Botswana Journal of African Studies, 14*: 186–95.
Chen, L. (2003), "The origins in overlap of place assimilation," paper presented at the XXIIth West Coast Conference of Formal Linguistics, San Diego, CA.

Cherry, E. C. (1953), "Some experiments on the recognition of speech, with one and two ears," *The Journal of the Acoustical Society of America, 25*: 975–9.

Cheung, H., H.-C. Chen, C. Y. Lai, O. C. Wong, and M. Hills (2001), "The development of phonological awareness: effects of spoken language experience and orthography," *Cognition, 81*: 227–41.

Cheung, C., L. Hamilton, K. Johnson, and C. F. Chang, (2015), "The auditory representation of speech sounds in the human motor cortex," *eLife* 2016;5, e12577.

Chitoran, I. (2016), "Relating the sonority hierarchy to articulatory timing patterns: a cross-linguistic perspective," in M. J. Ball and N. Müller (eds), *Challenging Sonority: Cross-linguistic Evidence*. Sheffield, UK and Bristol, CT: Equinox Publishing, pp. 45–62.

Chitoran, I., L. Goldstein, and D. Byrd (2002), "Gestural overlap and recoverability: articulatory evidence from Georgian," in C. Gussenhoven and N. Warner (eds), *Papers in Laboratory Phonology, 7*. Berlin: De Gruyter Mouton, pp. 419–47.

Cho, T. (2006), "Manifestation of prosodic structure in articulation: evidence from lip kinematics in English," in L. Goldstein, D. H. Whalen, and C. T. Best (eds), *Laboratory Phonology, 8*. Berlin: De Gruyter Mouton, pp. 519–48.

Cho, T. and P. Keating (2009), "Effects of initial position versus prominence in English," *Journal of Phonetics, 37*: 466–85.

Choi, J. D. (1991), "An acoustic study of Kabardian vowels," *Journal of the International Phonetic Association, 21*(1): 4–12.

Choi, J. D. (1992), "Phonetic underspecification and target-interpolation: an acoustic study of Marshallese vowel allophony," *UCLA Working Papers in Phonetics, 82*.

Chomsky, N. (1957), *Syntactic Structures*. The Hague, The Netherlands and Paris, France: Mouton.

Chomsky, N. (1966), *Cartesian Linguistics: A Chapter in the History of Rationalist Thought*. New York: Harper & Row.

Chomsky, N. (2001), "Derivation by phase," in M. Kenstowicz (ed.), *Ken Hale: A Life in Language*. Cambridge, MA: MIT Press, pp. 1–52.

Chomsky, N. and M. Halle (1968), *The Sound Pattern of English*. New York: Harper & Row.

Clark, M. (1990), *The Tonal System of Igbo*. Dordrecht: Foris.

Clements, G. N. (1981), "Akan vowel harmony: a non-linear analysis," *Harvard Studies in Phonology, 2*: 108–77.

Clements, G. N. (1985), "The geometry of phonological features," *Phonology Yearbook, 2*: 225–52.

Clements, G. N. (1990), "The role of the sonority cycle in core syllabification," in J. Kingston and M. Beckman (eds), *Papers in Laboratory Phonology I: Between the Grammar and Physics of Speech*. Cambridge, MA: Cambridge University Press, pp. 283–333.

Clements, G. N. (1992), "Phonological primes: features or gestures?" *Phonetica, 49*(3/4): 181–93.

Clements, G. N. (2001), "Representational economy in constraint-based phonology," in T. Alan Hall (ed.), *Distinctive Feature Theory*. Berlin: De Gruyter Mouton, pp. 71–146.

Clements, G. N. (2009a), "The role of features in phonological inventories," in E. Raimy and C. E. Cairns (eds), *Contemporary Views on Architecture and Representations in Phonology*. Cambridge, MA: MIT Press, pp. 19–68.

Clements, G. N. (2009b), "Does sonority have a phonetic basis?" in E. Raimy and C. E. Cairns, (eds), *Contemporary Views on Architecture and Representations in Phonological Theory*. Cambridge, MA: MIT Press, pp. 165–75.

Clements, G. and E. Sezer (1982), "Vowel and consonant disharmony in Turkish," in H. van der Hulst and N. Smith (eds), *The Structure of Phonological Representations (Part II)*, Dordrecht: Foris, pp. 213–55.

Clements, G. N. and S. J. Keyser (1983), *CV Phonology: A Generative Theory of the Syllable.* Cambridge, MA: MIT Press.

Clements, G. N. and E. V. Hume (1995), "The internal organization of speech sounds," in J. Goldsmith (ed.), *The Handbook of Phonological Theory.* Cambridge, MA and Oxford, UK: Blackwell, pp. 245–306.

Clements, G. N. and R. Ridouane (eds) (2011), *Where Do Phonological Features Come From? Cognitive, Physical and Developmental Bases of Distinctive Speech Categories.* Amsterdam, The Netherlands: John Benjamins Publishing Company.

Clements, G. N., A. Michaud, and C. Patin (2011), "Do we need tone features?" in J. A. Goldsmith, E. Hume, and W. L. Wetzels (eds), *Tones and Features: Phonetic and Phonological Perspectives.* Berlin: de Gruyter Mouton, pp. 3–24.

Coetzee, A. and P. S. Beddor (2014), "Emergent tonogenesis in Afrikaans," *The Journal of the Acoustical Society of America, 135*: 2421–2.

Coetzee, A. W. and D. Wissing (2007), "Global and local durational properties in three varieties of South African English," *The Linguistic Review, 24*: 263–89.

Cohn, A. C. (1990), "Phonetic and phonological rules of nasalization," *UCLA Working Papers in Phonetics, 76.*

Cohn, A. C. (1993), "Nasalisation in English: phonology or phonetics," *Phonology, 10*(1): 43–81.

Cohn, A. C. (2006), "Is there gradient phonology?" In G. Faneslow, C. Fery, R. Voegl, and M. Schlesewsky (eds), *Gradience in Grammar: Generative Perspectives.* Oxford: Oxford University Press, pp. 25–44.

Cohn, A. C. (2011), "Features, segments, and the sources of phonological primitives," in A. Cohn, G. N. Clements, and R. Ridouane (eds), *Where Do Features Come From?* Amsterdam, The Netherlands: John Benjamins, pp. 15–41.

Cole, J. (2009), "Emergent feature structures: harmony systems in exemplar models of phonology," *Language Sciences, 31*: 144–60.

Coleman, J. S. and J. Pierrehumbert (1997), "Stochastic phonological grammars and acceptability," Computational Phonology: Third meeting of the ACL special interest group in computational phonology. Somerset, NJ: Association for Computational Linguistics, pp. 49–56.

Collier, R. and J. t'Hart (1981), *Cursus Nederlandse Intonatie.* Louvain: Acco/de Horstink.

Collins, B. and I. M. Mees (1999), *The Real Professor Higgins: The Life and Career of Daniel Jones.* Berlin and New York: De Gruyter Mouton.

Comrie, B. (1981), *Language Universals and Linguistic Typology: Syntax and Morphology.* Chicago: University of Chicago Press.

Connell, B. (2000), "The perception of lexical tone in Mambila," *Language and Speech, 43*: 163–82.

Cooper, W. and J. Sorensen (1981), *Fundamental Frequency in Sentence Production.* Heidelberg: Springer.

Coulmas, F. (1989), *The Writing Systems of the World.* Oxford: Blackwell.

Coulmas, F. (1999), *The Blackwell Encyclopedia of Writing Systems.* Oxford: Blackwell.

Crasborn, O. (2012), "Phonetics," in R. Pfau, M. Steinbach, and B. Woll (eds), *Sign Language. An International Handbook.* Berlin: De Gruyter Mouton, pp. 4–20.

Crosswhite, K. (2001), *Vowel Reduction in Optimality Theory.* New York: Routledge.

Cruttenden, A. [1986] (1997), *Intonation.* New York: Cambridge University Press.

Crystal, D. (1969), *Prosodic Systems and Intonation in English.* Cambridge, UK: Cambridge University Press.

Cutler, A. (2005), "Lexical stress," in D. B. Pisoni and R. E. Remez (eds), *The Handbook of Speech Perception.* Oxford: Blackwell, pp. 264–89.

Cutler, A. (2012), *Native Listening: Language Experience and the Recognition of Spoken Words.* Cambridge, MA: MIT Press.

Cutler, A., and D. Norris (2002), "The role of strong syllables in segmentation for lexical access," in G. T. Altmann (ed.), *Psycholinguistics: Critical Concepts in Psychology* London: Routledge, pp. 157–77.

Cutler, A. and S. Butterfield (2003), "Rhythmic cues to speech segmentation: evidence from juncture misperception," in J. Field (ed.), *Psycholinguistics: A Resource Book for Students.* London: Routledge, pp. 185–9.

Cutler, A., D. Dahan, and W. van Donselaar (1997), "Prosody in the comprehension of spoken language: a literature review," *Language and Speech, 40*(2): 141–201.

Cutting, J. E. and B. S. Rosner (1974), "Categories and boundaries in speech and music," *Perception and Psychophysics, 16*: 564–70.

Daland, R., B. Hayes, J. White, M. Garallek, A. Davis, and I. Norrmann (2011), "Explaining sonority projection effects," *Phonology 28*(2): 197–234.

Daniels, P. T. and W. Bright (eds) (1996), *The World's Writing Systems.* New York: Oxford University Press.

Davidson, L. (2006), "Phonotactics and articulatory coordination interact in phonology: evidence from non-native production," *Cognitive Science, 30*(5): 837–62.

Davidson, L. (2018), "Phonation and laryngeal specification in American English voiceless obstruents," *Journal of the International Phonetic Association, 48*(3): 331–56.

Davidson, L. and C. Wilson (2016), "Processing nonnative consonant clusters in the classroom: perception and production of phonetic detail," *Second Language Research, 32*(4): 471–502.

Davidson, L., C. Wilson, and S. Martin (2015), "Stabilizing the production of nonnative consonant clusters with acoustic variability," *The Journal of the Acoustical Society of America, 137*: 856–72.

DeFrancis, J. (1989), *Visible Speech: The Diverse Oneness of Writing Systems.* Honolulu, HI: University of Hawaii Press.

de Jong, K. (2003), "Temporal constraints and characterizing syllable structure," in J. Local, R. Ogden, and R. Temple (eds), *Papers in Laboratory Phonology VI: Phonetic Interpretation.* Cambridge, UK: Cambridge University Press, pp. 253–68.

de Lacy, P. (1999), "Tone and prominence," University of Massachusetts, Amherst, <https: // www.pauldelacy.net/webpage/publications.html> (last accessed March 3, 2020).

de Lacy, P. (2002), *The formal expression of markedness.* Doctoral dissertation, University of Massachusetts, Amherst.

de Lacy, P. (2006), *Markedness: Reduction and Preservation in Phonology.* Cambridge, UK: Cambridge University Press.

de Lacy, P. (2007), "The interaction of tone, sonority and prosodic structure," in P. de Lacy (ed.), *The Cambridge Handbook of Phonology.* Cambridge, UK: Cambridge University Press, pp. 281–308.

de Lacy, P. and J. Kingston (2013), "Synchronic explanation," *Natural Language and Linguistic Theory, 31*(2): 287–355.

Delattre, P. C. (1966), "A comparison of syllable length conditioning among languages," *International Review of Applied Linguistics, 4*: 183–98.

Derbyshire, D. (1979), *Hixkaryana.* Amsterdam: North-Holland Publishing.

Diehl, R. (1991), "The role of phonetics within the study of language," *Phonetica, 48*: 120–34.

Diehl, R. and K. R. Kluender (1989), "On the objects of speech perception," *Ecological Psychology, 1*: 121–44.

Diehl, R. and B. Lindblom (2000), "Explaining the structure of feature and phoneme inventories," in S. Greenberg, W. A. Ainsworth, A. Popper, and R. Fay (eds), *Speech Processing in the Auditory System*. New York: Springer Verlag, pp. 101–62.

D'Imperio, M. (2012), "Tonal alignment," in A. C. Cohn, C. Fourgeron, and M. K. Huffman (eds), *The Oxford Handbook of Laboratory Phonology*. Oxford: Oxford University Press.

Disner, S. (1984), "Insights on vowel spacing," in I. Maddieson (ed.), *Patterns of Sounds*. Cambridge, MA: Cambridge University Press, pp. 136–55.

Donegan, P. J. and D. Stampe (1979), "The study of Natural Phonology," in D. A. Dinnsen (ed.), *Current Approaches to Phonological Theory*. Bloomington: Indiana University Press, pp. 126–73.

Donegan, P. J. and D. Stampe (2009), "Hypotheses of Natural Phonology," *Poznań Studies in Contemporary Linguistics, 45*: 1–31.

Downing, L. (1989), "The interaction of tone and intonation in Jita yes/no questions," *Studies in the Linguistic Sciences, 19*: 91–113.

Dressler, W. U. (1984), "Explaining Natural Phonology," *Phonology Yearbook, 1*: 29–51.

Duanmu, S. (1990), *A formal study of syllable, tone, stress and domain in Chinese languages*, Doctoral dissertation, Massachusetts Institute of Technology.

Duanmu, S. (1994), "Against contour tone units," *Linguistic Inquiry, 25*: 555–608.

Dunbar, E. and W. J. Idsardi (2016), "The acquisition of phonological inventories," in J. Lidz, W. Snyder, and J. Pater (eds), *The Oxford Handbook of Developmental Linguistics*. New York: Oxford University Press, pp. 7–26.

Dupoux, E., K. Kakehi, Y. Hirose, C. Pallier, and J. Mehler (1999), "Epenthetic vowels in Japanese: a perceptual illusion?" *Journal of Experimental Psychology: Human Perception and Performance, 25*(6): 1568–78.

Eccarius, P. and D. Brentari (2008), "Handshape coding made easier: a theoretically based notation for phonological transcription," *Sign Language and Linguistics, 11*: 69–101.

Elfner, E. (2015), "Recursion in prosodic phrasing: evidence from Connemara Irish," *Natural Language and Linguistic Theory*, 33: 1169–1208.

Elfner, E. (2018), "The syntax-prosody interface: current theoretical approaches and outstanding questions," *Linguistics Vanguard, 4*(1): 1–14.

Ellis, L. and W. J. Hardcastle (2002), "Categorical and gradient properties of assimilation in alveolar to velar sequences: evidence from EPG and EMA data," *Journal of Phonetics, 30*: 373–96.

Elordieta, G. and J. I. Hualde (2014), "Intonation in Basque," in S.-A. Jun (ed.), *Prosodic Typology II: The Phonology of Intonation and Phrasing*. Oxford: Oxford University Press, pp. 405–63.

Ernestus, M. (2006), "Statistically gradient generalizations for contrastive phonological features," *The Linguistic Review, 23*(3): 217–33.

Ernestus, M. (2011), "Gradience and categoricality in phonological theory," in M. van Oostendorp, C. J. Ewen, E. Hume, and K. Rice, (eds), *The Blackwell Companion to Phonology*. Oxford: Wiley-Blackwell, pp. 2115–36.

Errington, J. (2007), *Linguistics in a Colonial World: A Story of Language, Meaning, and Power*. Malden, MA and Oxford: Wiley-Blackwell.

Escudero, P. and P. Boersma (2004), "Bridging the gap between L2 speech perception research and phonological theory," *Studies in Second Language Acquisition, 26*: 551–85.

Ettlinger, M. (2008), *Input-driven opacity*. Doctoral dissertation, University of California, Berkeley.

Ewen, C. J. (1995), "Dependency relations in phonology," in J. Goldsmith (ed.), *The Handbook of Phonological Theory*. Oxford: Blackwell, pp. 570–85.

Fano, R. (1949), "The transmission of information," Technical Report Number 65. Cambridge, MA: The Research Laboratory of Electronics at MIT, pp. 1–34.

Fant. G. (2000), "Half a century in phonetics and speech research," in A. Botinis and N. Torstensson (eds), *Proceedings of fonetik 2000, the XIII Swedish phonetics conference*, Skövde, Sweden: Högskolan i Skövde Department of Languages, <http: //www.speech.kth.se/gunnarfant/halfcentury.pdf> (last accessed March 3, 2020).

Fear, B. D., A. Cutler, and S. Butterfield (1995), "The strong/weak syllable distinction in English," *The Journal of the Acoustical Society of America, 97*: 1893–1904.

Finley, S. (2017), "Learning metathesis: evidence for syllable structure constraints," *Journal of Memory and Language, 92*: 142–57.

Firth, J. R. (1948), "Sounds and prosodies," *Transactions of the Philological Society*, 127–52.

Flege, J. E. (1987), "The production of 'new' and 'similar' phones in a foreign language: evidence for the effect of equivalence classification," *Journal of Phonetics, 15*: 47–65.

Flege, J. E. (1995), "Second language speech learning: theory, findings, and problems," in W. Strange (ed.), *Speech Perception and Linguistic Experience: Issues in Cross-language Research*. Baltimore: York Press, pp. 233–77.

Flege, J. E. (2005), "Origins and development of the Speech Learning Model," First ASA workshop on L2 speech learning, <http: //www.jimflege.com/files/SLMvancouver_updated.pdf> (last accessed March 3, 2020).

Flemming, E. (1996), "Evidence for constraints on contrast: the Dispersion Theory of contrast," *UCLA Working Papers in Phonology, 1*: 86–106.

Flemming, E. (2002), *Auditory Representations in Phonology*. New York: Routledge.

Flemming, E. (2004), "Contrast and perceptual distinctiveness," in B. Hayes, R. Kirchner, and D. Steriade (eds), *Phonetically Based Phonology*. Cambridge, MA: Cambridge University Press, pp. 232–76.

Flemming, E. (2005), "Speech perception and phonological contrast," in D. B. Pisoni and R. E. Remez (eds), *The Handbook of Speech Perception*. Oxford: Blackwell, pp. 156–81.

Flemming, E. (2017), "Dispersion Theory and phonology," in M. Aronoff (ed.), *The Oxford Research Encyclopedia of Linguistics*, <https://oxfordre.com/linguistics/view/10.1093/acrefore/9780199384655.001.0001/acrefore-9780199384655-e-110?rskey=K1hMyI&result=1> (last accessed April 3, 2020).

Fletcher, J. (2010), "The prosody of speech: timing and rhythm," in W. J. Hardcastle, J. Laver, and F. E. Gibbon (eds), *The Handbook of Phonetic Sciences*, 2nd edn. Oxford: Blackwell, pp. 521–602.

Fletcher, J. (2014), "Intonation and prosody," in S.-A. Jun (ed.), *Prosodic Typology II: The Phonology of Intonation and Phrasing*. Oxford: Oxford University Press.

Fodor, J. A. (1983), *The Modularity of Mind: An Essay on Faculty Psychology*. Cambridge, MA: MIT Press.

Fougeron, C. and P. Keating (1997), "Articulatory strengthening at edges of prosodic domains," *The Journal of the Acoustical Society of America, 101*: 3728–40.

Fowler, C. A. (1990), "Calling a mirage a mirage: direct perception of speech produced without a tongue," *Journal of Phonetics, 18*: 529–41.

Fowler, C. A. (1996), "Listeners do hear sounds, not tongues," *The Journal of the Acoustical Society of America, 99*: 1730–41

Frisch, S. (2018), "Exemplar theories in phonology," in S. J. Hannahs and A. R. K. Bosch (eds), *The Routledge Handbook of Phonological Theory*. London: Routledge.

Frisch, S., J. Pierrehumbert, and M. Broe (2004), "Similarity avoidance and the OCP," *Natural Language and Linguistic Theory, 22*: 179–228.

Fromkin, V. (1971), "The non-anomalous nature of anomalous utterances," *Language 47*: 27–52.

Fromkin, V. (ed.) (1978), *Tone: A Linguistic Survey*. New York: Academic Press.
Frota, S. (2012), "Prosodic structure, constituents and their implementation," in A. C. Cohn, C. Fougeron, and M. K. Huffman (eds), *The Oxford Handbook of Laboratory Phonology*. Oxford University Press, pp. 255–65.
Frota, S. and P. Prieto (eds) (2007), "Special issue on prosodic phrasing," *The Linguistic Review 24*.
Fry, D. B. (1955), "Duration and intensity as physical correlates of linguistic stress," *The Journal of the Acoustical Society of America, 35*: 765–9.
Fry, D. B. (1958), "Experiments in the perception of stress," *Language and Speech, 1*: 120–52.
Fudge, E. C. (1967), "The nature of phonological primes," *Journal of Linguistics, 3*(1): 1–36.
Fujisaki, H. (1983), "Dynamic characteristics of voice fundamental frequency in speech and singing," in P. F. MacNeilage (ed.), *The Production of Speech*. Heidelberg: Springer-Verlag, pp. 39–55.
Fujisaki, H. (2004), "Information, prosody, and modeling, with emphasis on tonal features of speech," in B. Bel and I. Marlien (eds), *Speech prosody 2004*. Nara, Japan, <http://www.isca-speech.org/archive/sp2004> (last accessed April 7, 2020).
Fujisaki, H. and K. Hirose (1984), "Analysis of voice fundamental frequency contours for declarative sentences of Japanese," *Journal of the Acoustical Society of Japan (E), 5*(4): 233–42.
Gafos, A. (2002), "A grammar of gestural coordination," *Natural Language and Linguistic Theory, 20*(2): 269–337.
Gahl, S. (2008), "Time and thyme are not homophones: the effect of lemma frequency on word duration in spontaneous speech," *Language, 84*: 474–96.
Galantucci, B., C. A. Fowler, and M. T. Turvey (2006), "The Motor Theory of speech perception reviewed," *Psychonomic Bulletin and Review, 13*: 361–77.
Gallagher, G. (2011), "Acoustic and articulatory features in phonology—The case for [long VOT]," *The Linguistic Review, 28*(3): 281–313.
Gallagher, G. and J. Coon (2009), "Distinguishing total and partial identity: evidence from Chol," *Natural Language and Linguistic Theory, 27*: 545–82.
Gandour, J. (1974), "On the representation of tone in Siamese," in J. G. Harris and J. R. Chamberlain (eds), *Studies in Tai Linguistics in Honor of William J. Gedney*. Bangkok: Central Institute of English Language, pp. 170–95.
Gandour, J. (1978), "The perception of tone," in V. Fromkin (ed.), *Tone: A Linguistic Survey*. New York: Academic Press, pp. 41–76.
Gandour, J. (1981), "Perceptual dimensions of tone: evidence from Cantonese," *Journal of Chinese Linguistics, 9*: 20–36.
Gandour, J. (1983), "Tone perception in Far Eastern languages," *Journal of Phonetics, 11*: 149–75.
Gandour, J. and R. Harshman (1978), "Cross-language differences in tone perception: a multidimensional scaling investigation," *Language and Speech, 21*: 1–33.
Gandour, J., H. Li, B. Weinzapfel, D. van Lancker, and G. Hutchins (2000), "A cross-linguistic PET study of tone perception," *Journal of Cognitive Neuroscience, 12*: 207–22.
Gao, M. (2008), *Mandarin tones: an Articulatory Phonology account*. Doctoral dissertation, Yale University.
Gardiner, A. H. (1916), "The Egyptian origin of the Semitic alphabet," *Journal of Egyptian Archeology, 3*(1): 1–16.
Gårding, E. (1983), "A generative model of intonation," in A. Cutler and D. R. Ladd (eds), *Prosody: Models and Measurements*. Heidelberg: Springer, pp. 11–25.
Garellek, M. and P. Keating (2011), "The acoustic consequences of phonation and tone

interactions in Jalapa Mazatec," *Journal of the International Phonetic Association, 41*: 185–205.

Garellek, M., P. Keating, C. M. Esposito, and J. Kreiman (2013), "Voice quality and tone identification in White Hmong," *The Journal of the Acoustical Society of America, 133*: 1078–89.

Gelb, I. J. (1963), *A Study of Writing*. Chicago: University of Chicago Press.

Gick, B., F. Campbell, S. Oh, and L. Tamburri-Watt (2006), "Toward universals in the gestural organization of syllables: a cross-linguistic study of liquids," *Journal of Phonetics, 34*(1): 49–72.

Gilmour, R. (2006), *Grammars of Colonialism: Representing Languages in Colonial South Africa*. New York: Palgrave Macmillan.

Gimson, A. C. (1970), *An Introduction to the Pronunciation of English*. London: A. C. Arnold.

Gleason, H. A. (1955), *Workbook in Descriptive Linguistics*. New York: Henry Holt & Co.

Gnanadesikan, A. E. (2009), *The Writing Revolution: Cuneiform to the Internet*. Malden, MA and Oxford: Wiley-Blackwell.

Goldinger, S. D. (1996), "Words and voices: episodic traces in spoken word identification and recognition memory," *Journal of Experimental Psychology, Learning, Memory and Cognition, 22*: 1166–82.

Goldinger, S. D. (1997), "Words and voices: perception and production in an episodic lexicon," in K. Johnson and J. W. Mullennix (eds), *Talker Variability in Speech Processing*. San Diego: Academic Press, pp. 33–66.

Goldinger, S. D. (2000), "The role of perceptual episodes in lexical processing," paper presented at Spoken Word Access Processes, May, Max Planck Institute for Psycholinguistics, Nijmegen.

Goldrick, M. (2001), "Turbid output representations and the unity of opacity," in M. Hirotani, A. Coetzee, N. Hall, and J.-Y. Kim (eds), *Proceedings of the 30th annual meeting of the North East Linguistics Society*, volume I. Amherst, MA: GLSA, pp. 231–45.

Goldsmith, J. (1979), *Autosegmental Phonology*. New York: Garland.

Goldsmith, J. (1985), "Vowel harmony in Khalkha Mongolian, Yaka, Finnish and Hungarian", *Phonology Yearbook, 2* (1): 253–75.

Goldsmith, J. (2011), "The syllable," in J. Goldsmith, J. Riggle, and A. Yu (eds), *The Handbook of Phonological Theory*, 2nd edn. Malden, MA and Oxford: Wiley-Blackwell, pp. 164–96.

Goldsmith, J. and A. Xanthos (2009), "Learning phonological categories," *Language, 85* (1): 1–38.

Goldstein, E. B. and J. R. Brockmole (2017), *Sensation and Perception*. Boston: Cengage Learning.

Goldstein, L. (n.d.), Description of research: summary statement of research interests, <http://dornsife.usc.edu/cf/faculty-and-staff/faculty.cfm?pid=1016450> (last accessed March 3, 2020).

Goldstein, L. (2008), "A coupled oscillator model of speech production planning," paper presented at the 5th International Workshop on Language Production, Annapolis, MD.

Goldstein, L., D. Byrd, and E. Saltzman (2006), "The role of vocal tract gestural action units in understanding the evolution of phonology," in M. A. Arbib (ed.), *Action to Language via the Mirror Neuron System*. Cambridge, UK: Cambridge University Press, pp. 215–49.

Goldstein, L., I. Chitoran, and E. Selkirk (2007), "Syllable structure as coupled oscillator modes: evidence from Georgian vs. Tashlhiyt Berber," in W. J. Barry and J. Trouvain (eds), *Proceedings of the XVIth International Congress of Phonetic Sciences*, Saarbücken, Germany: Universität des Saarlandes, pp. 241–4.

Goldstein, L., H. Nam, E. Saltzman, and I. Chitoran (2009), "Coupled oscillator planning model of speech timing and syllable structure," in G. Fant, H. Fujisaki, and J. Shen (eds), *Frontiers in Phonetics and Speech Science*. Beijing: The Commercial Press, pp. 239–50.

Golston, C. and W. Kehrien (2015), "A prosodic theory of vocalic contrasts," in C. Cairns and E. Raimy (eds), *The Segment in Phonetics and Phonology*. Malden, MA and Oxford: Wiley-Blackwell, pp. 65–102.

Gordon, M. (2005), "A perceptually driven account of onset-sensitive stress," *Natural Language and Linguistics Theory, 23*: 595–653.

Gordon, M. (2007), "Functionalism in phonology," in P. de Lacy (ed.), *The Cambridge Handbook of Phonology*. Cambridge, UK: Cambridge University Press.

Gordon, M. and H. van der Hulst (2018), "Word prosody I: word-stress systems," in C. Gussenhoven and A. Chen (eds), *The Oxford Handbook of Language Prosody*. Oxford: Oxford University Press.

Gouskova, M. (2004), "Relational hierarchies in OT: the case of syllable contact," *Phonology, 21*(2): 201–50.

Gouskova, M. (2002), "Exceptions to sonority generalization," in *CLS 38: The Main Session. Papers from the 38th Meeting of the Chicago Linguistic Society*. Chicago: Chicago Linguistics Society, pp. 253–68.

Gouskova, M., E. Zsiga, and O. Tlale (2011), "Grounded constraints and the consonants of Setswana," *Lingua, 121*: 2120–52.

Grabe, E. and P. Warren (1995), "Stress shift: do speakers do it or do listeners hear it?" in B. Connell and A. Arvaniti (eds), *Phonology and Phonetic Evidence: Papers in Laboratory Phonology IV*. Cambridge, UK: Cambridge University Press, pp. 95–110.

Grabe, E. and E. L. Low (2002), "Durational variability in speech and the rhythm class hypothesis," in C. Gussenhoven and N. Warner (eds), *Laboratory Phonology VII*. Berlin: De Gruyter Mouton, pp. 515–46.

Grabe, E., G. Kochanski, and J. Coleman (2007), "Connecting intonation labels to mathematical descriptions of fundamental frequency," *Language and Speech, 50*: 281–310.

Greenberg, J. H. (1963), "Some universals of language with special reference to the order of meaningful elements," in J. H. Greenberg (ed.), *Universals of Language*. Cambridge, MA: MIT Press, pp. 73–113.

Grice, M. (1995), *The Intonation of Interrogation in Palermo Italian: Implications for Intonation Theory*. Tübingen: Niemeyer.

Grice, M., R. D. Ladd, and A. Arvaniti (2000), "On the place of 'phrase accents' in intonational phonology," *Phonology, 17*: 143–85.

Gruber, J. (2011), *An articulatory, acoustic, and auditory study of Burmese tone*. Doctoral dissertation, Georgetown University.

Gussenhoven, C. (1991), "The English rhythm rule as an accent deletion rule," *Phonology, 8*(1): 1–35.

Gussenhoven, C. (1992), "Dutch," *Journal of the International Phonetic Association, 22*(2): 45–7.

Gussenhoven, C. (2004), *The Phonology of Tone and Intonation*. Cambridge, UK: Cambridge University Press.

Gussenhoven, C. (2007), "Intonation," in P. de Lacy (ed.), *The Cambridge Handbook of Phonology*. Cambridge, UK: Cambridge University Press, pp. 253–80.

Gussenhoven, C. and T. Riad (eds) (2007a), *Tones and Tunes. Volume 1: Typological Studies in Word and Sentence Prosody*. Berlin: De Gruyter Mouton.

Gussenhoven, C. and T. Riad (eds) (2007b), *Tones and Tunes. Volume 2: Experimental Studies in Word and Sentence Prosody*. Berlin: De Gruyter Mouton.

Gussenhoven, C. and H. Jacobs (2011), *Understanding Phonology*, 3rd edn. New York: Routledge.

Gussenhoven, C. and A. Chen (eds) (2018), *The Oxford Handbook of Language Prosody*. Oxford: Oxford University Press.

Gussmann, E. (2002), *Phonology: Analysis and Theory*. Cambridge, UK: Cambridge University Press.

Haken, H., J. A. S. Kelso, and H. Bunz (1985), "A theoretical model of phase transitions in human hand movements," *Biological Cybernetics, 51*(5): 347–56.

Hale, M. and C. Reiss (2008), *The Phonological Enterprise*. New York: Oxford University Press.

Hall, N. (2010), "Articulatory Phonology," *Language and Linguistic Compass, 4*(9): 818-830.

Hall, N. (2018), "Articulatory Phonology," in S. J. Hannahs and A. R. K. Bosch (eds), *The Routledge Handbook of Phonological Theory*. London: Routledge, pp. 530–52.

Halle, M. (1954), "Why and how do we study the sounds of speech?" in H. J. Mueller (ed.), *Report of the Fifth Annual Round Table Meeting on Linguistics and Language Teaching*. Washington, DC: Georgetown University Press, pp. 73–80.

Halle, M. (1955), *The Russian Consonants, a Phonemic and Acoustical Investigation*. Doctoral dissertation, Harvard University.

Halle, M. (1959), *The Sound Pattern of Russian*. The Hague, The Netherlands: Mouton.

Halle, M. (1962), "Phonology in generative grammar," *Word 18*(1–3): 54–72.

Halle, M. (1971), "Remarks on Slavic accentology," *Linguistic Inquiry, 2*(1): 1–19.

Halle, M. (2009), "Two comments on 'The role of features in phonological inventories'," in E. Raimy and C. E. Cairns (eds), *Contemporary Views on Architecture and Representations in Phonology*. Cambridge, MA: MIT Press, pp. 69–73.

Halle, M. and K. Stevens (1971), "A note on laryngeal features," *Quarterly Progress Report, Research Laboratory Electronics MIT 101*: 198–213.

Halle, M. and J.-R. Vergnaud (1987), *An Essay on Stress*. Cambridge, MA: MIT Press.

Halliday, M. A. K. (1967), *Intonation and Grammar in British English*. The Hague: Mouton.

Halliday, M. A. K. (1970), *A Course in Spoken English: Intonation*. Oxford: Oxford University Press.

Hamann, S. (2011), "The phonetics-phonology interface," in N. C. Kula, B. Botma, and K. Nasukawa (eds), *The Bloomsbury Companion to Phonology*. London and New York: Bloomsbury, pp. 202–24.

Hamann, S. and L. J. Downing (2017), "*NT revisited again: an approach to postnasal laryngeal alternations with perceptual cue constraints," *Journal of Linguistics, 53*(1): 85–112.

Hamilton, H. (2018), "Phonotactic constraints and phonological categories in American Sign Language," manuscript, Gallaudet University.

Hammond, M. T. (1986), "The obligatory-branching parameter in metrical theory," *Natural Language and Linguistic Theory, 4*: 185–228.

Harms, R. (1968), *Introduction to Phonological Theory*. Englewood Cliffs: Prentice-Hall.

Harris, Z. S. (1944), "Simultaneous components in phonology," *Language, 20*: 181–205.

Harris, Z. S. (1951), *Methods in Structural Linguistics*. Chicago: University of Chicago Press.

Haspelmath, M. (2006), "Against markedness (and what to replace it with)," *Journal of Linguistics, 42*: 25–70.

Haugen, E. (1950), *First Grammatical Treatise: The Earliest Germanic Phonology*. London: Longman.

Havenhill, J. (2018), *Constraints on articulatory variability: audiovisual perception of lip rounding*. Doctoral dissertation, Georgetown University.

Hawkins, S. (1992), "An introduction to task dynamics," in G. J. Docherty and D. R. Ladd (eds), *Papers in Laboratory Phonology II: Gesture, Segment, Prosody*. Cambridge, UK: Cambridge University Press.

Hay, J. B., J. B. Pierrehumbert, and M. Beckman (2003), "Speech perception, well-formedness, and the statistics of the lexicon," in J. Local, R. Ogden, and R. Temple, (eds), *Phonetic*

Interpretation: Papers in Laboratory Phonology VI. Cambridge, UK: Cambridge University Press, pp. 58–74.

Hay, J. B., J. B. Pierrehumbert, A. J. Walker, and P. LaShell (2015), "Tracking word frequency effects through 130 years of sound change," *Cognition, 139*: 83–91.

Hayes, B. (1984), "The phonetics and phonology of Russian voicing assimilation," in M. Aronoff and R. T. Oehrle (eds), *Language Sound Structure*. Cambridge, MA: MIT Press, pp. 318–28.

Hayes, B. (1985), *A Metrical Theory of Stress Rules*. New York: Garland.

Hayes, B. (1995), *Metrical Stress Theory: Principles and Case Studies*. Chicago: University of Chicago Press.

Hayes, B. (1999), "Phonetically-driven phonology: the role of Optimality Theory and inductive grounding," in M. Darnell, E. Moravscik, M. Noonan, F. Newmeyer, and K. Wheatly (eds), *Functionalism and Formalism in Linguistics, Volume I: General Papers*. Amsterdam, The Netherlands: John Benjamins, pp. 243–85.

Hayes, B. and T. Stivers (1995), "The phonetics of post-nasal voicing," unpublished manuscript.

Hayes, B. and D. Steriade (2004), "Introduction: the phonetic basis of phonological markedness," in B. Hayes, R. Kirchner, and D. Steriade (eds), *Phonetically based Phonology*. Cambridge, MA: Cambridge University Press, pp. 1–32.

Hayes, B. and Z. C. Londe (2006), "Stochastic phonological knowledge: the case of Hungarian vowel harmony," *Phonology, 23*: 59–104.

Hayes, B. and C. Wilson (2008), "A maximum entropy model of phonotactics and phonotactic learning," *Linguistic Inquiry, 39*: 379–440.

Hermes, A., M. Grice, D. Mücke, and H. Niemann (2008), "Articulatory indicators of syllable affiliation in word initial consonant clusters in Italian," in *Proceedings of the eighth international seminar on speech production*, <https://issp2008.loria.fr/proceedings.html> (last accessed April 1, 2020).

Hermes, A., R. Ridouane, D. Mücke, and M. Grice (2011), "Kinematics of syllable structure in Tashlhyit Berber: the case of vocalic and consonantal nuclei," *Proceedings of the Ninth International Seminar on Speech Production*, Montréal, Canada, pp. 401–8, <https://issp2011.uqam.ca/upload/files/proceedings.pdf> (last accessed April 4, 2020).

Hermes, A., D. Mücke, and M. Grice (2013), "Gestural coordination of Italian word-initial clusters: the case of 'impure s'," *Phonology, 30*: 1–25.

Hirst, D. and A. Di Cristo (eds) (1998), *Intonation Systems: Survey of Twenty Languages*. Cambridge, UK: Cambridge University Press.

Hjelmslev, L. [1943] (1953), *Prolegomena to a Theory of Language*. Baltimore: Indiana University Publications in Anthropology and Linguistics (IJAL Memoir, 7).

Hochgesang, J. A. (2014), "Using design principles to consider representation of the hand in some notation systems," *Sign Language Studies, 14*(4): 488–542.

Hockett, C. (1942), "System of descriptive phonology," *Language, 18*(1): 3–21.

Hockett, C. (1958), *A Course in Modern Linguistics*. New York: Macmillan.

Hockett, C. (1960), "The origin of speech," *Scientific American, 203*: 88–111.

Hockett, C. (1961), "Grammar for the hearer," in R. Jakobson (ed.), *Structure of Language and Its Mathematical Aspects (Proceedings of Symposia in Applied Mathematics, vol. XII)*. Providence, RI: American Mathematical Society.

Holst, T. and F. Nolan (1995), "The influence of syntactic structure on [s] to [ʃ] assimilation," in B. Connell and A. Arvaniti (eds), *Phonology and Phonetic Evidence: Papers in Laboratory Phonology IV*. Cambridge, UK: Cambridge University Press, pp. 315–33.

Hombert, J.-M., J. J. Ohala, and W. G. Ewan (1979), "Phonetic explanations for the development of tones," *Language, 55*: 37–58.

Honeybone, P. (2005), "Diachronic evidence in segmental phonology: the case of obstruent laryngeal specifications," in M. van Oostendorp and J. van de Weijer (eds), *The Internal Organization of Phonological Segments*. Berlin: De Gruyter Mouton, pp. 319–54.

Honorof, D. N. (1999), *Articulatory gestures and Spanish nasal assimilation*. Doctoral dissertation, Yale University.

Hooper, J. B. (1972), "The syllable in phonological theory," *Language, 48*(3): 525–40.

Householder, F. (1952), "Review of *Methods in Structural Linguistics*, Zellig S. Harris," *International Journal of American Linguistics, 18*(4): 260–68.

Houston, D. M. (1999), *The role of talker variability in infant word representations*. Doctoral dissertation, The Johns Hopkins University.

Houston, D. M. (2005), "Speech perception in infants," in D. B. Pisoni and R. E. Remez (eds), *The Handbook of Speech Perception*. Oxford: Blackwell, pp. 417–48.

Hualde, J. I. and P. Prieto (2016), "Towards an International Prosodic Alphabet (IPrA)," *Laboratory Phonology: Journal of the Association for Laboratory Phonology, 7*(1), 5.

Huffman, M. K. (1989), "Implementation of nasal: timing and articulatory landmarks," *UCLA Working Papers in Phonetics, 75*.

Hume, E. (1994), *Front Vowels, Coronal Consonants and Their Interaction in Nonlinear Phonology*. New York: Garland.

Hume, E. (1998), "The role of perceptibility in consonant/consonant metathesis," in S. Blakem, E.-S. Kim, and K. Shahin (eds), *Proceedings of WCCFL XVII*. Stanford: CSLI, pp. 293–307.

Hume, E. (2001), "Metathesis: formal and functional considerations," in E. Hume, N. Smith, and J. van de Weijer (eds), *Surface Syllable Structure and Segment Sequencing*. Leiden: Holland Institute of Linguistics.

Hume, E. (2003), "Language specific markedness: the case of place of articulation," *Studies in Phonetics, Phonology and Morphology, 9*(2): 295–310.

Hume, E. and K. Johnson (2001), "A model of the interplay of speech perception and phonology," in E. Hume and K. Johnson (eds), *The Role of Speech Perception in Phonology*. New York: Academic Press.

Hyman, L. (1975), *Phonology: Theory and Analysis*. New York: Holt, Rinehard, Winston.

Hyman, L. (1977), "On the nature of linguistic stress," in L. Hyman (ed.), *Studies in Stress and Accent, Southern California Occasional Papers in Linguistics 4*. Los Angeles: University of Southern California.

Hyman, L. (2001), "On the limits of phonetic determinism in phonology: *NC revisited," in E. Hume and K. Johnson (eds), *The Role of Speech Perception Phenomena in Phonology*. San Diego: Academic Press, pp. 141–85.

Hyman, L. (2007), "Universals of tone rules: 30 years later," in T. Riad and C. Gussenhoven (eds), *Tone and Tunes, Volume 1: Typological Studies in Word and Sentence Prosody*. Berlin: De Gruyter Mouton, pp. 1–34.

Hyman, L. (2008), "Universals in phonology," *The Linguistic Review, 25*: 83–137.

Hyman, L. (2010), "Do tones have features?" in J. Goldsmith, E. Hume, and W. L. Wetzels (eds), *Tones and Features: Phonetic and Phonological Perspectives*. Berlin: de Gruyter Mouton, pp. 50–80.

Hyman, L. (2011a), "Tone: is it different?" in J. Goldsmith, J. Riggle, and A. Yu (eds), *The Handbook of Phonological Theory, 2nd Edition*. New York: Blackwell, pp. 197–239.

Hyman, L. (2011b), "Does Gokana really have no syllables? Or: what's so great about being universal?" *Phonology, 28*(1): 55–85.

Hyman, L. (2014), "How autosegmental is phonology?" *The Linguistic Review, 31*(2): 363–400.

Hyslop, G. (2009), "Kurtop tone: a tonogenetic case study," *Lingua, 119*(6): 827–45.

Inkelas, S. (2014), *The Interplay of Phonology and Morphology*. Oxford: Oxford University Press.

Inkelas, S, and W. Leben (1990), "Where phonology and phonetics intersect: the case of Hausa intonation," in J. Kingston and M. Beckman (eds), *Papers in Laboratory Phonology I*. Cambridge, UK: Cambridge University Press.

Inkelas, S. and D. Zec (eds) (1990), *The Phonology-syntax Connection*. Chicago: University of Chicago Press.

Inkelas, S. and D. Zec (1995), "The phonology/syntax interface," in J. Goldsmith (ed.), *Handbook of Phonological Theory*. Oxford: Blackwell Publishing.

Inkelas, S. and S. Shih (2016), "Re-representing phonology: consequences of Q Theory," in C. Hammerly and B. Prickett (eds), *Proceedings of the 46th Annual Meeting of the North East Linguistics Society, Volume II*. Amherst, MA: University of Massachusetts Graduate Linguistics Student Association, pp. 161–74.

International Phonetic Association (1912), "The principles of the International Phonetic Association [Supplement]," *Le Maître Phonétique, 27*: 1–40.

Itô, J. (1986), *Syllable theory in prosodic phonology*. Doctoral dissertation, University of Massachusetts, Amherst.

Itô, J. and A. Mester (2003), "Lexical and postlexical phonology in Optimality Theory: evidence from Japanese," in G. Fanselow and C. Féry (eds), *Linguistische Berichte. Sonderheft 11: Resolving conflicts in grammars*. Hamburg: Buske, pp. 183–207.

Itô, J. and A. Mester (2013), "Prosodic subcategories in Japanese," *Lingua, 124*: 20–40.

Jakobson, R. [1941] (1968), *Child Language, Aphasia and Phonological Universals* (R. A. Keiler, trans.). The Hague, The Netherlands: Mouton.

Jakobson, R. [1942] (1990), "The concept of phoneme," in Linda R. Waugh and Monique Moville-Burston (eds), *On language*. Cambridge, MA: Harvard University Press, pp. 218–41.

Jakobson, R., G. Fant, and M. Halle (1952), *Preliminaries to Speech Analysis: The Distinctive Features and Their Correlates*. (Technical Report 13). Cambridge, MA: Acoustics Laboratory, MIT.

Jespersen, O. (1904), *Lehrbuch der Phonetik*. Leipzig: Teubner.

Johnson, K. (1989), "On the perceptual representation of vowel categories," *Research on Speech Perception*. Progress Report No. 15. Department of Psychology, Indiana University, Bloomington, pp. 343–58.

Johnson, K. (1997), "Speech perception without speaker normalization: an exemplar model," in K. Johnson and J. Mullenix, (eds), *Talker Variability in Speech Processing*, San Diego: Academic Press, pp. 145–66.

Johnson, K. (2005), "Speaker normalization in speech perception," in D. B. Pisoni and R. E. Remez (eds), *The Handbook of Speech Perception*. Oxford: Blackwell, pp. 363–89.

Johnson, K. (2007), "Decisions and mechanisms in exemplar-based phonology," in M. J. Solé, P. Beddor, and M. Ohala (eds), *Experimental Approaches to Phonology. In Honor of John Ohala*. Oxford: Oxford University Press, pp. 25–40.

Johnson, K. (2013a), *Acoustic and Auditory Phonetics*. Malden, MA and Oxford: Blackwell.

Johnson, K. (2013b), "Basic audition and speech perception," in *Acoustic and Auditory Phonetics*. Malden, MA and Oxford: Blackwell.

Johnson, K. and M. Babel (2010), "On the perceptual basis of distinctive features: evidence from the perception of fricatives by Dutch and English speakers," *Journal of Phonetics, 38*: 127–36.

Johnson, R. E. and S. K. Liddell (2010), "Toward a phonetic representation of signs: sequentiality and contrast," *Sign Language Studies, 11*(2): 241–74.

Jones, D. (1910), "Albanian (Gheg dialect) [from dictation of M. E. Durham]," *Le Maître Phonétique, 2*(25): 139–41.

Jones, D. (1911), "Chinese (Standard Cantonese dialect) [from dictation of Kwing Tong Woo]," *Le Maître Phonétique, 2*(26): 80–4.

Jones, D. (1912), "Mono (Western Solomon Islands Melanesian) [from dictation of G. C. Wheeler]," *Le Maître Phonétique, 2*(27): 15.

Jones, D. (1914), "Manx [pronounced by W. Radcliffe]," *Le Maître Phonétique, 2*(29): 76.

Jones, D. (1924), "Korean," *Le Maître Phonétique, 3*(1): 4–5.

Jones, D. (1925), "Gã [pronounced by K. K. Lokko]," *Le Maître Phonétique, 3*(3): 6–9.

Jones, D. (1950), *The Phoneme: Its Nature and Use*. Cambridge, UK: Cambridge University Press.

Jones, D. [1909] (1966), *The Pronunciation of English*, 4th edn. Cambridge, MA: Cambridge University Press.

Jones, D. and S. T. Plaatje (1916), *A Sechuana Reader*. London: London University Press.

Joos, M. (1948), *Acoustic Phonetics (Language monograph no. 23)*. Baltimore: Waverly Press.

Jun, J. (1996), "Place assimilation is not the result of gestural overlap: evidence from Korean and English," *Phonology, 13*(3): 377–407.

Jun, S.-A. (1995), "Asymmetrical prosodic effects on the laryngeal gesture in Korean," in B. Connell and A. Arvaniti (eds), *Phonology and Phonetic Evidence: Papers in Laboratory Phonology IV*. Cambridge, UK: Cambridge University Press, pp. 235–53.

Jun, S.-A. (ed.) (2005), *Prosodic Typology: The Phonology of Intonation and Phrasing*. Oxford: Oxford University Press.

Jun, S.-A. (ed.) (2014), *Prosodic Typology II: The Phonology of Intonation and Phrasing*. Oxford: Oxford University Press.

Jun, S.-A. and Janet Fletcher (2014), "Methodology of studying intonation: from data collection to data analysis", in Sun-Ah Jun (ed.), *Prosodic Typology II: The Phonology of Intonation and Phrasing*. Oxford: Oxford University Press. pp. 493–519.

Jusczyk, P. (1997), *The Discovery of Spoken Language*. Cambridge, MA: MIT Press.

Jusczyk, P. and R. N. Aslin (1995), "Infants' detection of the sound patterns of words in fluent speech," *Cognitive Psychology, 29*(1): 1–23.

Jusczyk, P. W. (1992), "Developing phonological categories from the speech signal," in C. A. Ferguson, L. Menn, and C. Stoel-Gammon (eds), *Phonological Development: Models, Research, Implications*. Timonium, MD: York Press, pp. 17–64.

Jusczyk, P. W. (1999), "How infants begin to extract words from speech," *Trends in Cognitive Sciences, 3*: 323–8.

Jusczyk, P. W., P. Smolensky, and T. Allocco (2002), "How English-learning infants respond to markedness and faithfulness constraints," *Language Acquisition, 10*(1): 31–73.

Kabak, B, and J. Grijzenhout (eds) (2009), *Phonological Domains: Universals and Deviations*. Berlin and New York: De Gruyter Mouton, pp. 135–94.

Kager, R. (1989), *A Metrical Theory of Stress and Destressing in English and Dutch*. Dordrecht: Foris.

Kager, R. (1999), *Optimality Theory*. Cambridge, UK: Cambridge University Press,

Kager, R. (2007), "Feet and metrical stress," in Paul de Lacy (ed.), *The Cambridge Handbook of Phonology*. Cambridge, UK: Cambridge University Press, pp. 161–94.

Kahn, D. (1976), *Syllable-based generalizations in English phonology*. Doctoral dissertation, Massachusetts Institute of Technology.

Kaisse, E. (1982), "Review of fundamental frequency in sentence production by William Cooper and John M. Sorensen," *Language, 58*(2): 478–9.

Kaisse, E. M. (1985), *Connected Speech: The Interaction of Syntax and Phonology*. New York: Academic Press.

Kaisse, E. M. (1990), "Toward a typology of postlexical rules," in S. Inkelas and D. Zec (eds), *The Phonology-syntax Connection*. Chicago: The University of Chicago Press, pp. 127–43.
Kaisse, E. M. and P. A. Shaw (1985), "On the theory of lexical phonology," *Phonology Yearbook, 2*: 1–30.
Kang, Y. and S. Han (2013), "Tonogenesis in early contemporary Seoul Korean: a longitudinal case study," *Lingua 134*: 62–74.
Karlin, R. (2018), *Tone gestures and tone-bearing units: acoustic studies*. Doctoral dissertation, Cornell University
Karlin, R. and S. Tilsen (2014), "The articulatory tone-bearing unit: gestural coordination of lexical tone in Thai," *The Journal of the Acoustical Society of America, 136*(4): 2176.
Kaye, J., J. Lowenstamm, and J.-R.Vergnaud (1985), "The internal structure of phonological elements: a theory of charm and government," *Phonology Yearbook 2*: 305–28.
Kaye, J., J. Lowenstamm, and J.-R.Vergnaud (1990), "Constituent structure and government in phonology," *Phonology 7*: 193–231.
Kaye, J. J. (1989), *Phonology: A Cognitive View*. Hillsdale, NJ: Erlbaum.
Keane, E. (2014), "The intonational phonology of Tamil," in S.-A. Jun (ed.), *Prosodic Typology II: The Phonology of Intonation and Phrasing*. Oxford: Oxford University Press.
Keating, P. and A. C. Cohn (1988), "Cross-language effects of vowels on consonant onsets," paper presented at the Acoustical Society of America, November, Honolulu, HI.
Keating, P. and A. Lahiri (1993), "Fronted velars, palatalized velars, and palatals," *Phonetica, 50*: 73–101.
Keating, P., T. Cho, C. Fougeron, and C.-S. Hsu (2003), "Domain-initial strengthening in four languages," in J. Local, R. Ogden, and R. Temple (eds), *Phonetic Interpretation: Papers in Laboratory Phonology VI*. Cambridge, UK: Cambridge University Press, pp. 143–61.
Keating, P. A. (1985), "Universal phonetics and the organization of grammars," in V. Fromkin (ed.), *Phonetic Linguistics: Essays in Honor of Peter Ladefoged*. Orlando: Academic Press, pp. 115–32.
Keating, P. A. (1990a), "Phonetic representations in a generative grammar," *Journal of Phonetics, 18*: 321–34.
Keating, P. A. (1990b), "The window model of coarticulation: articulatory evidence," in J. Kingston and M. Beckman (eds), *Papers in Laboratory Phonology I*. Cambridge, UK: Cambridge University Press, pp. 451–70.
Keating, P. A. (1996), "The phonology-phonetics interface," in U. Kleinhenz (ed.), *Interfaces in Phonology*. Berlin: Akademie Verlag, pp. 262–78.
Keating, P. A. (2006), "Phonetic encoding of prosodic structure," in J. Harrington and M. Tabain (eds), *Speech Production: Models, Phonetic Processes, and Techniques*, Macquarie Monographs in Cognitive Science. New York: Psychology Press, pp. 167–86.
Kelso, J. A. S. (1982), *Human Motor Behavior: An Introduction*. Hillsdale, NJ: Erlbaum.
Kelso, J. A. S., E. L. Saltzman, and B. Tuller (1986), "The dynamical perspective on speech production: data and theory," *Journal of Phonetics, 14*: 29–59.
Kenstowicz, M. (1994), *Phonology in Generative Grammar*. Cambridge, MA and Oxford: Blackwell.
Kenstowicz, M. and C. Kisseberth (1977), *Generative Phonology*. New York: Academic Press.
Kim, M.-R. (2013), "Tonogenesis in contemporary Korean with special reference to the onset-tone interaction and the loss of a consonant opposition," *The Journal of the Acoustical Society of America*, 133: 3570.
Kingston, J. (1985), *The phonetics and phonology of the timing of oral and glottal events*. Doctoral dissertation, University of California, Berkeley.

Kingston, J. (2005), "The phonetics of Athabaskan tonogenesis," in S. Hargus and K. Rice (eds), *Athabaskan Prosody*. Amsterdam: John Benjamins. pp. 137–84.
Kingston, J. (2007), "The phonetics-phonology interface," in P. de Lacy (ed.), *Handbook of Phonology*. Cambridge, UK: Cambridge University Press, pp. 435–56.
Kingston, J. (2011), "Tonogenesis," in M. van Oostendorp, C. J. Ewen, E. Hume, and K. Rice (eds), *Blackwell Companion to Phonology*, vol. 4, Oxford: Blackwell Publishing, pp. 2304–34.
Kingston, J. and R. L. Diehl (1994), "Phonetic knowledge," *Language, 70*: 419–54.
Kingston, J. and R. L. Diehl (1995), "Intermediate properties in the perception of distinctive feature values," in B. Connell and A. Arvaniti (eds), *Phonology and Phonetic Evidence: Papers in Laboratory Phonology IV*, Cambridge, UK: Cambridge University Press, pp. 7–27.
Kingston, J., S. Kawahara, D. Chambless, M. Key, D. Mash, and S. Watsky (2014), "Context effects as auditory contrast," *Attention, Perception, & Psychophysics, 76*: 1437–72.
Kiparsky, P. (1982), "Lexical morphology and phonology," in the Linguistic Society of Korea (ed.), *Linguistics in the Morning Calm*. Seoul, South Korea: Hanshin, pp. 3–91.
Kiparsky, P. (1985), "Some consequences of lexical phonology," *Phonology Yearbook, 2*: 85–138.
Kiparsky, P. (1994), "Pāṇinian linguistics," in R. E. Asher (ed.), *Encyclopedia of Language and Linguistics*. Oxford and New York: Pergamon Press, pp. 2918–23.
Kiparsky, P. (2000), "Opacity and cyclicity," *The Linguistic Review, 17*: 351–65.
Kiparsky, P. (2002), "On the architecture of Panini's grammar," January and March, <https: // web.stanford.edu/~kiparsky/Papers/hyderabad.pdf> (last accessed March 3, 2020).
Kiparsky, P. (2006), "Amphichronic linguistics vs. Evolutionary Phonology," *Theoretical Linguistics, 32*: 217–36.
Kiparsky, P. (2008), "Universals constrain change; change results in typological generalizations," in J. Good (ed.), *Language Universals and Language Change*. Oxford: Oxford University Press, pp. 23–53.
Kiparsky, P. (in press), "Panini," in E. Dresher and H. v. d. Hulst (eds), *Handbook of the History of Phonology*. Oxford: Oxford University Press.
Kirby, J. (2014), "Incipient tonogenesis in Phnom Penh Khmer: acoustic and perceptual studies," *Journal of Phonetics, 43*: 69–85.
Kirchner, R. (1998), *An effort based approach to consonant lenition*. Doctoral dissertation, University of California Los Angeles.
Kirchner, R. (2004), "Consonant lenition," in B. Hayes, R. Kirchner, and D. Steriade (eds), *Phonetically Based Phonology*. Cambridge, MA: Cambridge University Press, pp. 313–45.
Kisseberth, C. (1994), "On domains," in J. Cole and C. Kisseberth (eds), *Perspectives in Phonology*. Stanford: CSLI Publications, pp. 133–66.
Klatt, D. H. (1976), "Linguistic uses of segmental duration in English: acoustic and perceptual evidence," *The Journal of the Acoustical Society of America, 59*(5): 1208–21.
Klatt, D. H. (1979), "Speech perception: a model of acoustic-phonetic analysis and lexical access," *Journal of Phonetics, 7*: 279–312.
Klatt, D. H. (1989), "Review of selected models of speech perception," in W. Marslen-Wilson (ed.), *Lexical Representation and Process*. Cambridge, MA: MIT Press, pp. 169–226.
Klima, E. S. and U. Bellugi (1979), *The Signs of Language*. Cambridge, MA: Harvard University Press.
Kluender, K. R., R. L. Diehl, and P. R. Killeen (1987), "Japanese quail can learn phonetic categories," *Science, 237*: 1195–7.
Kochetov, A. and M. Pouplier (2008), "Phonetic variability and grammatical knowledge: an articulatory study of Korean place assimilation," *Phonology, 25*: 399–431.

Kochetov, A. and L. Colantoni (2011), "Spanish nasal assimilation revisited: a cross-dialect electropalatographic study," *Journal of Laboratory Phonology, 2*: 487–523.

Kochetov, A., M. Pouplier, and M. Son (2007), "Cross-language differences in overlap and assimilation patterns in Korean and Russian," in W. J. Barry and J. Trouvain (eds), *Proceedings of the XVIth International Congress of Phonetic Sciences,* Saarbücken, Germany: Universität des Saarlandes, pp. 1361–4.

Köhnlein, B. (2016), "Contrastive foot structure in Franconian tone-accent dialects," *Phonology, 33*: 87–123.

Kopp, G. A. and H. C. Green (1946), "Basic phonetic principles of Visible Speech," *The Journal of the Acoustical Society of America, 18*: 74–89.

Krakow, R. A. (1989), *The articulatory organization of syllables: a kinematic analysis of labial and velar gestures.* Doctoral dissertation, New Haven: Yale University.

Krakow, R. A. (1993), "Nonsegmental influences on velum movement patterns: syllables, sentences, stress, and speaking rate," in M. A. Huffman and R. A. Krakow (eds), *Nasals, Nasalization, and the Velum (Phonetics and Phonology V).* New York: Academic Press, pp. 87–116.

Krakow, R. A. (1999), "Physiological organization of syllables: a review," *Journal of Phonetics, 27*(1): 23–54.

Krämer, M. and D. Zec (2020), "Nasal consonants, sonority and syllable phonotactics: the dual nasal hypothesis," *Phonology, 37*: 1–37.

Kristoffersen, G. (2000), *The Phonology of Norwegian.* Oxford: Oxford University Press.

Krueger, J. (1962), *A Grammar of Yakut.* Bloomington: Indiana University Press.

Kuang, J. (2017), "Covariation between voice quality and pitch: revisiting the case of Mandarin creaky voice," *The Journal of the Acoustical Society of America, 142*: 1693.

Kuhl, P. K. (1991), "Human adults and human infants show a 'perceptual magnet effect' for the prototypes of speech categories, monkeys do not," *Perception and Psychophysics 50*: 93–107.

Kuhl, P. K. and J. D. Miller (1975), "Speech perception by the chinchilla: voiced–voiceless distinction in alveolar plosive consonants," *Science, 190*: 69–72.

Kuhl, P. K. and D. M. Padden (1983), "Enhanced discriminability at the phonetic boundaries for voicing feature in macaques," *The Journal of the Acoustical Society of America, 73*: 1003–10.

Kuhl, P. K. and P. Iverson (1995), "Linguistic experience and the 'perceptual magnet effect'," in W. Strange (ed.), *Speech Perception and Linguistic Experience: Issues in Cross-language Research.* Baltimore: York Press, pp. 121–54.

Kuhl, P. K., K. A. Williams, F. Lacerda, K. N. Stevens, and B. Lindblom (1992), "Linguistic experience alters phonetic perception in infants by 6 months of age," *Science, 255*: 606–8.

Kuhl, P. K., E. Stevens, A. Hayashi, T. Deguchi, S. Kiritani, and P. Iverson (2006), "Infants show a facilitation effect for native language phonetic perception between 6 and 12 months," *Developmental Science, 9*: 13–21.

Kuhl, P. K., B. T. Conboy, S. Coffey-Corina, D. Padden, M. Rivera-Gaxiola, and T. Nelson (2008), "Phonetic learning as a pathway to language: new data and Native Language Magnet theory expanded (NLM-e)," *Philosophical Transactions of the Royal Society B, 363*: 979–1000.

Kühnert, B., P. Hoole, and C. Mooshammer (2006), "Gestural overlap and C-center in selected French consonant clusters," in H. C. Yehia, D. Demolin, and R. Labossière (eds), *Proceedings of the 7th International Seminar on Speech Production (ISSP),* December, Ubatuba, Brazil, pp. 327–34, <https://univ-paris3.academia.edu/BarbaraKuhnert> (last accessed April 1, 2020).

Ladd, D. R. (1992), "An introduction to intonational phonology," in G. J. Docherty and D. R.

Ladd (eds), *Papers in Laboratory Phonology II: Gesture, Segment, Prosody*, Cambridge, UK: Cambridge University Press, pp. 321–34.

Ladd, D. R. (2008), *Intonational Phonology*. Cambridge, UK: Cambridge University Press.

Ladd, D. R. (2014), *Simultaneous Structure in Phonology*. Oxford: Oxford University Press.

Ladd, D. R. (in press), "Mid-century American phonology: the post-Bloomfieldians," in E. Dresher and H. van der Hulst (eds), *Handbook of the History of Phonology*. Oxford: Oxford University Press.

Ladd, D. R., and A. Schepman (2003), "'Sagging transitions' between high accent peaks in English: experimental evidence," *Journal of Phonetics, 31*: 81–112.

Ladd, D. R. and J. M. Scobbie (2003), "External sandhi as gestural overlap? Counter-evidence from Sardinian," in J. Local, R. Ogden, and R. Temple (eds), *Papers in Laboratory Phonology VI: Phonetic Interpretation*. Cambridge, UK: Cambridge University Press, pp. 164–82.

Ladd, D. R., A. Schepman, L. White, L. M. Quarmby, and R. Stackhouse (2009), "Structural and dialectal effects on pitch peak alignment in two varieties of British English," *Journal of Phonetics, 37*(2): 154–61.

Ladefoged, P. (1967), *Three Areas of Experimental Phonetics*. Oxford: Oxford University Press.

Ladefoged, P. (1980), "What are linguistic sounds made of?" *Language, 56*, 485–502.

Ladefoged, P. (1993), *A Course in Phonetics (Third Edition)*. New York: Harcourt Brace.

Ladefoged, P. (2004), "Phonetics and phonology in the last 50 years," *UCLA Working Papers in Phonetics, 103*: 1–11.

Ladefoged, P. and D. E. Broadbent (1957), "Information conveyed by vowels," *The Journal of the Acoustical Society of America*, 29: 98–104.

Lado, R. (1957), *Linguistics across Cultures*. Ann Arbor: University of Michigan.

Lahiri, A. (2000), "Phonology: structure, representation, and process," in L. Wheeldon (ed.), *Aspects of Language Production*. Hove: Psychology Press, pp. 165–225.

Lahiri, A. and W. D. Marslen-Wilson (1991), "The mental representation of lexical form: a phonological approach to the recognition lexicon," *Cognition, 38*: 245–94.

Lahiri, A. and W. D. Marslen-Wilson (1992), "Lexical processing and phonological representation," in G. J. Docherty and D. R. Ladd (eds), *Papers in Laboratory Phonology II: Gesture, Segment, Prosody*. Cambridge, UK: Cambridge University Press, pp. 229–54.

Lahiri, A. and H. Reetz (2002), "Underspecified recognition," in C. Gussenhoven and N. Warner (eds.), *Laboratory Phonology VII*. Berlin: de Gruyter Mouton, pp. 637–77.

Lahiri, A. and H. Reetz (2010), "Distinctive features: phonological underspecification in processing," *Journal of Phonetics, 38*: 44–59.

Lavoie, L. (2001), *Consonant Strength: Phonological Patterns and Phonetic Manifestations*. London: Routledge.

Leben, W. R. (1973), *Suprasegmental phonology*. Doctoral dissertation, Massachusetts Institute of Technology.

Leben, W. R. (1978), "The representation of tone," in V. Fromkin (ed.), *Tone: A Linguistic Survey*. New York: Academic Press, pp. 173–219.

Lees, R. (1957), "Review of *Syntactic Structures*," *Language, 33*(3): 375–408.

Lehiste, I. (1970), *Suprasegmentals*. Cambridge, MA: MIT Press.

Liberman, A. M. (1970), "Discrimination in speech and nonspeech modes," *Cognitive Psychology, 2*: 131–57.

Liberman, A. M. (1996), *Speech: A Special Code*. Cambridge, MA: MIT Press

Liberman, A. M. and I. G. Mattingly (1985), "The Motor Theory of speech perception revised," *Cognition, 21*: 1–36.

Liberman, A. M. and D. H. Whalen (2000), "On the relation of speech to language," *Trends in Cognitive Sciences, 4*: 187–96.
Liberman, A. M., K. S. Harris, H. S. Hoffman, and B. Griffith (1957), "The discrimination of speech sounds with and across phoneme boundaries," *Journal of Experimental Psychology, 53*: 358–68.
Liberman A. M., F. S. Cooper, D. P. Shankweiler, and M. Studdert-Kennedy (1967), "Perception of the speech code," *Psychological Review, 74*: 476.
Liberman, M. and A. Prince (1977), "On stress and linguistic rhythm," *Linguistic Inquiry, 8*: 249–336.
Liberman, M. and J. Pierrehumbert (1984), "Intonational invariance under changes in pitch range and length," in M. Aronoff, R. Oehrle, F. Kelley, and B. Wilker Stephens (eds), *Language Sound Structure*. Cambridge, MA: MIT Press, pp. 157–233.
Liberman, M. Y. (1985), *The Intonation System of English*. New York: Garland.
Lieberman, P. (1967), *Intonation, Perception, and Language*. Cambridge, MA: MIT Press.
Liljencrants, J. and B. Lindblom (1972), "Numerical simulation of vowel quality systems: the role of perceptual contrast," *Language, 48*(4): 839–62.
Lindau, M. and P. Ladefoged (1986), "Variability of feature specifications," in J. S. Perkell and D. H. Klatt (eds), *Invariance and Variability of Speech Processes*. Hillsdale, NJ: Lawrence Erlbaum, pp. 464–78.
Lindblom, B. (1967), "Vowel duration and a model of lip-mandible coordination," *Speech Transmission Laboratories Progress Status Report, 4*: 1–29.
Lindblom, B. (1990), "On the notion of 'possible speech sound'," *Journal of Phonetics, 18*: 135–52.
Lindblom, B. (1986), "Phonetic universals in vowel systems," in J. Ohala and J. Jaeger (eds), *Experimental Phonology*. Orlando: Academic Press, pp. 13–44.
Lisker, L. and A. S. Abramson (1964), "A cross-language study of voicing in initial stops: acoustical measurements," *Word, 20*(3): 384–422.
Lisker, L. and A. S. Abramson (1970), "The voicing dimension: some experiments in comparative phonetics," in H. Bohuslav, M. Romportl, and P. Janota (eds), *Proceedings of the 6th International Conference of Phonetic Sciences*. Prague: Academia Publishing House of the Czechoslovak Academy of Sciences, pp. 563–7.
Löfqvist, A. and V. L. Gracco (1999), "Interarticulator programming in VCV sequences: lip and tongue movements," *The Journal of the Acoustical Society of America, 105*(3): 1864–76.
Lombardi, L. (1991), *Laryngeal Features and Laryngeal Neutralization*. New York: Garland.
Luce, P. A. and C. T. McLennan (2005), "Spoken word recognition: the challenge of variation," in D. B. Pisoni and R. E. Remez (eds), *The Handbook of Speech Perception*. Oxford: Blackwell, pp. 591–609.
McCarthy, J. J. (1979), "Stress and syllabification," *Linguistic Inquiry, 10*: 443–66.
McCarthy, J. J. (1988), "Feature geometry and dependency: a review," *Phonetica, 45* : 84–108.
McCarthy, J. J. (2005), "Optimal paradigms," in L. J. Downing, T. A. Hall, and R. Raffelsiefen (eds), *Paradigms in Phonological Theory*. Oxford and New York: Oxford University Press, pp. 170–210.
McCarthy, J. J. (2007), *Hidden Generalizations: Phonological Opacity in Optimality Theory*. London: Equinox.
McCarthy, J. J. (2008), *Doing Optimality Theory*. Malden, MA: Wiley-Blackwell.
McCarthy, J. J. (2009), "Harmony in harmonic serialism," *Linguistics Department Faculty Publications, 41*, <http: //scholarworks.umass.edu/linguist_faculty_pubs/41> (last accessed March 3, 2020).
McCarthy, J. J. (2011), "Autosegmental spreading in Optimality Theory," in J. Goldsmith,

E. Hume, and W. L. Wetzels (eds), *Tones and Features: Phonetic And Phonological Perspectives*. Berlin: De Gruyter Mouton, pp. 195–222.
McCarthy, J. J. and A. Prince (1993), "Generalized alignment," in G. Booij and J. Van Marle (eds), *Yearbook of Morphology 1993*. Dordrecht: Springer, pp. 79–153.
McCawley, J. (1978), "What is a tone language?" in V. Fromkin (ed.), *Tone: A linguistic survey*. New York: Academic Press.
McClelland, J. L. and J. L. Elman (1986), "The TRACE model of speech perception," *Cognitive Psychology, 18*: 1–86.
McDermott, J. (2009), "The cocktail party problem," *Current Biology, 19*(22): R1024–7.
McGurk, H. and J. W. MacDonald (1976), "Hearing lips and seeing voices," *Nature, 264*: 746–8.
Macken, M. (1978), "Permitted complexity in phonological development: one child's acquisition of Spanish consonants," *Lingua, 44*: 219–53.
McMurray, B., R. N. Aslin, and J. C. Toscano (2009), "Statistical learning of phonetic categories: insights from a computational approach," *Developmental Science, 12*(3): 369–78.
McQueen, J. M. and A. Cutler (1997), "Cognitive processes in speech perception," in W. J. Hardcastle and J. Laver (eds), *The Handbook of Phonetic Sciences*. Oxford: Blackwell, pp. 566–85.
Maddieson, I. (1978), "Universals of tone," in J. Greenberg (ed.), *Universals of Human Language, 2*. Stanford: Stanford University Press, pp. 337–465.
Maddieson, I. (1984), *Patterns of Sounds*. Cambridge, UK: Cambridge University Press.
Maddieson, I. (2013), "Consonant inventories," in M. S. Dryer and M. Haspelmath (eds), *The World Atlas of Language Structures Online*. Leipzig: Max Planck Institute for Evolutionary Anthropology, <http: //wals.info/chapter/1> (last accessed March 3, 2020).
Marin, S. (2013), "The temporal organization of complex onsets and codas in Romanian: a gestural approach," *Journal of Phonetics, 41*(3/4): 211–27.
Marin, S. and M. Pouplier (2010), "Temporal organization of complex onsets and codas in American English: testing the predictions of a gestural coupling model," *Motor Control, 14*(3): 380–407.
Martinet, A. (1955), *Économie des changements phonétiques*. Berne, Switzerland: Francke.
Mascaró, J. (1976), *Catalan phonology and the phonological cycle*. Doctoral dissertation, Massachusetts Institute of Technology.
Massaro, D. W. (1987), "Psychophysics vs. specialized processes in speech perception: an alternative perspective," *Psychological Review, 84*: 452–71.
Massaro, D. W. and T. H. Chen (2008), "The Motor Theory of speech perception revisited," *Psychonomic Bulletin and Review, 15*(2): 453–62.
Maye, J. C. (2000). *Learning speech sound categories from statistical information*. Doctoral dissertation, Department of Linguistics, University of Arizona.
Mielke, J. (2008), *The Emergence of Distinctive Features*. Oxford: Oxford University Press.
Mielke, J. (2013), "Phonologization and the typology of feature behaviour," in A. Yu (ed.), *Origins of Sound Change: Approaches to Phonologization*. Oxford: Oxford University Press, pp. 165–80.
Miller, B. (2012), "Sonority and the larynx," in S. Parker (ed.), *The Sonority Controversy*. Berlin and Boston: Walter de Gruyter, pp. 257–88.
Mixdorff, H., S. Luksaneeyanawin, H. Fujisaki, and P. Charnavit, (2002), "Perception of tone and vowel quality in Thai," paper presented at the 7th International Congress on Spoken Language Processing, September, Denver, Colorado.
Mohanan, K. P. (1982), *Lexical phonology*. Doctoral dissertation, Massachusetts Institute of Technology.

Moore, B. C. J. (1997), "Aspects of auditory processing related to speech perception," in W. J. Hardcastle and J. Laver (eds), *The Handbook of Phonetic Sciences*. Oxford: Blackwell, pp. 539–65.

Morén, B. and E. Zsiga (2006), "The lexical and post-lexical phonology of Thai tones," *Natural Language and Linguistic Theory, 24*: 113–78.

Morén-Duolljá, B. (2013), "The prosody of Swedish underived nouns: no lexical tones required," *Nordlyd 40(1)*: 196–248

Moreton, E. (2002), "Structural constraints in the perception of English stop-sonorant clusters," *Cognition, 84*, 55–71.

Moreton, E. and J. Pater (2012), "Structure and substance in artificial-phonology learning. Part I, Structure. Part II, Substance," *Language and Linguistics Compass* 6(11): 686–701, 702–18.

Morton, E. W. (1977), "On the occurrence and significance of motivation-structural rules in some bird and mammal sounds," *The American Naturalist, 111*: 855–69.

Mowrey, R. A. and I. R. MacKay (1990), "Phonological primitives: electromyographic speech error evidence," *The Journal of the Acoustical Society of America, 88*(3): 1299–1312.

Myers, S. (1990), *Tone and the Structure of Words in Shona*. New York: Garland.

Myers, S. (1996), "Boundary tones and the phonetic implementation of tone in Chichewa," *Studies in African Linguistics, 25*: 29–60.

Myers, S. (2000), "Boundary disputes: the distinction between phonetic and phonological sound patterns," in N. Burton-Roberts, P. Carr, and G. Docherty (eds), *Phonological Knowledge: Conceptual and Empirical Issues*. Oxford: Oxford University Press, pp. 245–72.

Myers, S. (2003), "F0 timing in Kinyarwanda," *Phonetica, 60*: 71–97.

Nakatani, L. H. and C. H. Aston (1978), "Perceiving the stress pattern of words in sentences," *Journal of the Acoustical Society of America, 63*: S55.

Nam, H. (2007), "Syllable-level intergestural timing model: split-gesture dynamics focusing on positional asymmetry and moraic structure," in J. Cole and J. I Hualde (eds), *Laboratory Phonology, 9*. Berlin: De Gruyter Mouton, pp. 483–506.

Nam, H. and E. Saltzman (2003), "A competitive, coupled oscillator model of syllable structure," in M. J. Sole, D. Recasens, and J. Romero (eds), *Proceedings of the 15th International Conference on Phonetic Sciences*, Barcelona, pp. 2253–6, <http://www.internationalphoneticassociation.org/icphs/icphs2003> (last accessed April 8, 2020).

Nam, H., L. Goldstein, and E. Saltzman (2009), "Self-organization of syllable structure: a coupled oscillator model," in F. Pellegrino, E. Marsico, I. Chitoran, and C. Coupé (eds), *Approaches to Phonological Complexity*. Berlin: De Gruyter Mouton, pp. 299–328.

Nava, E., L. Goldstein, E. Saltzman, H. Nam, and M. Proctor (2008), "Modeling prosodic rhythm: evidence from L2 speech," poster presented at the 156th meeting of the Acoustical Society of America, November, Miami, FL.

Nespor, M. and I. Vogel (1986), *Prosodic Phonology*. Dordrecht: Foris Publications.

Nespor, M. and I. Vogel (1989), "On clashes and lapses," *Phonology, 6*(1): 69–116.

Newport, E. L. (2016), "Statistical language learning: computational, maturational and linguistic constraints," *Language and Cognition, 8*: 447–61.

Nitisaroj, R. (2006), *Effects of stress and speaking rate on Thai tones*. Doctoral dissertation, Georgetown University.

Nolan, F. (1992), "The descriptive role of segments: evidence from assimilation," in G. J. Docherty and D. R. Ladd (eds), *Papers in Laboratory Phonology II: Gesture, Segment, Prosody*. Cambridge, UK: Cambridge University Press, pp. 261–79.

Norris, D., J. M. McQueen, and A. Cutler (2000), "Merging information in speech recognition: feedback is never necessary," *Behavioral and Brain Sciences, 23*: 299–370.

Nosofsky, R. M. (2014), "The generalized context model: an exemplar model of classification," in M. Pothos and A. Wills (eds), *Formal Approaches in Categorization*. New York: Cambridge University Press, pp. 18–39.

Nycz, J. (2018), "Stylistic variation among mobile speakers: using old and new regional variables to construct complex place identity," *Language Variation and Change, 30*(20): 175–202.

Nygaard, L. C. (2005), "Perceptual integration of linguistic and nonlinguistic properties in speech," in D. B. Pisoni and R. E. Remez (eds), *The Handbook of Speech Perception*. Oxford: Blackwell, pp. 390–413.

O'Connor, J. and G. Arnold (1973), *Intonation of Colloquial English*. London: Longman.

Odden, D. (1981), *Problems in tone assignment in Shona*. Doctoral dissertation: University of Illinois at Urbana-Champaign.

Odden, D. (1996), "Reviewed work(s): Grounded Phonology by Diana Archangeli and Douglas Pulleyblank," *Language, 72*(1): 153–5.

Ohala, J. J. (1973), "The physiology of tone," in L. M. Hyman (ed.), *Consonant Types and Tone: Southern California Occasional Papers in Linguistics 1*. Los Angeles: University of Southern California, pp. 1–14.

Ohala, J. J. (1977), "The physiology of stress," in L. Hyman (ed.), *Studies in Stress and Accent. Southern California Occasional Papers in Linguistics 4*. Los Angeles: University of Southern California.

Ohala, J. J. (1978), "The production of tone," in V. Fromkin (ed.), *Tone: A Linguistic Survey*. New York: Academic Press, pp. 5–40.

Ohala, J. J. (1981), "The listener as a source of sound change," in C. S. Masek, R. A. Hendrick, and M. F. Miller (eds), *Papers from the Parasession on Language and Behavior*. Chicago: Chicago Linguistic Society, pp. 178–203.

Ohala, J. J. (1983), "Cross-language use of pitch: an ethological view," *Phonetica, 40*: 1–18.

Ohala, J. J. (1990), "There is no interface between phonology and phonetics," *Journal of Phonetics, 18*: 153–71.

Ohala, J. J. (1992), "The segment: primitive or derived?" in G. J. Docherty and D. R. Ladd (eds), *Papers in Laboratory Phonology II: Gesture, Segment, Prosody*. Cambridge, UK: Cambridge University Press, pp. 166–83.

Ohala, J. J. (1996), "The frequency code underlies the sound symbolic use of voice pitch," in L. Hinton, J. Nichols, and J. J. Ohala (eds), *Sound Symbolism*. Cambridge, UK: Cambridge University Press, pp. 325–47.

Ohala, J. J. and W. G. Ewan (1972), "Speed of pitch change," *The Journal of the Acoustical Society of America, 53*: 345.

Orr, C. (1962), "Ecuador Quichua phonology," in B. Elson (ed.), *Studies in Ecuadorian Indian Languages*. Norman, OK: Summer Institute in Linguistics, pp. 60–77.

Padgett, J. (1994), "Stricture and nasal place assimilation," *Natural Language and Linguistic Theory, 12*(3): 465–513.

Padgett, J. (2001), "Contrast dispersion and Russian palatalization," in E. V. Hume and K. Johnson (eds), *The Role of Speech Perception Phenomena in Phonology*. San Diego: Academic Press, pp. 187–218.

Padgett, J. (2002), "Feature classes in phonology," *Language, 78*(1): 81–110.

Padgett, J. (2011), "Consonant-vowel place feature interactions," in M. van Oostendorp, C. Ewen, E. Hume, and K. Rice (eds), *The Blackwell Companion to Phonology*. Malden, MA: Wiley-Blackwell, pp. 1761–86.

Palmer, H. E. (1922), *English Intonation with Systemic Exercises*. Cambridge, UK: Heffner.

Pan, H. (2007), "Initial strengthening of lexical tones in Taiwanese Min," in C. Gussenhoven and T. Riad (eds), *Tones and Tunes, Volume 2: Experimental Studies in Word and Sentence Prosody*. Berlin: De Gruyter Mouton.

Parker, S. and A. Lahiri (2012), *The Sonority Controversy*. Berlin and Boston: de Gruyter Mouton.

Passy, P. (1890), *Étude sur les changements phonétiques et leurs caractères généraux*. Paris: Firmin-Didot.

Pater, J. (1999), "Austronesian nasal substitution and other NC effects," in R. Kager, H. van der Hulst, and W. Zonneveld (eds), *The Prosody-morphology Interface*. Cambridge, UK: Cambridge University Press, pp. 310–43.

Pater, J. (2010), "Morpheme-specific phonology: constraint indexation and inconsistency resolution," in S. Parker (ed.), *Phonological Argumentation: Essays on Evidence and Motivation*. London: Equinox, pp. 123–54.

Pearl, L. and S. Goldwater (2016), "Statistical learning, inductive bias, and Bayesian inference in language acquisition," in J. Lidz, W. Snyder, and J. Pater (eds), *Oxford Handbook of Developmental Linguistics*. Oxford: Oxford University Press. pp. 664–95.

Peperkamp, S. and E. Dupoux (2002), "A typological study of stress deafness," in C. Gussenhoven and N. Warner (eds), *Laboratory Phonology 7*. Berlin: de Gruyter Mouton, pp. 203–40.

Peperkamp, S., I. Vendelin, and E. Dupoux (2010), "Perception of predictable stress: a cross-linguistics investigation," *Journal of Phonetics, 38*: 422–30.

Perera, H. S. and D. Jones (1919), *A Colloquial Sinhalese Reader*. Manchester: Manchester University Press.

Petersen, S. (2018), *Accounting for dipthongs: duration as contrast in vowel Dispersion Theory*. Doctoral dissertation, Georgetown University.

Pfiffner, A. (2019), *(Incomplete) neutralization: the loss and maintenance of contrast and its trajectory over time*. Doctoral dissertation, Georgetown University.

Picanço, G. (2005), *Munduruku: phonetics, phonology, synchrony, diachrony*. Doctoral dissertation, University of British Columbia.

Pierrehumbert, J. (1990b), "Phonological and phonetic representation," *Journal of Phonetics, 18*: 375–94.

Pierrehumbert, J. (1994), "Syllable structure and word structure: a study of triconsonantal clusters in English," in P. Keating (ed.), *Papers in Laboratory Phonology III*. Cambridge, UK: Cambridge University Press. pp. 168–88.

Pierrehumbert, J. (2000), "The phonetic grounding of phonology," *Bulletin de la Communication Parlée, 5*: 7–23.

Pierrehumbert, J. B. (1990a), *The Phonology and Phonetics of English Intonation*. New York: Garland Press.

Pierrehumbert, J. B. (2002), "Word-specific phonetics," in C. Gussenhoven and N. Warner (eds), *Laboratory Phonology VII*. Berlin: De Gruyter Mouton, pp. 101–39.

Pierrehumbert, J. B. (2003), "Phonetic diversity, statistical learning, and acquisition of phonology," *Language and Speech, 46*: 115–54.

Pierrehumbert, J. B. (2012), "The dynamic lexicon," in A. Cohn, M. Huffman, and C. Fougeron (eds), *Handbook of Laboratory Phonology*. Oxford: Oxford University Press, pp. 173–83.

Pierrehumbert, J. B. (2016), "Phonological representation: beyond abstract vs. episodic," *Annual Review of Linguistics, 2*: 33–52.

Pierrehumbert, J. B. and M. E. Beckman (1988), *Japanese Tone Structure*. Cambridge, MA: MIT Press.

Pierrehumbert, J. B. and S. Steele (1989), "Categories of tonal alignment in English," *Phonetica, 46*: 181–96.

Pierrehumbert, J. and D. Talkin (1992), "Lenition of /h/ and glottal stop," in G. Docherty and D. R. Ladd (eds), *Gesture, Segment, Prosody: Papers in Laboratory Phonology II*. Cambridge, UK: Cambridge University Press, pp. 90–116.

Pierrehumbert, J., M. E. Beckman, and D. R. Ladd (2000), "Conceptual foundations of phonology as a laboratory science," in N. Burton-Roberts, P. Carr, and G. Docherty (eds), *Phonological Knowledge: Conceptual and Empirical Issues*. Oxford: Oxford University Press, pp. 273–303.

Pike, K. (1945), *The Intonation of American English*. Ann Arbor: University of Michigan Press.

Pike, K. (1948), *Tone Languages*. Ann Arbor: University of Michigan Press.

Pisoni, D. and P. A. Luce (1986), "Speech perception: research, theory, and the principal issues," in E. C. Schwab and L. C. Nusbaum (eds), *Perception of Speech and Visual Form*. New York: Academic Press. pp. 1–50.

Pluymaekers, M., M. Ernestus, and R. H. Baayen (2005), "Lexical frequency and acoustic reduction in spoken Dutch," *The Journal of the Acoustical Society of America, 118*: 2561–9.

Polivanov, E. D. (1931), "La perception des sons d'une langage étrangère," *Travaux du Cercle Linguistique de Prague, 4*: 79–96.

Port, R. F. and A. Leary (2005), "Against formal phonology," *Language, 81*: 927–64.

Poser, W. J. (1982), "Phonological representation and action at a distance," in H. van der Hulst and N. Smith (eds), *The Structure of Phonological Representations (part II)*. Dordrecht: Foris, pp. 121–58.

Poser, W. J. (1992), "The structural typology of phonological writing," paper presented at the Berkeley Linguistics Society Meeting, February, Berkeley, CA.

Potisuk, S., J. Gandour, and M. P. Harper (1996), "Acoustic correlates of stress in Thai," *Phonetica, 53*: 200–20.

Pouplier, M. (2003), *Units of phonological encoding: empirical evidence*. Doctoral dissertation, Yale University.

Pouplier, M. and W. Hardcastle (2005), "A re-evaluation of the nature of speech errors in normal and disordered speakers," *Phonetica, 62*, 227–43.

Pouplier, M. and Š. Beňuš (2011), "On the phonetic status of syllabic consonants: evidence from Slovak," *Laboratory Phonology, 2*(2): 243–73.

Prieto, P. (2011), "Tonal alignment," in M. van Oostendrop, C. Ewen, B. Hume, and K. Rice (eds), *The Blackwell Companion to Phonology*. Malden, MA: Wiley Blackwell, pp. 1185–1203.

Prieto, P. (1997), "Prosodic manifestation of syntactic structure in Catalan," in F. Martínez-Gil and A. Morales-Front (eds), *Issues in the Phonology of the Iberian Languages*. Washington DC: Georgetown University Press, pp. 179–99.

Prieto, P., M. D'Imperio, and B. Gili Fivela (2005), "Pitch accent alignment in Romance: primary and secondary associations with metrical structure," *Language and Speech, 48*(4): 359–96.

Prince, A. (1980), "A metrical theory for Estonian quantity," *Linguistic Inquiry, 11*(3): 511–62.

Prince, A. and Smolensky, P. (2004), *Optimality Theory: Constraint Interaction in Generative Grammar*. Oxford: Blackwell.

Pulleyblank, D. (1986), *Tone in Lexical Phonology*. Dordrecht: Reidel.

Pullum, G. K. (1976), "The Duke of York gambit," *Journal of Linguistics, 12*: 83–102.

Pye, C., D. Ingram, and H. List (1987), "A comparison of initial consonant acquisition in English and Quiché," *Children's Language, 6*: 175–90.

Raimy, E. and C. E. Cairns (2009), "Architecture and representation in phonology," in E. Raimy and C. E. Cairns (eds), *Contemporary Views on Architecture and Representation in Phonology*. Cambridge, MA: MIT Press, pp. 1–16.

Raimy, E. and C. E. Cairns (eds) (2015a), *The Segment in Phonetics and Phonology*. Malden, MA and Oxford: Wiley-Blackwell.

Raimy, E. and C. E. Cairns (2015b), "Introduction," in *The Segment in Phonetics and Phonology*. Malden, MA and Oxford: Wiley-Blackwell.

Raphael, L. J. (2005), "Acoustic cues to the perception of segmental phonemes," in D. B. Pisoni and R. E. Remez (eds), *The Handbook of Speech Perception*. Oxford: Blackwell. pp. 182–206.

Read, C., Y.-F. Zhang, N. Hong-Yin, and D. Bao-Qing (1986), "The ability to manipulate speech sounds depends on knowing alphabetic writing," *Cognition, 24*: 31–44.

Reiss, C. (2007), "Modularity in the 'sound' domain: implications for the purview of universal grammar," in G. Ramchand and C. Reiss (eds), *The Oxford Handbook of Linguistic Interfaces*. New York: Oxford University Press, pp. 53–80.

Reiss, C. (2018), "Substance-free Phonology," in S. J. Hannahs and A. R. K. Bosch (eds), *The Routledge Handbook of Phonological Theory*. London: Routledge, pp. 425–52.

Remez, R. (2001), "The interplay of phonology and perception considered from the perspective of perceptual organization," in E. Hume and K. Johnson (eds), *The Role of Speech Perception in Phonology*. New York: Academic Press, pp. 27–52.

Remez, R. E., P. E. Rubin, D. B. Pisoni, and T. D. Carrell (1981), "Speech perception without traditional speech cues," *Science, 212*: 947–9

Repp, B. H. and H. Lin (1989), "Acoustic properties and perception of stop consonant release transients," *The Journal of the Acoustical Society of America, 85*(1): 379–96.

Rice, K. (2007), "Markedness in phonology," in P. de Lacy (ed.), *The Cambridge Handbook of Phonology*. Cambridge, UK: Cambridge University Press, pp. 79–97.

Ridouane, R. (2008), "Syllables without vowels: phonetic and phonological evidence from Tashlhyit Berber," *Phonology, 25*: 321–59.

Ridouane, R. (2017), "La syllabe en tachlhit: une cause célèbre en phonologie et en phonétique," *Bulletin de la Societe de Linguistique de Paris, 112*: 301–30. doi: 10.2143/BSL.112.1.3271883.

Ridouane, R. and C. Fougeron (2011), "Schwa elements in Tashlhiyt word-initial clusters," *Laboratory Phonology, 2*(2): 275–300.

Ridouane, R., A. Hermes, and P. Hallé (2014), "Tashlhyit's ban of complex syllable onsets: phonetic and perceptual evidence," *STUF–Language Typology and Universals, 67*(1): 7–20.

Rizzolatti G. and L. Craighero (2004), "The mirror-neuron system," *Annual Review of Neuroscience, 27*: 169–92.

Robins, R. H. (1957), "Vowel nasality in Sundanese: a phonological and grammatical study," in J. R. Firth (ed.), *Studies in Linguistic Analysis (Philological Society)*. Oxford: Blackwell, pp. 87–103.

Roengpitya, R. (2007), "The variations, quantification, and generalizations of Standard Thai tones," in M.-J. Solé, P. Beddor, and M. Ohala (eds), *Experimental Approaches to Phonology*. Oxford: Oxford University Press. pp. 270–301.

Rogers, H. (2004), *Writing Systems: A Linguistic Approach*. Malden, MA and Oxford: Wiley-Blackwell.

Rose, S. (2018), "ATR vowel harmony: new patterns and diagnostics," in G. Gallagher, M. Gouskova, and S. Yin (eds), *Proceedings of the 2017 Annual Meeting on Phonology*. Washington, DC: Linguistic Society of America, <https://journals.linguisticsociety.org/proceedings/index.php/amphonology/article/view/4254/3889> (last accessed April 8, 2020).

Rosenblum, L. D. (2005), "Primacy of multimodal speech perception," in D. B. Pisoni and R. E. Remez (eds), *The Handbook of Speech Perception*. Oxford: Blackwell, pp. 51–78.

Roussel, N. (1996), "Speech perception: a psychoacoustic perspective," *National Student Speech Hearing Association Journal, 23*: 46–54.

Rubach, J. (1981), *Cyclic Phonology and Palatalization in Polish and English*. Warsaw, Poland: Wydawnictwa Uniwersytetu Warszawskiego.

Rudy, S. (1990), *Roman Jakobson, 1896–1982: A Complete Bibliography of His Writings*. New York: De Gruyter Mouton.

Saffran, J., R. Aslin, and E. L. Newport (1996), "Statistical learning by 8-month-old infants," *Science, 274*: 1926–8.

Sagey, E. C. (1986), *The representation of features and relations in non-linear phonology*. Doctoral dissertation, Massachusetts Institute of Technology.

Sagey, E. C. (1988), "On the ill-formedness of crossing association lines," *Linguistic Inquiry, 19*(1): 109–18.

Saltzman, E. L. (1986), "Task dynamic coordination of the speech articulators: a preliminary model. Generation and modulation of action patterns," in H. Heuer and C. Fromm (eds), *Experimental Brain Research, Series, 15*. New York: Springer-Verlag, pp. 129–44.

Saltzman, E. L. and K. G. Munhall (1989), "A dynamical approach to gestural patterning in speech production," *Ecological Psychology, 1*(4): 333–82.

Saltzman, E., H. Nam, J. Krivokapic, and L. Goldstein (2008), "A task-dynamic toolkit for modeling the effects of prosodic structure on articulation," in P. A. Barbosa, S. Madureira, and C. Reis (eds), *Proceedings of the 4th International Conference on Speech Prosody (Speech Prosody 2008)*. Campinas, Brazil, pp. 175–84, <http://www.isca-speech.org/archive/sp2008> (last accessed April 8, 2020).

Sampson, G. (2015), *Writing Systems*, 2nd edn. Sheffield: Equinox Publishing.

Sandler, W. and D. Lillo-Martin (2006), *Sign Language and Linguistic Universals*. Cambridge, UK and New York: Cambridge University Press.

Sapir, E. (1921), *Language: An to the Study of Speech*. New York: Harcourt, Brace & Company.

Sapir, E. [1933] (1949), "The psychological reality of phonemes," in D. G. Mandelbaum (ed.), *Selected Writings of Edward Sapir in Language, Culture, and Personality*. Berkeley and Los Angeles: University of California Press, pp. 45–60.

Sapir, S. (1989), "The intrinsic pitch of vowels: theoretical, physiological, and clinical considerations," *Journal of Voice, 3*: 44–51.

de Saussure, F. [1915] (1959), *Course in General Linguistics*. New York: McGraw-Hill.

Schane, S. A. (1984), "The fundamentals of Particle Phonology," *Phonology Yearbook 1*: 129–55.

Scheer, T. (2011), *A Guide to Morphosyntax-phonology Interface Theories: How Extra-Phonological Information is Treated in Phonology since Trubetzkoy's Grenzsignale*. Berlin: De Gruyter Mouton.

Scheer, T. and N. C. Kula (2018), "Government Phonology: element theory, conceptual issues, and introduction," in S. J. Hannahs and A. R K. Bosch (eds), *The Routledge Handbook of Phonological Theory*. New York: Routledge, pp. 226–61.

Scheer, T. and E. Cyran (2018), "Interfaces in Government Phonology," in S. J. Hannahs and A. R K. Bosch (eds), *The Routledge Handbook of Phonological Theory*. New York: Routledge, pp. 262–92.

Scherer, K. R. (2000), "Cross-cultural investigations of emotion inferred from voice and speech: implications for speech technology," in *Proceedings of the Sixth International Conference on the Processing of Spoken Language*, 2: 379–82, <https://www.isca-speech.org/archive/icslp_2000> (last accessed April 8, 2020).

Scobbie, J. M. (2007), "Interface and overlap in phonetics and phonology," in G. Ramchand and C. Reiss (eds), *The Oxford Handbook of Linguistic Interfaces*. Oxford: Oxford University Press, pp. 17–52.

Scobbie, J. M. and A. A. Wrench (2003), "An articulatory investigation of word final /l/ and /l/-sandhi in three dialects of English," in M. J. Sole, D. Recasens, and J. Romero (eds), *Proceedings of the 15th International Conference on Phonetic Sciences*, Barcelona, pp. 1871–4, <http://www.internationalphoneticassociation.org/icphs/icphs2003> (last accessed April 8, 2020).

Seidl, A. (2001), *Minimal Indirect Reference: A Theory of the Syntax-phonology Interface (Outstanding Dissertations in Linguistics)*. London and New York: Routledge.

Selkirk, E. (1984b), *Phonology and Syntax: The Relation between Sound and Structure*. Cambridge, MA: MIT Press.

Selkirk, E. (1995), "Sentence prosody: intonation, stress and phrasing," in J. Goldsmith (ed.), *The Handbook of Phonological Theory*. London: Blackwell, pp. 550–69.

Selkirk, E. (2011), "The syntax-phonology interface," in J. Goldsmith, J. Riggle, and A. Yu (eds), *The Handbook of Phonological Theory*. Oxford: Blackwell.

Selkirk, E. O. (1978), "On prosodic structure and its relation to syntactic structure," in T. Fretheim (ed.), *Nordic Prosody II*. Trondheim: Tapir, pp. 111–40.

Selkirk, E. O. (1980), "Prosodic domains in phonology: Sanskrit revisited," *Juncture, 7*: 107–29.

Selkirk, E. O. (1982), "The syllable," in H. van der Hulst and N. Smith (eds), *The Structure of Phonological Representations*. Dordrecht: Foris, pp. 337–84.

Selkirk, E.O. (1984a), "On the major class features and syllable theory," in M. Aronoff and R. Oehrl (eds), *Language Sound Structure*. Cambridge, MA: MIT Press, pp. 107–36.

Selkirk, E. O. (1986), "On derived domains in sentence phonology," *Phonology Yearbook 3*: 371–405.

Shannon, C. E. (1948), "A mathematical theory of communication," *The Bell System Technical Journal, 27*: 379–423, 623–56.

Shaw, J., A. I. Gafos, P. Hoole, and C. Zeroual (2011), "Dynamic invariance in the phonetic expression of syllable structure: a case study of Moroccan Arabic consonant clusters," *Phonology, 28*(3): 455–90.

Shih, S. and S. Inkelas (2019), "Autosegmental aims in surface optimizing phonology," *Linguistic Inquiry, 50*: 137–96.

Silva, D. (1992), *The phonetics and phonology of stop lenition in Korean*. Doctoral dissertation, Cornell University.

Silverman, D. (1997), *Phasing and Recoverability*. New York: Garland.

Silverman, D. (2006), *A Critical Introduction to Phonology: Of Sound, Mind, and Body*. London and New York: Continuum Press.

Silverman, K. and J. Pierrehumbert (1990), "The timing of pre-nuclear high accents in English," in J. Kingston and M. Beckman (eds), *Papers in Laboratory Phonology I: Between the Grammar and Physics of Speech*. Cambridge, MA: Cambridge University Press, pp. 72–106.

Skinner, B. F. (1957), *Verbal Behavior*. Acton, MA: Copley Publishing Group.

Smiljanić, R. (2004), *Lexical, Pragmatic, and Positional Effects on Prosody in Two Dialects of Croatian and Serbian: An Acoustic Study*. New York: Routledge.

Smith, J. (2002), *Phonological Augmentation in Prominent Positions*. PhD dissertation, Linguistics, University of Massachusetts Amherst.

Smolensky, P. and G. Legendre (2006), *The Harmonic Mind: From Neural Computation to Optimality-theoretic Grammar*. Cambridge, MA: MIT Press.

Smolensky, P. and M. Goldrick (2016), "Gradient symbolic representations in grammar: the case of French liaison," ROA-1286, Rutgers Optimality Archive, <http://roa.rutgers.edu/content/article/files/1552_smolensky_1.pdf> (last accessed April 9, 2020).

Sohoglu, E., J. E. Peelle, R. P. Carlyon, and M. H. Davis (2012), "Predictive top-down integration of prior knowledge during speech perception," *Journal of Neuroscience 32*(25): 8443–53.

Son, M., A. Kochetov, and M. Pouplier (2007), "The role of gestural overlap in perceptual place assimilation: evidence from Korean," in J. Cole and J. I. Hualde (eds), *Papers in Laboratory Phonology IX*. Berlin: De Gruyter Mouton, pp. 507–34.

Sproat, R. (2006), *A Computational Theory of Writing Systems*. Cambridge, UK: Cambridge University Press.

Sproat, R. and O. Fujimura (1993), "Allophonic variation in English /l/ and its implications for phonetic implementation," *Journal of Phonetics, 21*: 291–311.

Stampe, D. (1979), *A Dissertation on Natural Phonology*. New York: Garland.

Stankiewicz, E. (1972a), "Baudouin de Courtenay: his life and work," in E. Stankiewicz (trans. and ed.), *A Baudouin de Courtenay Anthology: The Beginnings of Structural Linguistics*. Bloomington: Indiana University Press, pp. 3–48.

Stankiewicz, E. (trans. and ed.) (1972b), *A Baudouin de Courtenay Anthology: The Beginnings of Structural Linguistics*. Bloomington: Indiana University Press.

Steriade, D. (1982), *Greek prosodies and the nature of syllabification*. Doctoral dissertation, Massachusetts Institute of Technology.

Steriade, D. (1994), "Positional neutralization and the expression of contrast," unpublished manuscript, University of California, Los Angeles.

Steriade, D. (1997), "Phonetics in phonology: the case of laryngeal neutralization," unpublished manuscript, University of California, Los Angeles.

Steriade, D. (2001), "Directional asymmetries in place assimilation: a perceptual account," in E. Hume and K. Johnson (eds), *The Role of Speech Perception in Phonology*. New York: Academic Press, pp. 219–50.

Stetson, R. H. (1928), *Motor Phonetics*. Amsterdam: North Holland.

Stevens, K. N. (1960), "Towards a model for speech perception," *The Journal of the Acoustical Society of America, 32*: 47–55.

Stevens, K. N. (1989), "On the quantal nature of speech," *Journal of Phonetics, 17*: 3–45.

Stevens, K. N. (2000a), *Acoustic Phonetics*. Cambridge, MA: MIT Press.

Stevens, K. N. (2000b), "Diverse acoustic cues at consonantal landmarks," *Phonetica 57*(2–4): 139–51.

Stevens, K. N. (2002), "Toward a model for lexical access based on acoustic landmarks and distinctive features," *The Journal of the Acoustical Society of America, 111*: 1871–91.

Stevens, K. N. (2005), "Features in speech perception and lexical access," in D. B. Pisoni and R. E. Remez (eds), *The Handbook of Speech Perception*. Oxford: Blackwell, pp. 125–55.

Stokoe, W. C. (1960), "Sign language structure: an outline of the visual communication system of the American Deaf," *Studies in Linguistics: Occasional Papers 8*. Buffalo: University of Buffalo Department of Anthropology and Linguistics.

Strand, E. A. (2000), *Gender stereotype effects in speech processing*. Doctoral dissertation, Ohio State University, Columbus.

Sumby, W. H. (1954), "Visual contribution to speech intelligibility in noise," *The Journal of the Acoustical Society of America, 26*: 212–15.

Svantesson, J.-O. and D. House (2006), "Tone production, tone perception, and Kammu tonogenesis," *Phonology, 23*: 309–33.

Sweet, H. (1877), *A Handbook of Phonetics: Including a Popular of the Principles of Spelling Reform*. Oxford: Clarendon Press.

Teeranon, P. (2007), "The plausibility of tonal evolution in the Malay dialect spoken in Thailand: evidence from an acoustic study," *Taiwan Journal of Linguistics, 5*(2): 45–64.

t'Hart, J. and R. Collier (1975), "Integrating different levels of intonation analysis," *Journal of Phonetics, 3*: 235–55.

t'Hart, J., R. Collier, and A Cohen (1990), *A Perceptual Study of Intonation: An Experimental-phonetic Approach*. Cambridge, UK: Cambridge University Press.

Tilsen, S. (2016), "Selection and coordination: the articulatory basis for the emergence of phonological structure," *Journal of Phonetics, 55*: 53–77.

Tkachman, O., K. Hall, A. Xavier, and B. Gick (2016), "Sign Language Phonetic Annotation meets Phonological CorpusTools: Towards a sign language toolset for phonetic notation and phonological analysis," in G. O. Hansson, A. Farris-Trimble, K. McMullin, and D. Pulleyblank (eds), *Supplemental Proceedings of the 2015 Annual Meeting on Phonology*. Washington, DC: Linguistic Society of America, <http://journals.linguisticsociety.org/proceedings/index.php/amphonology/article/view/3667> (last accessed April 9, 2020).

Tomlinson Jr., J. M., Q. Liu, and J. E. F. Tree (2014), "The perceptual nature of stress shifts," *Language Cognition and Neuroscience, 29*(9): 1046–58. doi: 10.1080/01690965.2013.813561.

Trager, G. L. and H. L. Smith Jr. (1951), "An outline of English structure," in *Studies in Linguistics: Occasional Papers 3*. Washington, DC: American Council of Learned Societies.

Traunmüller, H. (1990), "Analytical expressions for the tonotopic sensory scale," *The Journal of the Acoustical Society of America, 88*: 97.

Trofimov, M. and D. Jones (1923), *The Pronunciation of Russian*. Cambridge, UK: Cambridge University Press.

Trubetzkoy, N. S. [1939] (1969), *Foundations of Phonology*, (C. A. M. Baltaxe, trans.), Berkeley: University of California Press.

Trudell, B. and C. T. Adger (2015), "Early reading success in Africa: the language factor," in E. C. Zsiga, O. T. Boyer, and R. Kramer (eds), *Languages in Africa*. Washington, DC: Georgetown University Press, pp. 12–20.

Tserdanelis, G. (2001), "A perceptual account of manner dissimilation in Greek," *Ohio State Working Papers in Linguistics, 55*: 172–99, <http://hdl.handle.net/1811/80836> (last accessed April 8, 2020).

Turk, A. (2012), "The temporal implementation of prosodic structure," in A. C. Cohn, C. Fougeron, and M. K. Huffman (eds), *The Oxford Handbook of Laboratory Phonology*. Oxford: Oxford University Press, pp. 242–53.

Turk, A. and S. Shattuck-Hufnagel (2000), "Word-boundary-related duration patterns in English," *Journal of Phonetics, 28*(4): 397–440.

Turk, A. and S. Shattuck-Hufnagel (2007), "Multiple targets of phrase-final lengthening in American English words," *Journal of Phonetics*, 35(4): 445–72.

Turk, A. and S. Shattuck-Hufnagel (2014), "Timing in talking: what is it used for, and how is it controlled?" *Philosophical Transactions of the Royal Society B: Biological Sciences, 369*(1658), 20130395. doi: 10.1098/rstb.2013.0395.

Turk, A. and S. Shattuck-Hufnagel (2020), *Speech Timing: Implications for Theories of Phonology, Phonetics, and Speech Motor Control*. Oxford: Oxford University Press.

Turvey, M. T. (1990), "Coordination," *American Psychologist, 45*(8): 938–53.

van der Hulst, H. (2016). "Monovalent 'features' in phonology," *Language and Linguistics Compass, 10*: 83–102.

van der Hulst, H. and J. van de Weijer (2018), "Dependency Phonology," in S. J. Hannahs and A. R K. Bosch (eds), *The Routledge Handbook of Phonological Theory*. New York: Routledge, pp. 325–59.

van Heuven, V. J. and A. Turk (2018), "Phonetic correlates of word and sentence stress," in C. Gussenhoven and A. Chen (eds), *The Oxford Handbook of Language Prosody*. Oxford: Oxford University Press.

Vaux, B. and B. Samuels (2015), "Explaining vowel systems: Dispersion Theory vs. natural selection," *The Linguistic Review, 32*(3): 573–99.

Veilleux, N., S. Shattuck-Hufnagel, and A. Brugos (2006), "Transcribing prosodic structure of spoken utterances with ToBI," Massachusetts Institute of Technology: MIT OpenCourseWare, <https://ocw.mit.edu/courses/electrical-engineering-and-computer-science/6-911-transcribing-prosodic-structure-of-spoken-utterances-with-tobi-january-iap-2006/index.htm> (last accessed April 9, 2020).

Vennemann, T. (1972), "On the theory of syllabic phonology," *Linguistische Berichte, 18*: 1–18.

Vicenik, C. and S. A. Jun (2014), "An autosegmental-metrical analysis of Georgian intonation," in S.-A. Jun (ed.), *Prosodic Typology II: The Phonology of Intonation and Phrasing.* Oxford: Oxford University Press.

Vitevitch, M. S. and P. A. Luce (1998), "When words compete: levels of processing in spoken word perception," *Psychological Science, 9*: 325–9.

Vitevitch, M. S. and P. A. Luce (1999), "Probabilistic phonotactics and neighborhood activation in spoken word recognition," *Journal of Memory and Language, 40*: 374–408.

Vogel, I. (1977), *The syllable in phonological theory with special reference to Italian.* Doctoral dissertation, Stanford University.

Vogel, I., H. T. Bunnell, and S. Hoskins (1995), "The phonology and phonetics of the rhythm rule," in B. Connell and A. Arvaniti (eds), *Phonology and Phonetic Evidence: Papers in Laboratory Phonology IV.* Cambridge, UK: Cambridge University Press.

Walker, S., V. Bruce, and C. O'Malley (1995), "Facial identity and facial speech processing: familiar faces and voices in the McGurk effect," *Perception and Psychophysics, 57*: 1124–33.

Waugh, L. R. and M. Monville-Burston (1990), "Introduction: the life, work, and influence of Roman Jakobson," in L. Waugh and M. Monville-Burston (eds), *Roman Jakobson: On Language*, Cambridge, MA: Harvard University Press, pp. 1–45.

Wedekind, K. (1983), "A six-tone language of Ethiopia," *Journal of Ethiopian Studies, 16*: 129–56.

Wedel, A. B. (2006), "Exemplar models, evolution and language change," *The Linguistic Review, 23*(3): 247–74.

Weiner, N. (1950), *The Human Use of Human Beings: Cybernetics and Society.* Boston: Houghton Mifflin.

Wells, R. (1947), "Review of Pike 1945," *Language, 23*: 255–73.

Welmers, W. (1973), *African Language Structures.* Berkeley: University of California Press.

Wetzels, W. L. and J. Mascaró (2001), "The typology of voicing and devoicing," *Language, 77*(2): 207–44.

Whalen, D. H. (1990), "Coarticulation is largely planned," *Journal of Phonetics, 18*: 3–35.

Whalen, D. H. (2019), "The Motor Theory of speech perception," in M. Aronoff (ed.), Oxford Research Encyclopedia of Linguistics. Oxford University Press. doi. 10.1093/acrefore/9780199384655.013.404.

Wightman, C. W., S. Shattuck-Hufnagel, M. Ostendorf, and P. J. Price (1992), "Segmental durations in the vicinity of prosodic phrase boundaries," *The Journal of the Acoustical Society of America, 91*(3): 1707–17.

Wilson, C. (2006), "Learning phonology with substantive bias: an experimental and computational study of velar palatalization," *Cognitive Science, 30*: 945–82.

Wilson, C., L. Davidson, and S. Martin (2014), "Effects of acoustic-phonetic detail on cross-language speech production," *Journal of Memory and Language, 77*, 1–24.

Wilson, S. M., A. P. Saygin, M. I. Sereno, and M. Iacoboni (2004), "Listening to speech activates motor areas involved in speech production," *Nature Neuroscience, 7*: 701–2.

Woo, N. H. (1969), *Prosody and phonology*. Doctoral dissertation, Massachusetts Institute of Technology.

Wright, D., J. H. Hebrank, and B. Wilson, (1974), "Pinna reflections as cues for localization," *The Journal of the Acoustical Society of America, 56*: 957–62.

Wright, R. (1996), *Consonant clusters and cue preservation in Tsou*. Doctoral dissertation, University of California, Los Angeles.

Wright, R. (2001), "Perceptual cues in contrast maintenance," in E. Hume and K. Johnson (eds), *The Role of Speech Perception in Phonology*. New York: Academic Press, pp. 251–77.

Xu, Y. (1997), "Contextual tonal variations in Mandarin," *Journal of Phonetics, 25*: 61–83.

Xu, Y. (1998), "Consistency of tone-syllable alignment across different syllable structures and speaking rates," *Phonetica, 55*: 179–203.

Xu, Y. (1999), "F0 peak delay: where, when and why it occurs," in J. J. Ohala, Y. Hasegawa, M. Ohala, D. Granville, and A. C. Bailey (eds), *Proceedings of the 14th International Congress of Phonetic Sciences*, San Francisco, CA, pp. 1881–4, <https://www.internationalphoneticassociation.org/icphs-proceedings/ICPhS1999/p14_1881.html> (last accessed April 9, 2020).

Xu, Y. (2004), "Understanding tone from the perspective of production and perception," *Language and Linguistics, 5*: 757–97.

Xu, Y. (2005), "Speech melody as articulatorily implemented communicative functions," *Speech Communication, 46*: 220–51.

Xu, Y. (2011), "Speech prosody: a methodological review," *Journal of Speech Sciences, 1*: 85–115.

Xu, Y. and Xu, C. X. (2005), "Phonetic realization of focus in English declarative intonation," *Journal of Phonetics, 33*: 159–97.

Xu, Y., A. Lee, S. Prom-on, and F. Liu (2015), "Explaining the PENTA model: a reply to Arvaniti and Ladd," *Phonology, 32*: 505–35.

Yi, H. (2017), *Lexical tone gestures*. Doctoral dissertation, Cornell University.

Yi, H. and S. Tilsen (2014), "A gestural account of mandarin tone sandhi," *The Journal of the Acoustical Society of America, 136*: 2144.

Yip, M. (1989), "Contour tones," *Phonology, 6*: 149–74.

Yip, M. (1995), "Tone in East Asian languages," in J. Goldsmith (ed.), *Handbook of Phonological Theory*. Oxford: Blackwell. pp. 476–94.

Yip, M. (2001), "The complex interaction of tone and prominence," in M. Kim and U. Strauss (eds), *Proceedings of NELS 31*, Amherst, MA: GLSA.

Yip, M. (2002), *Tone*. Cambridge, UK: Cambridge University Press.

Yip, M. (2007). "Tone," in P. de Lacy (ed.), *The Cambridge Handbook of Phonology*. Cambridge, UK: Cambridge University Press, pp. 229–52.

Yu, A. C. L. (2004), "Explaining final obstruent voicing in Lezgian: Phonetics and history," *Language, 80*(1): 73–97.

Yu, A. C. L. (2006), "Tonal effects on perceived vowel duration," presentation at the 10th Conference on Laboratory Phonology, Paris.

Yu, A. C. L. (2011), "On measuring phonetic precursor robustness: a response to Moreton," *Phonology, 28*(3) 491–518.

Yu, A. C. L. (ed.) (2013), *Origins of Sound Change: Approaches to Phonologization*. Oxford: Oxford University Press.

Yu, K. M. and H. W. Lam (2014), "The role of creaky voice in Cantonese tonal perception," *The Journal of the Acoustical Society of America, 136*(3): 1320–33.

Zec, D. (1988), *Sonority constraints on prosodic structure*. Doctoral dissertation, Stanford University.

Zec, D. (1995), "Sonority constraints on syllable structure," *Phonology, 12*: 85–129.

Zec, D. (2007), "The syllable," in P. de Lacy (ed.), *The Cambridge Handbook of Phonology*. Cambridge, UK: Cambridge University Press, pp. 161–94.

Zec, D. and E. C. Zsiga (2019), "Tone and stress as agents of cross-dialectal variation: the case of Serbian," talk presented at the 6th NINJAL International Conference on Phonetics and Phonology, December, National Institute for Japanese Language and Linguistics, Tachikawa, Japan.

Zhang, J. (2002), *The Effects of Duration and Sonority on Contour Tone Distribution: A Typological Survey and Formal Analysis*. New York: Routledge.

Zimmerman, K. and B. Kellermeier-Rehbein (eds) (2015), *Colonialism and Missionary Linguistics*. Berlin: De Gruyter Mouton.

Zipf, G. K. (1935), *The Psycho-biology of Language*. Boston: Houghton Mifflin Co.

Zsiga, E. C. (1992), "A mismatch between morphological and prosodic domains: evidence from two Igbo rules," *Phonology*, *9*: 101–35.

Zsiga, E. C. (1993), *Features, gestures, and the temporal aspects of phonological organization*. Doctoral dissertation, Yale University.

Zsiga, E. C. (1994), "Acoustic evidence for gestural overlap in consonant sequences," *Journal of Phonetics, 22*: 121–40.

Zsiga, E. C. (1995), "An acoustic and electropalatographic study of lexical and post-lexical palatalization in American English," in B. Connell and A. Arvaniti (eds), *Phonology and Phonetic Evidence: Papers in Laboratory Phonology IV*. Cambridge, UK: Cambridge University Press, pp. 282–302.

Zsiga, E. C. (1997), "Features, gestures, and Igbo vowels: an approach to the phonology/phonetics interface," *Language, 73*(2): 227–74.

Zsiga, E. C. (2000), "Phonetic alignment constraints: consonant overlap and palatalization in English and Russian," *Journal of Phonetics, 28*: 69–102.

Zsiga, E. C. (2003), "Articulatory timing in a second language: evidence from English and Russian," *Studies in Second Language Acquisition, 25*(3), 399–432.

Zsiga, E. C. (2011), "External sandhi in a second language: the phonetics and phonology of obstruent nasalization in Korean and Korean-accented English," *Language, 87*: 289–345.

Zsiga, E. C. (2012), "Contrastive tone and its implementation," in A. C. Cohn, C. Fourgeron, and M. K. Huffman (eds), *The Oxford Handbook of Laboratory Phonology*. Oxford: Oxford University Press, pp. 196–206.

Zsiga, E. C. (2013a). *The Sounds of Language: An Introduction to Phonetics and Phonology*. Malden, MA and Oxford: Wiley-Blackwell.

Zsiga, E. C. (2013b), "What is a possible language?" in *The Sounds of Language: An Introduction to Phonetics and Phonology*. Oxford: Wiley Blackwell, Chapter 12.

Zsiga, E. C. (2018a), "In favor of [Fortis]: evidence from Setswana and Sebirwa," in G. Gallagher, M. Gouskova, and S. Yin (eds), *Proceedings of the 2017 Annual Meeting on Phonology*. Washington, DC: Linguistic Society of America, <https: //journals.linguisticsociety.org/proceedings/index.php/amphonology/article/view/4252/3896> (last accessed March 3, 2020).

Zsiga, E. C. (2018b), "Tonal timing and vowel reduction in Igbo: implications for the representation of prosodic structure in Articulatory Phonology," poster presented at LabPhon 16, June, Lisbon.

Zsiga, E. C. and R. Nitisaroj (2007), "Tone features, tone perception, and peak alignment in Thai," *Language and Speech, 50*: 343–83.

Zsiga, E. C. and D. Zec (2013), "Contextual evidence for the representation of pitch accents in Standard Serbian," *Language and Speech, 56*(1): 69–104.

Zsiga, E. C. and O. T. Boyer (2017), "A natural experiment in learning an unnatural alternation: Sebirwa in contact with Setswana," in J. Kandybowicz and H. Torrence (eds), *Africa's Endangered Languages: Documentary and Theoretical Approaches*. Oxford: Oxford University Press, pp. 343–66.

INDEX

Figures and tables are shown by fig and tab next to the relevant locators.